INNOVATION
IN GLOBAL INDUSTRIES

U.S. FIRMS COMPETING IN A NEW WORLD

C O L L E C T E D   S T U D I E S

Jeffrey T. Macher and David C. Mowery, Editors

Committee on the Competitiveness and Workforce Needs of U.S. Industry

Board on Science, Technology, and Economic Policy

Policy and Global Affairs

)NAL RESEARCH COUNCIL
OF THE NATIONAL ACADEMIES

HC
79
.T4
I 54
2008

THE NATIONAL ACADEMIES PRESS
Washington, D.C.
www.nap.edu

THE NATIONAL ACADEMIES PRESS   500 Fifth Street, N.W.   Washington, DC 20001

NOTICE: The project that is the subject of this report was approved by the Governing Board of the National Research Council, whose members are drawn from the councils of the National Academy of Sciences, the National Academy of Engineering, and the Institute of Medicine. The members of the committee responsible for the report were chosen for their special competences and with regard for appropriate balance.

This study was supported by Contract/Grant No. SB 1341-06-Z-0011, TO #2 between the National Academy of Sciences and the Technology Administration of the U.S. Department of Commerce; Contract/Grant No. SLON 2005-10-18 between the National Academy of Sciences and the Alfred P. Sloan Foundation; and Contract/Grant No. P116Z05283 between the National Academy of Sciences and the U. S. Department of Education. Any opinions, findings, conclusions, or recommendations expressed in this publication are those of the author(s) and do not necessarily reflect the views of the organizations or agencies that provided support for the project.

International Standard Book Number-13: 978-0-309-11631-2
International Standard Book Number-10: 0-309-11631-7
Library of Congress Catalog Card Number: 2008926521

Limited copies are available from:
Board on Science, Technology, and Economic Policy
National Research Council
500 Fifth Street, N.W., Keck Center 574, Washington, D.C. 20001
Phone: (202) 334-2200
Fax: (202) 334-1505
E-mail: **step@nas.edu**

Cover: The cover design depicts a possible far distant future configuration bringing nearly all of the earth's land mass into a single supercontinent, including merging the Americas into Asia.

THE NATIONAL ACADEMIES
Advisers to the Nation on Science, Engineering, and Medicine

The **National Academy of Sciences** is a private, nonprofit, self-perpetuating society of distinguished scholars engaged in scientific and engineering research, dedicated to the furtherance of science and technology and to their use for the general welfare. Upon the authority of the charter granted to it by the Congress in 1863, the Academy has a mandate that requires it to advise the federal government on scientific and technical matters. Dr. Ralph J. Cicerone is president of the National Academy of Sciences.

The **National Academy of Engineering** was established in 1964, under the charter of the National Academy of Sciences, as a parallel organization of outstanding engineers. It is autonomous in its administration and in the selection of its members, sharing with the National Academy of Sciences the responsibility for advising the federal government. The National Academy of Engineering also sponsors engineering programs aimed at meeting national needs, encourages education and research, and recognizes the superior achievements of engineers. Dr. Charles M. Vest is president of the National Academy of Engineering.

The **Institute of Medicine** was established in 1970 by the National Academy of Sciences to secure the services of eminent members of appropriate professions in the examination of policy matters pertaining to the health of the public. The Institute acts under the responsibility given to the National Academy of Sciences by its congressional charter to be an adviser to the federal government and, upon its own initiative, to identify issues of medical care, research, and education. Dr. Harvey V. Fineberg is president of the Institute of Medicine.

The **National Research Council** was organized by the National Academy of Sciences in 1916 to associate the broad community of science and technology with the Academy's purposes of furthering knowledge and advising the federal government. Functioning in accordance with general policies determined by the Academy, the Council has become the principal operating agency of both the National Academy of Sciences and the National Academy of Engineering in providing services to the government, the public, and the scientific and engineering communities. The Council is administered jointly by both Academies and the Institute of Medicine. Dr. Ralph J. Cicerone and Dr. Charles M. Vest are chair and vice chair, respectively, of the National Research Council.

www.national-academies.org

STEP Staff

Stephen A. Merrill
Executive Director

Charles W. Wessner
Program Director

Sujai J. Shivakumar
Senior Program Officer

David E. Dierksheide
Program Officer

Mahendra Shunmoogam
Program Associate

Jeffrey C. McCullough
Program Associate

Cynthia A. Getner
Financial Associate

Preface and Acknowledgments

In 1999 the National Academies' Board on Science, Technology, and Economic Policy (STEP) released a series of industry studies analyzing the sources of competitive resurgence from the 1980s to the 1990s of many U.S.-based firms in a variety of manufacturing and service sectors. These studies, published under the title *U.S. Industry in 2000: Studies in Competitive Performance*, included steel, chemicals, metal working, trucking, grocery retailing, retail banking, computing, semiconductors, hard disk drives, apparel, pharmaceuticals, and biotechnology.

The general picture of stronger performance in the mid-to-late 1990s than in the early 1990s was attributed to a variety of factors including heavy investment in applications of information technology, supportive public policies, openness to innovation, and changes in supplier and customer relationships. Vigorous foreign competition forced cost-cutting changes in manufacturing processes, organization, and strategy but then receded, making the performance of U.S. industries look even better. As none of these favorable conditions could be assumed to be permanent, the collected studies persuasively made the point that U.S. industries' superior performance is not guaranteed to continue.

In late 2005 the STEP Board decided to reprise the study, focusing on the acceleration in global sourcing of innovation and emergence of new locations of research capacity, new sources of skilled technical workers, and the implications of these developments for U.S. businesses and workforce. Although the current study involves several of the same industries—in particular, semiconductors, personal computing, financial services, pharmaceuticals, and biotechnology—the overall selection shifted markedly toward technology-intensive producing, supporting, or using sectors to include software, flat panel displays, solid state lighting, logistics, and venture capital finance. The group of industries examined does not represent a carefully selected sample representative of the economy as a whole. Rather, it reflects a decision to again capitalize on the work of university-based multidisciplinary research teams studying economic performance and technological change at the industry level. Most of these groups were formed and supported under the Industry Centers Program of the Alfred P. Sloan Foundation.

To help integrate this work, the Board again asked **David C. Mowery**, Professor at the Haas School of Business at the University of California at Berkeley, to develop a general framework for analyzing changes in the structure of innovation over the past 10 to 15 years. Mowery in turn recruited **Jeffrey T. Macher**, Associate Professor, McDonough School of Business, Georgetown University, to assist in this effort and co-edit the resulting volume. The chapters in this volume were drafted independently by individual authors, and their findings and any policy recommendations do not represent a consensus among all of the contributors to the volume. They also do not necessarily represent the opinions and views of the Committee on Competitiveness and Workforce Needs of U.S. Industry, the STEP Board, the National Academies, or the sponsoring organizations.

In the course of their work, the editors and chapter authors participated in two public workshops in Washington, D.C. The first, on April 19, 2006, reviewed their preliminary findings with industry representatives and other analysts including **Irving Wladawsky-Berger**, IBM Corporation; **Jack Gill**, Vanguard Ventures and Harvard Medical School; **Richard S. Golaszewski**, GRA, Inc.; **Jeffrey D. Tew**, General Motors; **Jerome H. Grossman**, LionGate Corporation and Harvard University; **Gordon W. Day**, Optoelectronic Industry Development Association; **Timothy J. Sturgeon**, Massachusetts Institute of Technology; **Charles W. Wade**, Technology Forecasters, Inc.; **Richard B. Freeman**, Harvard University; **Nancy Hauge**, K12; **Harold Salzman**, the Urban Institute; and **Navi Radjou**, Forrester Research, Inc.

A year later a second workshop was held, on April 20, 2007, to try to anticipate trends over the next several years in three broad sectors encompassing most of the industries being studied—information and computing technology, biopharmaceuticals, and finance. Speakers in addition to committee members and authors included **Undersecretary Robert C. Cresanti**, Commerce Department's Technology Administration; **Barry Jaruzelski**, Booz Allen Hamilton; **Robert D. Atkinson**, Information Technology and Innovation Foundation; **Alex Soojung-Kim Pang**, Institute for the Future; **Bhaskar Chakravorti**, McKinsey and Company; **David Moschella**, Leading Edge Forum; **Michael E. Fawkes**, Hewlett-Packard Company; **Anna D. Barker**, National Cancer Institute; **Thomas R. Cech**, Howard Hughes Medical Institute; **Joseph Jasinski**, Health Care Life Science, IBM; **Andy Lee**, Pfizer Inc.; **T. L. Stebbins**, Canaccord Adams, Inc.; **Karen G. Mills**, Solera Capital; and **Alex J. Pollock**, American Enterprise Institute.

As the editors state in their summary introduction to this collection, despite the emergence of robust R&D and innovative capabilities in East, Southeast, and

South Asia, and concerted efforts to develop them in other parts of the world, patterns of innovation are highly variable across industries and across firms within industries. Many industries and some firms within nearly all industries retain leading-edge capacity in the United States. The flat panel display sector, in which innovative activity for the most part has followed production abroad, is not as yet the norm. This is no reason for complacency about the outlook for the future, however. Empirically-based analyses such as those in this volume are inevitably backward-looking. Even recently issued patents generally represent filings two to five years back and R&D investments considerably earlier. Although not pessimistic overall, our authors compellingly document the rapidity of contemporary industrial change and shifts in competitive advantage. For that reason alone, innovation deserves more sustained public policy attention than it has been receiving.

The STEP Board is grateful to the authors, the editors, and the workshop participants as well as to the sponsors of this activity—**the Alfred P. Sloan Foundation**, the **U.S. Department of Education**, and the **Technology Administration of the U.S. Department of Commerce**.

This collection has been reviewed in draft from by individuals chosen for their diverse perspectives and technical expertise, in accordance with procedures approved by the National Academies' Report Review Committee. The purpose of this independent review is to provide candid and critical comments that will assist the institution in making the published report as sound as possible and to ensure that the report meets institutional standards for objectivity, evidence, and responsiveness to the study charge. The review comments and draft manuscript remain confidential to protect the integrity of the deliberative process.

We wish to thank the following individuals for their review of this report: **Suma Athreye**, Brunel University; **MaryAnn Feldman**, University of Toronto; **Jeffrey Furman**, Boston University; **Bronwyn Hall**, University of California at Berkeley; **Megan MacGarvie**, Boston University; **Deepak Somaya**, University of Maryland; **Jerry Thursby**, Emory University; and **Philip Webre**, Congressional Budget Office.

Although the reviewers listed above have provided many constructive comments and suggestions, they were not asked to endorse the content of the report, nor did they see the final draft of the report before its release. Responsibility for the final content of this report rests entirely with the individual authors.

David T. Morgenthaler, *Chair*
Stephen A. Merrill, *Study Director*

Contents

Introduction

JEFFREY T. MACHER
Georgetown University
DAVID C. MOWERY
University of California at Berkeley

The causes and consequences for U.S. competitiveness and living standards of innovation by foreign nations and firms have been long-standing topics of scholarly and policy debate within the United States. During the 1960s and 1970s, much of this debate focused on U.S. multinational firms' investment in offshore research and development (R&D) and production facilities, most of which were located in other industrial economies. Concern was expressed that the transfer of technology by U.S. corporations through their offshore R&D and manufacturing investments would contribute to the growth of foreign competitors in these and other industries and reduce domestic employment opportunities. The debates of the 1980s and early 1990s adopted a slightly different tone, emphasizing the growth of foreign competitors in industries such as automobiles and semiconductors whose innovative performance and high-quality products threatened the viability of U.S. firms (including multinational U.S. firms) and entire industrial sectors. These debates were concerned less with the offshore transfer of U.S. technological capabilities than with the threat posed by foreign firms' competitive strengths.

A previous volume (*U.S. Industry in 2000: Studies in Competitive Performance*) released by the National Academies' Board on Science, Technology, and Economic Policy in 1999 argued that much of the pessimism of the 1980s and 1990s over U.S. industrial competitiveness proved to be exaggerated or misplaced. U.S. firms in a number of industries developed new business models and new products, which enabled them to address competitive threats with considerable success. In some cases, the responses of U.S. firms relied on their position within a large domestic market of innovative users who proved to be important sources for new ideas and products. In other cases, improvements in U.S. competitive and innovative performance relied on the robust domestic "R&D infra-structure" comprising industrial, governmental, and university research facilities, much of which had benefited from large federal investments spanning the post-1945 period. U.S. firms and consumers alike also benefited from low-cost imports of some products, such as personal computers and components that were critical inputs for the innovation and restructuring processes described in this volume.

A more recent wave of concern over U.S. competitive prospects in the 21st century combines elements of all of these previous debates. The actions of many U.S. firms (not all of which can be considered multinational by any conventional definition) to "outsource" activities formerly undertaken by U.S.-based professional, scientific, and engineering employees have raised widespread popular concerns over the erosion of employment opportunities in occupations and industries (including many service industries) that formerly were minimally exposed to foreign competition. At the same time, the growth of innovative and manufacturing capabilities in countries such as China, India, South Korea, and Taiwan has raised concerns over new sources of competition for U.S. firms. A 2006 study of U.S. competitiveness and innovative performance phrased these concerns as follows:

> ... the committee is deeply concerned that the scientific and technological building blocks critical to our economic leadership are eroding at a time when many other nations are gathering strength. We strongly believe that a worldwide strengthening will benefit the world's economy—particularly in the creation of jobs in countries that are far less well-off than the United States. But we are worried about the future prosperity of the United States. Although many people assume that the United States will always be a world leader in science and technology, this may not continue to be the case inasmuch as great minds and ideas exist throughout the world. We fear the abruptness with which a lead in science and technology can be lost—and the difficulty of recovering a lead once lost, if indeed it can be regained at all. (National Research Council, 2006, Executive Summary, p. 2)[1]

A central issue in the long debate over U.S. competitiveness that is briefly summarized above is the processes through which industrial firms in the United States and other economies create innovative new products and processes. What has changed in the global and U.S. domestic economies to transform the near euphoria in popular evaluations of U.S. economic and innovative performance during the "New Economy" of the late 1990s to the concerns expressed by

---

[1]A similar sentiment may be found in Freeman (2005): "But the US will also face economic dif-ficulties as its technological superiority erodes. What is good for the world is not inevitably good for the U.S. The group facing the biggest danger from the loss of America's technological edge are work-ers whose living standards depend critically on America's technological superiority. The decline in monopoly rents from being the lead country will make it harder for the US to raise wages and benefits to workers. The big winners from the spread of technology will be workers in developing countries, and the firms that employ them, including many U.S. multinational corporations" (p. 27).

the blue-ribbon panel cited above? In particular, what characteristics of the innovation-related activities of U.S. and non-U.S. firms have changed since the late 1990s so significantly to trigger these concerns? On the other hand, recognizing that many elements of this debate are not new but have been widely shared since the 1960s, what aspects have not changed in the innovation-related competition in which U.S. and foreign firms find themselves engaged? How well are scholars or policy makers able to measure any such change at a sufficiently fine-grained level of analysis to inform such debates? Finally, what are the implications for public policy of change in the structure (especially globalization) of U.S. firms' innovation-related activities? This volume examines these questions by providing detailed studies of structural change in the innovation process in 10 manufacturing and service industries.

Any study of issues related to innovation and competitiveness must address the widely held view that "firms compete, nations don't." In other words, the innovation-related and employment consequences of global competition are the result of private-sector investment and management decisions—public policy is of little importance. Just as it distorts reality to claim that international economic competition is solely a matter of competition among governments, the claim that private managers' decisions are all that matter is also an oversimplification, particularly in light of the evidence presented in many of the chapters in this volume. The international performance of firms, including multinationals, is affected by policy and other economic conditions in their home countries. And this link is especially strong for firms' innovation-related activities, which rely on a complex infrastructure of public and private institutions devoted to knowledge creation and transmission, personnel training, and other activities. Indeed, one of the most striking findings in many of the chapters in this volume is the extent to which the inventive activities of firms in many knowledge-intensive industries remain concentrated in their home countries. Simply put, both firms and nations matter.

THE STEP BOARD STUDY

Recognizing that the debate over the international transfer of technological and innovative capabilities and potential loss of U.S. competitiveness is a long-running one, the Science, Technology, and Economic Policy (STEP) Board of the National Academies undertook a study in 2006-2007 of the changes in the structure of the innovation process that are associated with shifting perceptions of the competitive outlook for U.S. firms and domestic employment, especially in professional and engineering occupations.

The STEP Board study examined 10 industries: personal computers, software, semiconductors, flat panel displays, lighting, pharmaceuticals, biotechnology, logistics, venture capital, and financial services. The choice of industries reflected several factors: (1) coverage of knowledge-intensive industries that have been the focus of many expressions of concern over waning U.S. technological strength; (2) inclusion of service industries, which historically have received little attention in debates over foreign competition and innovation; and (3) the willingness of scholars who have conducted extensive research on these industries to examine the issues of change in the structure and globalization of innovation-related activities. Reflecting the study organizers' interest in highlighting similarities and differences across industry-level studies, we drew on scholars affiliated with the Industry Centers established with initial financial support from the Alfred P. Sloan Foundation.

The following is a summary of the findings of each industry study.

Personal Computing

The personal computer (PC) industry now operates as a global network of independent suppliers of systems, components, peripherals, and software. Although the pace of innovation in the industry is rapid, its character now is largely incremental because of the dominance of the "Wintel" PC architectural standard. One important future challenge is the integration of the PC with the proliferating array of consumer devices that "orbit the user" and provide computing and communication capability (e.g., PDAs, phones, music players).

The global division of innovation-related activities within the industry is characterized as follows: component-level R&D (concept design and product planning) is performed in the United States and Japan; applied R&D and development of new platforms (particularly notebook computers) take place in Taiwan; and product development for mature products (mainly desktop computers) and a majority of production and sustaining engineering are performed in China.

U.S. PC firms have benefited from the international division of labor in innovation that has supported rapid innovation and quicker integration of new technologies into new products. The growing demand for smaller, more mobile products plays to U.S. firms' strengths in product architecture and early stage development. The shift in production activities away from the United States has pulled new product development activities to Asia, but design jobs, which are relatively few in number, are expected to remain largely in the United States.

Software

U.S. firms dominate global trade in both packaged software products and software services, although their leadership position is weakening in software services. Important non-U.S. providers of software services are located in India (software services), Ireland (software logistics, localization, and development), and Israel (product development and R&D). Despite some change in the location of leading providers of software services, there has been relatively little change in the location of new software product development. Inventive software development activity (at least as measured by patents) is concentrated in the United States and is controlled by U.S. firms. Some inventive activity by U.S. firms has shifted abroad but represents a relatively small share of the overall inventive activity of U.S. firms.

The importance of repeated interaction between software developers and users is especially important in the early stages of complex projects, and the enormous size and sophistication of the U.S. market for software and services means that the United States is likely to retain its dominant role in this industry for some time to come. Nonetheless, some software development activities will continue to move to offshore sites characterized by lower labor costs and high-quality manpower. The chapter highlights the importance for the future U.S. software industry of federal support for computer science R&D in industry and academia in the face of continued upgrading of the capabilities of offshore sites for R&D and product development.

Semiconductors

Significant structural change occurred in the U.S. semiconductor industry during the 1990s. Among the most important changes were shifts in markets for semiconductor component applications, changes in the location of production and geographic structure of markets for semiconductor components, and increased vertical specialization in industry structure. Despite these changes, the contributions of "offshore" sites to U.S. semiconductor firms' innovation-related activities remain surprisingly modest. For example, process R&D remains predominantly "homebound"—concentrated in global semiconductor firms' home countries. The patenting activity of U.S. and non-U.S. semiconductor firms alike is similarly dominated by domestic inventive activity. The innovation-related activities of global firms in this industry remain remarkably "nonglobalized," even in the face of greater international flows of capital and technology, far-reaching change in the structure of semiconductor manufacturing, and significant shifts in the structure of demand.

The vertically specialized industry structure that now characterizes the semiconductor industry has enabled U.S. firms specializing in design and marketing of semiconductor components to access global production networks and grow rapidly. Nevertheless, continued growth in production capacity and design capabilities in Southeast Asia is likely to result in expanded offshore product design and development activity by U.S. firms and the entry of new firms based in this region.

Flat Panel Displays

The flat panel display industry originated in the United States, but production-related activities quickly migrated to Japan, followed by Korea and Taiwan, and now are expanding in China. Innovation in the flat panel display industry has been driven by periodic shifts to larger "form factors" (i.e., larger screens), which affect the design of new products and new manufacturing processes. Innovative activity has tended to follow production investment in the industry because of the high demand for process innovation. The migration of production away from the United States means that U.S. firms play a limited but important role in the innovation process, primarily as suppliers of specialized components (e.g., glass substrates). The innovation-related activities of these U.S. firms remain concentrated in the United States, although all of these firms have invested in offshore R&D and related activities that are located near major customers.

Lighting

The structure, characteristics, and location of innovation in lighting contrast with those of predecessor lighting technologies. The traditional lighting industry, which relied on incandescent, gas-discharge, and fluorescent technologies, was dominated by a three-firm oligopoly (GE, Philips, and OSRAM) based in the United States and Europe. All three firms still operate global production networks for the manufacture of traditional lighting products, but they face increased competition from low-cost Southeast Asian producers in traditional lighting products. Moreover, the growth of markets for lighting technologies based on light-emitting diodes (LEDs) has begun to transform the competitive balance of the industry. The three dominant firms were slow to enter the LED market and have subsequently utilized joint ventures and acquisitions for technological catch-up, but they have failed to achieve market dominance in this technology.

Lighting innovation, which historically was dominated by the U.S. and European R&D facilities of the leading firms, has shifted to Southeast Asia. An analysis of patenting in lighting indicates that Japanese and U.S. firms hold modest leadership positions in inventive activity, but their lead is shrinking as Taiwanese and other Southeast Asian companies improve their R&D capabilities. Most global firms have established Asian-based manufacturing, engineering, and R&D operations, principally in Japan and Taiwan. The lighting industry also has developed a vertically specialized structure, with firms specializing in R&D, production, packaging, and other functions within the value chain.

Pharmaceuticals

The structure of the U.S. pharmaceutical industry has been transformed since 1990 by the rise of biotechnology and intensified competition from global generic manufacturers. Innovative activity in the industry (measured in terms of industry-financed R&D investment and patenting) is concentrated in the United States, some European Union countries, and Japan. U.S. pharmaceutical firms have been leading investors in offshore R&D, which has been concentrated in high-income economies, since the 1980s. Important economic factors, such as localized knowledge spillovers, intellectual property protection, and government policies related to price regulation, state procurement of drugs, and health and safety regulation, have helped to reinforce this geographic concentration of innovative activity. Nonetheless, increasing vertical specialization within the industry, as well as improvements in the scientific and engineering capabilities of nations such as India, has been associated with shifting some innovation-related activities to offshore sites. Thus far, the innovation-related activities that have experienced some movement include manufacturing process innovation and clinical trials management.

In some respects, the trends in offshore movement of pharmaceutical R&D resemble those in the U.S. software industry in the early 1990s. U.S. firms continue to dominate the innovative efforts of the industry, a position that has been reinforced considerably by large public investments in biomedical R&D in academic and government research facilities. The U.S. market is also an attractive site for product innovations, given the minimal price controls that characterize it at present.

Biotechnology

Although it has attracted a great deal of attention from policy makers, investors, and entrepreneurs, the U.S. biotechnology industry employs a relatively small number of individuals overall, and even fewer scientists and engineers. The industry consists of several distinct segments that span biomedical, industrial, and agricultural biotechnology. Development of the industry has been dominated by biomedical applications, for which prices and profits tend to be greatest. The biomedical segment is concentrated in particular regions in the United States and Western Europe, reflecting the importance of the interaction of biomedical biotechnology firms with science-based university research. Although many countries around the world now "host" a biotechnology industry (of varying importance), biotechnology activity within most of these nations is often centered in a single metropolitan area. Nevertheless, an increasing number of distinct locations in the United States and an increasing number of countries support modest to significant biotechnology activity.

Vertical specialization has played an important role in the development of the biomedical segment in particular, since many new firms in this segment serve as "research boutiques," conducting R&D in new drugs that are subsequently developed for commercial purposes by larger pharmaceutical firms. The biotechnology industry itself has experienced considerable vertical specialization, and (as in other industries examined in this volume) the development of a vertically specialized structure has tended to support the globalization of innovation-related activities. Based on its large academic and public biomedical R&D infrastructure, however, the United States remains the dominant location for advanced R&D and product development in the industry, and the growth of offshore R&D and related activities is likely to have a minor impact on U.S. employment in this industry for the foreseeable future.

Logistics

The logistics industry manages the planning and control of the flow and storage of goods, services, and related information between the point of production and the point of consumption. The industry has expanded rapidly since 1990, in parallel with the globalization of manufacturing, and now includes a large number of specialized firms. Innovation in logistics frequently occurs in response to new customer requirements, and this process in many respects resembles the "co-invention" activity that typifies innovation in computer software. Advances in supporting technologies, such as information technology and communications, are another important source of innovation in logistics.

Close interaction between the developers and users of logistics services is essential to innovation, and the global spread of logistics networks has been associated with growth in the offshore innovation-related activities of U.S. and non-U.S. logistics firms. Nevertheless, analysis of logistics-related patents indicates that U.S. logistics firms specializing in information technology (IT)-related software and services remain dominant within the industry. As the logistics industry develops a more global structure, the role of governments in creating and enforcing intellectual property protection, in reducing trade barriers and standardizing import rules, and in supporting the training of managers, engineers, and technicians capable of furthering innovation will grow in importance.

Venture Capital

The venture capital (VC) "industry" in its modern form emerged in the United States during the post-1945 period. Although the industry now operates globally, U.S. firms remain dominant. Globalization of the VC industry has occurred through cross-border partnerships, the establishment by U.S. VC firms of overseas offices, and expansion within the United States by foreign VC firms. This process is likely to continue as countries develop clusters of technological expertise that attract the attention and investments of VC firms throughout the global economy, as multinational corporations acquire more foreign startups, and as financial markets throughout the world develop sufficiently to support the liquidation by venture capitalists of their investments. Indeed, in some important respects, globalization of innovation in VC reflects the growth of innovation-related entrepreneurship in other economies.

The primary focus of investments by U.S. and non-U.S. VC firms has been the IT sector, including semiconductor, computer software, computer hardware, and related industries. Indeed, the historic strength and innovative dynamism of the U.S. IT sector is one factor behind U.S. VC firms' dominant position in the industry. Although the globalization of VC has not had negative consequences for the U.S. innovation system, U.S. VC firms will continue to expand their offshore activities and support the creation of foreign startups that compete directly against U.S. firms in the IT and other high-technology sectors. The effect of these investments will be strongly influenced by most economies' evolving regulatory and legal systems.

Financial Services

The rapid expansion in offshore investment in business process support by U.S. financial services firms has led to increased innovation within these offshore locations by subsidiaries of U.S. firms as well as independent providers of specialized services. The growth of vertical specialization within the global financial services industry also has affected the structure of innovation-related activities. There are strong complementarities between process and product innovation in this industry, but the offshore movement (mainly to Asian countries) of "back-office" functions has supported increased offshore innovation in these processing functions. Firms based in high-income markets such as the United States and Europe remain the primary sources of product-oriented "customer-facing" innovations, but market-mediated interaction between end users and providers is less important to the process innovation activities of many of these offshore sites.

WHAT HAS CHANGED SINCE 1990?

These summaries of structural change in the innovation-related activities of the industries examined in the volume highlight four broader trends: (1) the growth of innovative capabilities in a number of foreign nations that 30 years ago were classified as "developing economies," (2) the growth of sophisticated manufacturing and services-production activities in these and other economies, (3) the growth of demand for cutting-edge technologies (particularly in IT) in markets outside of the United States, and (4) the growth of "vertical specialization" in many knowledge-intensive industries. A discussion of each of these trends follows.

Improved Innovative Capabilities in New Regions of the Global Economy

The first and perhaps most important of these trends is the growth of innovative capabilities in countries such as China, India, Taiwan, and South Korea, none of which were active in R&D or product development for global markets during the 1960s and 1970s. In some of these countries, indigenous firms or subsidiaries of foreign firms are performing fundamental research. In most of them, improvements in innovative capabilities have enhanced the ability of these countries to contribute to the design and development of advanced products, including those in service-based industries such as financial services and logistics. Particularly in India and China, advances in regional innovative capabilities have been associated with growth in domestic scientific and engineering workforces.

With the important exception of India, whose role in the production of software and services has assuredly expanded, the transformation of these countries' innovative capabilities has been linked to growth in the domestic manufacture (in some cases, in foreign-owned facilities) of products for global markets in industries ranging from PCs to automobiles. And just as has been true of innovation in the industrial economies, the growth in innovation-related activities within countries such as India, China, and Taiwan has been associated with regional concentration and agglomeration—Bangalore, Shanghai, and Hsinchu are examples of regional "high-technology" agglomerations in India, China, and Taiwan, respectively.

Expansion of Production Activities Outside of the United States

A second factor in the transformation of the innovation processes in the industries discussed in this volume is the expansion of production activities outside of the United States in these and other regions. The extent and timing of this expansion of offshore production vary among industries (e.g., offshore production is hardly a new feature of the automobile industry, but is less important in biotechnology). In a number of industries, however, ranging from semiconductors to flat panel displays and PCs, U.S. firms rely on sites outside of the United States (through ownership or contracts) for a growing share of their production requirements. Much of this offshore expansion in manufacturing activity has occurred in Asia and Southeast Asia, particularly in China, Taiwan, and South Korea. In the flat panel display industry, growth in Asian production by U.S. firms and the entry into production by Asian firms have "pulled" many innovation-related activities (e.g., process innovation) to Asian sites. Increased offshore manufacturing by U.S. semiconductor firms, by contrast, has had more modest consequences for the location of innovation-related activities. There is little evidence of shifts to offshore locations in the patenting activities of U.S. (or non-U.S.) semiconductor firms, and no evidence of offshore shifts in the location of process-innovation activities of U.S. semiconductor firms.

Growth in global "production networks" in many of the industries discussed in this volume has provided a powerful impetus for the expansion of logistics that has in turn spurred and depended on significant innovation in the logistics industry. Expanding offshore production and product-development networks in industries such as semiconductors and software also has accelerated growth in foreign nations' VC industries.

The Changing Profile of Demand for Advanced
Products in Foreign Markets

Yet another influence on the movement of product design and development activities away from the United States in industries such as software, semiconductors, and PCs is interregional shifts in the scope and sophistication of consumer demand. Consumer markets for wireless and digital devices in countries such as South Korea, for example, are growing more rapidly than are similar markets in the United States. Equally important is the fact that many consumers in these markets (including firms producing advanced electronic-systems products) demand more advanced applications than is true of consumers elsewhere in the global economy. Users play a crucial role in demanding and in some cases developing or "co-inventing" new applications in the aforementioned industries, as well as in logistics. Firms seeking to exploit and develop new applications for these dynamic user-driven markets typically must locate a portion of their product development and design activities within these markets. In industries such as semiconductors, U.S. firms' offshore design activities rely on close contacts with local firms who design and produce the new consumer products that incorporate advanced semiconductor components.

The "product cycle" model that influenced academic analysis of U.S. firms' offshore manufacturing and R&D activities during the 1960s (Vernon, 1966) posited that U.S. firms developed and introduced their most advanced products within their domestic market before marketing and (eventually) manufacturing these products offshore. Although product demand in a number of the industries examined in this volume remains important, several of the most advanced markets in these industries now are located in foreign economies and, therefore, attract increased investment by U.S. firms seeking to develop advanced products. In effect, the product cycle has been reversed, with important implications for the location by U.S. firms of their product development activities.

Increased "Vertical Specialization"

Structural change in the industries examined in this volume has influenced the shifting structure and location of innovation-related activities. Perhaps the most pervasive and important type of structural change, one that is observed in industries ranging from PCs to pharmaceuticals and biotechnology, is vertical specialization—the development of an industry structure populated by firms that specialize in one or a limited set of activities who contract with other firms that specialize in different activities within the industry. For example, one group of firms in the pharmaceutical industry now focuses on drug discovery and contracts with other firms for drug development (e.g., clinical trials) and post-approval marketing. In semiconductors, manufacturing "foundries" collaborate on a contractual basis with "fabless" semiconductor firms that specialize in design and marketing of semiconductor components. This type of contract-based collaboration among specialized firms differs considerably from the operations of firms that are vertically integrated in all functions ranging from R&D through manufacturing to marketing.

In many industries, vertical specialization has developed in parallel with (and in many cases has accelerated) global shifts in production activities. The manufacturing specialists in semiconductors are largely located in Asia, whereas fabless semiconductor firms remain largely based in the United States. Similarly, the systems architecture, software operating systems, and semiconductor components within PCs are designed in the United States, but almost all production activity is located offshore and managed by firms not affiliated with the U.S. semiconductor or software enterprises. Moreover, the offshore manufacturers of PCs rely on specialized suppliers of components ranging from disk drives to displays.

Vertical specialization thus far has had varied effects on the location of innovation-related activities in the industries discussed in this volume. Although flat panel display production is located almost entirely outside of the United States, U.S. firms retain important roles in technology development (including investing in U.S.-based R&D) as suppliers of specialized inputs and equipment. In PCs, vertical specialization has been associated with the geographic separation of manufacturing from high-level design activities. Although some semiconductor design activities have migrated to the East Asian sites where the bulk of specialized semiconductor producers are located, U.S. sites retain an important role in advanced design activities. The location of U.S. pharmaceuticals and biotechnology R&D does not appear to have shifted in response to growing offshore drug production and marketing activities.

In some industries, the factors determining the location of advanced R&D activities seem to differ significantly from those influencing the location of manufacturing. In semiconductors and pharmaceuticals, vertical specialization has supported the formation of new U.S.-based firms whose business models rely on collaboration with offshore manufacturers. Vertical specialization also has aided the growth of innovation and globalization in financial services and logistics by facilitating the complex web of transactions that underpin the structure of these industries. But in other industries, such as lighting, shifts in the location of production have had significant implications for the location of innovation-related activities.

WHAT HAS NOT CHANGED?

Although many aspects of the innovation process in the industries examined in this volume (as well as many others) have undergone significant change since 1990, the broad economic and policy challenges associated with such structural change have changed little. In most of these industries, U.S.-based firms continue to perform the majority of their (most advanced) R&D within the United States. Inventive activity, as measured by the location of inventors for U.S. patents filed by U.S.-based firms, remains remarkably "homebound" in industries such as biotechnology, pharmaceuticals, semiconductors, and software.[2]

---

[2]Indeed, patent-based indicators suggest that the inventive activity of foreign-based firms in these industries also remains concentrated in their home countries.

Nevertheless, in other industries, a growing share of the R&D and inventive activity at the technological frontier now appears to be located outside of the United States. In the lighting and flat panel display industries, non-U.S. firms and nations have become more prominent innovators since 1990. Moreover, even within industries such as software or semiconductors, in which the inventive activity of U.S.-based firms appears to be concentrated in the United States, a substantial (and poorly measured) portion of the design and development of new products has moved offshore, either to exploit lower labor costs or to collaborate more closely with innovative users. As we noted earlier, however, the ability to exploit offshore innovative talent has supported the entry and growth of numerous U.S. firms pursuing new business models and technology strategies.

Thus, economic change has affected the structure of the innovation process in all of the industries studied. The characteristics of structural change in virtually all of these industries resemble those emphasized in the analyses of U.S. competitiveness highlighted earlier: Industries and activities in which U.S. workers (defined in this case to include scientists and engineers) add less value are the most vulnerable to foreign competition and the most likely ones to move to foreign sites. The improved capabilities of scientists and engineers in many of these foreign locations, the identity of these locations themselves, and the changing outlook of demand and growth in the U.S. and foreign markets, however, may be causing more rapid shifts in competitive advantage and affecting a broader range of activities, including innovation-related activities, than in earlier decades. Nevertheless, the fundamental conclusion remains unchanged: For U.S. firms, consumers, and workers to profit from the expanding opportunities in the global economy, their innovative and productivity performance must continue to improve; the U.S. economy must remain open to inflows of goods, technology, and capital; and the infrastructure underpinning the domestic U.S. R&D "system" must remain highly innovative and attractive as a site for investment by U.S. and non-U.S. firms alike.

Another important element of continuity that contemporary analyses of innovation and globalization share with earlier discussions of this topic is the poor quality of the data on which they rely. As the previous STEP Board study (*U.S. Industry in 2000*) noted, restructuring in the domestic and international R&D systems means that conventional R&D investment data are less reliable as a guide to structural change in the innovation process. The R&D investment data collected by the National Science Foundation (NSF) and other public statistical agencies in the United States and other industrial economies arguably do not include a number of the activities (e.g., product design, or spending by firms on acquisitions as a means of gaining access to new technologies or capabilities) that play a central role in the innovation process of the 21st century. Moreover, the NSF data provide limited information on the international dimensions of R&D investment by U.S. and non-U.S. firms. These problems with the R&D investment data have been the subject of a number of studies by the STEP Board, the National Research Council, and other expert panels, but the fact remains that much of the analysis of globalization in the innovation process is hampered by limited, dated, and imperfect data and indicators. These data limitations are especially serious for service-based industries, which have expanded their investment in R&D and offshore production significantly since 1990.

A number of the chapters in this volume rely on patent data to supplement the limited R&D investment data available for their industry. Patent data have a number of advantages, including their disaggregation into specific technology classes, and their reporting of both the assignment and geographic location of the patent owner(s). Nevertheless, patents have important disadvantages as well. They measure inventive activity, which is an important input to the overall process of commercial innovation, but do not measure the output of the innovation process. The coverage by patents of even inventive activity within different industries and technology classes also varies, as does their commercial and economic value among fields of invention. Equally important is the fact that the grant of a patent follows a period of review of the patent application that typically takes at least 18 months and frequently requires 3 to 5 years. Therefore, patent data provide a "retrospective" measure of inventive activity occurring as many as 5 years ago, and this inventive activity itself results from investments in R&D and other activities made still earlier. Although patent data represent a valuable additional set of indicators of innovation-related activities in a much more complex global economic environment, their limitations must be kept in mind.

Yet another area in which the quality of available data makes it difficult to draw definitive conclusions is the effects on domestic scientific and engineering employment from the globalization of innovation-related activities that is occurring in many of these industries. Data on industry-wide employment trends for scientists and engineers in many of these industries (e.g., logistics, venture capital, PCs, software) do not exist, reflecting the complex structure of the industries and the outdated structure of publicly available data on industry employment. Moreover, the central topic of these chapters is not shifts in the location of these industries but shifts in the location of specific functions within these industries. And many of the trends described in these chapters (e.g., greater reliance on advanced information and communications technologies, vertical specialization) facilitate the geographic separation of different activities within industries, rather than the relocation of entire industries.[3]

---

[3]The distinction is an important one, since the gloomy predictions made by Freeman (2005) and others assume that the United States will lose its historic dominance in knowledge-intensive industries as a result of the growing technological and scientific capabilities and workforce in nations such as India and China. What these chapters indicate is that some specific functions (e.g., product manufacture, software coding, product development) may shift to offshore locations. But these shifts need not pull other knowledge-intensive activities in their wake, and in some cases (as in semiconductors) these shifts in location create opportunities for the growth of new firms in the United States. The Freeman predictions cannot be dismissed, although Branstetter and Foley (2007) present a more skeptical view of the current level of MNE R&D and innovation within China. Nevertheless, the trends described in

The chapters in this volume on semiconductors, software, and PCs all conclude that the employment consequences for scientists and engineers from the restructuring of innovation-related activities thus far are modest and not clearly negative or positive. Indeed, leading U.S. firms in these IT-related industries consistently complain about the lack of sufficient immigration visas to hire foreign-born engineers needed to address shortages (Lohr, 2007).[4] It is also difficult if not impossible to separate the "contributions," negative or positive, to scientific and engineering employment of the globalization of an industry's innovation-related activities from myriad other factors.

Overall, therefore, the data underpinning the conclusions of all of the chapters in this volume provide a clearer understanding of the past than they do of the future. Although the nature of the innovation process is such that the near-term future is not likely to differ radically from the recent past, the fact remains that the data underpinning detailed industry studies such as these provide a limited foundation for forecasts.

POLICY ISSUES AND CHALLENGES

The fundamental challenges for policy created by the processes of globalization described in this volume have changed little from those described in studies of this topic that date back to the 1960s and 1970s. To preserve and expand employment in the functions and professions that benefit from the globalization of innovation, the United States must sustain the high levels of innovative performance that have supported the competitiveness of U.S. industry and have made the United States a major destination for R&D investment from foreign firms. Among other things, this goal means that support for the "R&D infrastructure" that decades of public and private investment have created must be strengthened.

---

this volume reflect a different process of economic change. Indeed, it is highly plausible that stronger scientific and engineering capabilities in India and China will produce effects similar to those observed after Japanese and European "convergence" with U.S. levels of innovation-related expertise, as Bhagwati et al. (2004) point out: "When the revival of Europe and Japan brought their skill levels closer to those of the United States, the gains from trade induced by 'factor endowment differences' were increasingly replaced by gains from 'intraindustry' trade; for example, the United States now specializes in high-end chips such as Pentium, while leaving more standard semiconductor chips to foreign producers. Similarly, we can confidently expect 'intraservice' and 'intraindustry' trade to grow between the United States on the one hand and India and China on the other as the latter acquire more skills" (p. 108). Alternatively (and equally plausibly), one may observe growth in "intrafunction" trade within such activities as new product development, based on the same factors.

[4] H-1B visas are given to foreign workers with high-technology skills or in specialty occupations by the Citizenship and Immigration Services Agency. U.S. companies seek H-1B visas on behalf of foreign scientists and engineers to fill hiring shortfalls, but Congress mandates that the Agency cap the number of visas granted to 65,000. Some claim that U.S. H-1B visa policies are counterproductive and detrimental to U.S. technological and economic competitiveness, while others see them as critical to protecting domestic workers.

This infrastructure has supported investment and innovation by U.S. and foreign firms in the industries examined in this volume, as well as others, and has yielded great benefits to U.S. consumers.

Other developing and developed countries now recognize the importance of such an infrastructure and, in many respects, are emulating U.S. policies by making similar public and private investments. A failure by the United States to maintain its commitment to the strength and quality of its public and private R&D infrastructure could limit the benefits for U.S. citizens of the globalization of innovation-related activities described in this volume. Here, as elsewhere, the competitive dynamics should not be seen as a "zero-sum" competition—U.S. citizens benefit from higher levels of R&D investment by foreign governments, just as foreign citizens have benefited from U.S. public R&D investment. But the mobility of innovation-related activities means that the United States must remain an attractive site for these activities by U.S. and non-U.S. firms in order to maintain employment opportunities for skilled personnel.

Beyond sustaining this infrastructure, however, public policies must ensure that government R&D investments yield the highest possible public returns. Achieving this goal means that university-industry research collaboration should be supported by public policy, without imperiling the critical role of U.S. universities as educational institutions that produce world-class scientists and engineers. Any obstacles to such collaboration imposed by shortsighted university patenting and licensing policies also should be reviewed critically by university administrators, industry managers, and policy makers.[5] As Thursby and Thursby (2006) note, one of the most important influences on the location of multinational corporations' advanced scientific research facilities is proximity and access to university researchers. U.S. universities, like U.S. firms, face growing competition from foreign institutions for industry-supported collaborative R&D and must adjust their policies toward intellectual property management accordingly. Policies that limit federal support for academic research on politically sensitive topics such as embryonic stem cells also reduce the attractiveness of U.S. universities as research collaborators and therefore weaken the "magnetic force" of these important institutional assets for R&D investment in the United States from foreign sources.

Users of advanced technologies play an important role in innovation in many of the industries examined in this volume, especially those in the IT sector. Prox-

---

[5]"Largely as a result of the lack of federal funding for research, American Universities have become extremely aggressive in their attempts to raise funding from large corporations. . . . Large US based corporations have become so disheartened and disgusted with the situation they are now working with foreign universities, especially the elite institutions in France, Russia and China, which are more than willing to offer extremely favorable intellectual property terms" (testimony before the Subcommittee on Science, Technology, and Space, U.S. Senate Commerce Committee of R. Stanley Williams, September 17, 2002; statement reproduced at http://www.memagazine.org/contents/current/webonly/webex319.html; accessed April 2, 2005).

imity to sophisticated users is an important factor in the decisions of U.S. firms to locate a portion of their innovation-related activities offshore. An important part of the R&D infrastructure that attracts (or retains) investment in innovation-related activities supports user-driven applications in advanced technologies. One of the most celebrated recent examples of investment in such infrastructure was the public investment in the computer-networking infrastructure (originally referred to as ARPANET) that laid the foundations for the Internet. U.S. policy supported public and private investments in the networking technology and infrastructure, and U.S. trade policy encouraged widespread imports and adoption by users of low-cost desktop computing hardware. These policies helped create a large domestic "testbed" for demanding users of computing technology to develop new applications, which in turn helped propel the explosive growth during the 1990s of commercial investment in Internet-related firms (Mowery and Simcoe, 2002).

One contemporary (and closely related) equivalent to the computer-networking infrastructure of the ARPANET and NSFNET is broadband communications technology, which remains less widely available in the United States than in other (notably Nordic) nations (Turner, 2006). Moreover, differences in such access between urban and rural users depress the size of the domestic market for advanced applications developed on this testbed by innovative users. Broadband access is an indispensable foundation for continued growth in the user-driven innovation that now is prominent in many of these industries. In this area, as well as others affecting the viability of user-led innovation, public policy and private investment should support the development of widely accessible testbeds for sophisticated users to develop new applications and business models. Such an infrastructure could support the development of new firms and industries from domestic sources, investments in related fields from foreign firms, and continued innovation and growth in the U.S. economy.

The broader process of economic globalization, of which the restructuring of innovation-related activities is one part, is on the whole beneficial for the United States. Consumers benefit from higher-quality, lower-cost, and more innovative products; employees benefit from the ability to exploit their skills in a global rather than a domestic market; firms benefit from lower costs and economies of specialization through vertical specialization and increased collaboration; and the processes of trade liberalization can have beneficial political consequences for international relations as well. In addition, of course, literally millions of non-U.S. citizens benefit from the expanded economic opportunities in their home nations provided by the process of economic globalization.

Nevertheless, the distributional consequences of trade liberalization and globalization are significant, and, in a democracy, the political effects of worker displacement and flat or declining wages can intensify resistance to trade liberalization. These concerns are affected much more by relocation of manufacturing and services employment, rather than by change in the structure of innovation-related activities, and raise issues that go far beyond the focus of the studies in this volume. Nevertheless, in the absence of more effective policies within the United States to address the legitimate concerns and needs of the domestic economic "losers" from globalization, political resistance to policies seeking to further liberalize international flows of trade and investment seems likely to grow. And such political resistance has the potential to undercut the globalization in innovation-related activities that has proven highly beneficial to U.S. and non-U.S. citizens alike.

ACKNOWLEDGMENTS

Research for this chapter was supported by the Alfred P. Sloan Foundation, the Ewing M. Kauffman Foundation, and the Andrew W. Mellon Foundation. A portion of David Mowery's research was supported by the National Science Foundation (SES-0531184).

REFERENCES

Bhagwati, J., A. Panagariya, and T. N. Srinavasan. (2004). The muddles over outsourcing. *Journal of Economic Perspectives* 18(4):93-114.

Branstetter, L., and C. F. Foley. (2007). Facts and Fallacies about U.S. FDI in China. National Bureau of Economic Research Working Paper #13470.

Freeman, R. B. (2005). Does Globalization of the Scientific/Engineering Workforce Threaten U.S. Economic Leadership. National Bureau of Economics Working Paper 11457, pp. 1-47.

Lohr, S. (2007). Parsing the truths about visas for tech workers. *New York Times*. April 15.

Mowery, D. C., and T. S. Simcoe. (2002). Is the Internet a U.S. invention?—An economic and technological history of computer networking. *Research Policy* 31:1369-1387.

National Academy of Sciences, National Academy of Engineering, Institute of Medicine. Committee on Prospering in the Global Economy of the 21st Century. (2006). *Rising above the Gathering Storm: Energizing and Employing America for a Brighter Economic Future.* Washington, D.C.: The National Academies Press.

Thursby, J., and M. Thursby. (2006). Where is the new science in corporate R&D. *Science* 314: 1547-1548.

Turner, S. D. (2006). Broadband reality check II: The truth behind America's digital decline. *Consumers Union* 1-44.

Vernon, R. (1966). International investment and international trade in the product cycle. *Quarterly Journal of Economics* 80:190-207.

1

Personal Computing

JASON DEDRICK
University of California at Irvine
KENNETH L. KRAEMER
University of California at Irvine

INTRODUCTION

August 2006 marked the 25th anniversary of the release of the original IBM personal computer (PC), the product that defined the standards around which a vast new industry formed. Unlike the vertically integrated mainframe industry, the PC industry consisted of a global network of independent suppliers of systems, components, peripherals, and software (Grove, 1999; Dedrick and Kraemer, 1998). The key factor shaping the industry's structure was the design of the IBM PC as a modular, open system with standard interfaces, which allowed many newcomers to enter the market by specializing in one industry segment and developing innovations that could be integrated into any IBM-compatible system. It also permitted producers of parts, components, and systems to achieve global economies of scale as most of the world adopted the IBM standard. In time, desktop PCs were joined by portable laptop/notebook PCs and PC servers as the industry innovated on this common standard.

Today, the core personal computing industry includes not only traditional desktop and laptop PCs and PC servers but also smart handheld devices such as personal digital assistants (PDAs) and smart phones. This core industry is supported by a large number of component suppliers, manufacturing services and logistics providers, distributors, retailers, service specialists, and others. These companies also support other segments of the electronics industry, and so are counted here not as part of the PC industry but as part of its overall production and innovation network. This network not only supports innovation in the core industry segments but also provides the necessary infrastructure for innovations in newer product categories such as ultramobile PCs, MP3 players (e.g., the iPod), and smart phones.

Worldwide revenues for the core PC industry totaled $235 billion in 2005: $191 billion in desktop and portable PCs, $28 billion in PC servers, and $16 billion in smart handheld devices (IDC, 2006a). In addition, PC software accounts for about half of the packaged software industry, whose 2006 sales were $225 billion, and PC use also drives sales of information technology (IT) services and of other hardware such as storage, peripherals, and networking equipment (IDC, 2006c).

The PC has undergone considerable innovation and change since it was first introduced. The traditional PC is no longer expected to be the sole locus of innovation in the future, but simply one of many devices "orbiting the user" (*Economist*, 2006). Communications devices (phones, PDAs) have acquired computing capabilities and people now send e-mail with a BlackBerry or download music on a mobile phone. Digital photos can be transferred from a camera to a PC and uploaded to a website, transferred directly to a printer, or shot and e-mailed with a mobile phone. And although the traditional desktop and laptop PC is becoming less central to all computing activities, over 225 million PCs were sold in 2006 and the PC is often the first place to find innovations that may migrate later to other devices.

As important as product innovation has been, equally important is the steady price declines in recent years, which have brought PCs within the reach of more of the world's population. Emerging markets such as China and India are growing much faster than the more mature developed markets, and PC makers have begun to focus on innovation that addresses the needs of those markets at low prices. Globalization of production has been credited for making computer hardware 10 to 30 percent cheaper than it would be otherwise (Mann, 2003). The availability of ever cheaper, smaller, and more powerful hardware has continued to expand the market and has stimulated ongoing innovation in hardware, software, and services.

Although globalization has been a major factor in the growth and innovation of the PC industry, it raises issues for U.S. companies, government and other institutions, and workers. U.S. PC makers are struggling to eke out a profit in an environment of falling prices and intense international competition. Government policy issues include tax incentives, antitrust, immigration, and market access. Universities must ensure that they are training people with the skills that industry needs, and workers must invest their own time and money to acquire those skills even as more highly skilled knowledge work is moved offshore.

The impacts of globalization have been debated extensively. An optimistic view is that U.S. firms are outsourcing and offshoring lower-end manufacturing and routine engineering work, freeing resources to focus on more dynamic innovation that will sustain profitability and create new jobs in the United States.

A more pessimistic view is that innovation will follow manufacturing offshore, leaving U.S. firms uncompetitive and draining the United States of the innovation that drives growth and employment (Kotkin and Friedman, 2004).

While macro-level data can be useful in analyzing the impacts of globalization, trends and impacts can be easier to spot at the industry level, especially when looking at more dynamic industries where change is happening faster. Personal computing is one such industry. Therefore, this chapter examines the globalization of innovation in the PC industry, its causes, its impacts, and its strategy and policy implications. The focus is mainly on innovation-related activities in U.S.-branded PC companies set in their global context; it is not an analysis of PC companies in other economies such as Japan, Taiwan, or China, although it brings them in as part of the global supply chain and the competitive context.

This chapter is a fact-based analysis grounded in over 200 personal interviews with industry executives in the United States and Asia, data from the International Data Corporation (IDC), Taiwan's Market Intelligence Center, Reed Electronics Research and other sources, published empirical research, and our study of the industry for over 20 years.

We find that the global division of innovation-related activities can be characterized as follows: component-level research and development (R&D), concept design, and product planning are performed mostly in the United States and Japan; applied R&D and development of new platforms mostly take place in Taiwan; and product development for mature products and a majority of production and sustaining engineering are performed in China.

U.S. PC firms have benefited from this international division of labor, which has supported rapid innovation and quicker integration of new technologies into their products. The growing demand for smaller, more mobile products plays to U.S. firms' strengths in product architecture and early-stage development. Their bigger problem is earning profits from innovation in an industry dominated by Microsoft and Intel, who capture very high profit margins thanks to their control of key standards. From the perspective of U.S. knowledge workers, the situation is more mixed. The shift in production away from the United States has pulled many new product development jobs to Asia, whereas design and early-stage development work has remained largely in the United States. Still, the new jobs created by the industry's growth are largely outside of the United States. Finally, consumers in the United States have been clear beneficiaries of the very low cost structure that globalization has produced in PCs as average selling prices have been reduced continually.

Following this Introduction, the structure of this chapter is as follows. The section "Innovation in the Industry" analyzes the nature of innovation in PCs and how production and innovation are organized across the value network. "Changing International Structure of Demand and Supply" describes international trends in PC demand and production. The fourth section, "Globalization of Innovation," reviews the global structure of innovation in the PC industry and the factors driving globalization. "Implications of Globalization of Innovation" considers the implications of the foregoing trends for firm strategy and U.S. national policy.

INNOVATION IN THE INDUSTRY

The PC industry has introduced many innovations in its 25-year history. Product innovation includes the creation of new product categories such as notebook PCs and PDAs, as well as the creation of new product platforms such as multimedia PCs and wireless "mobility" notebooks. The scope and outcome of product innovation in PCs is shaped by the presence of global architectural standards set originally by IBM and now largely controlled by Microsoft and Intel. Common interface standards enable innovators to reach a global market with standard product lines; thus, economies of scale can be achieved to support investments in product development and manufacturing capacity. This is different from other industries, such as mobile phones or video games, in which multiple incompatible standards exist. An example of the benefits of standardization is the acceptance of 802.11 as a common standard, which spurred the introduction of wireless networking as a standard feature on notebook PCs. On the other hand, standardization battles can constrain innovation because PC makers are reluctant to incorporate technologies before a standard is set, as is the case with second-generation DVD technology.

When PC makers do innovate, they face hard choices in trying to capture profits from their innovations. One alternative is to incorporate the innovation only in their own products to differentiate their PCs from those of competitors, but there is a question of whether they can convince customers to pay for the differentiation and also whether customers will want to adopt a nonstandard technology. Another is to license the technology broadly, which might bring in license fees and even establish the technology as an industry standard, but which will eliminate product differentiation. One current example is Hewlett-Packard's (HP's) Personal Media Drive (PMD), a portable hard drive that slides into a special slot in HP Media Center PCs. HP incorporated the special slot into some of its own products, while letting customers connect the PMD to competitors' PCs using a slower USB connection, thus differentiating HP's PCs. By contrast, HP has licensed its LightScribe technology, for labeling DVDs and CDs, to other PC makers. In either case, it can be difficult to translate innovation into profits sufficient to justify the R&D effort.

Despite these challenges, which may discourage more fundamental product innovation, PC makers are pushed to incremental innovation by component makers (such as for semiconductors, storage, or power supply) who introduce frequent changes in their products (faster speed, greater capacity, smaller form factor, longer life) in efforts to gain greater market share within their industry sector. They also are pushed by consumers who want the latest technologies. PC makers feel they have to adopt these often-incremental changes rather than risk being left behind by a competitor that does adopt.

As a result, PC makers have tended to concentrate on operational efficiency, marketing, and distribution rather than trying to use product differentiation as a source of sustainable competitive advantage (Porter, 1996). Product innovation at the system level tends to be incremental and emphasizes developing slightly different products for narrowly defined market niches, such as PC gamers who demand high performance or business travelers who desire ultralight notebooks, rather than more distinctively innovative products.[1] Instead, most product innovation occurs upstream in components and software, which are then incorporated by PC makers.

Consistent with the emphasis on efficiency and distribution, the industry has introduced business process innovations such as outsourced manufacturing, using the Internet as a direct sales channel, vendor-managed inventory, third-party logistics, and build-to-order (BTO) production. At the plant level, some firms have replaced assembly lines with small production cells to facilitate BTO production and have adopted process improvements such as reducing the number of steps and improving quality in final assembly. They also have employed a range of information technologies such as shop floor management systems, bar coding, and automated software downloads to improve manufacturing performance (Kraemer et al., 2000). However, while early adoption of these innovations benefited some companies, particularly Dell Inc., competing PC makers have since adopted these and other process innovations and closed the gap on key measures such as inventory turnover and time to market for new products (Dedrick and Kraemer, 2005). Today, most companies use a mix of build-to-forecast and BTO processes that is optimal for their targeted markets. The result is greater efficiency in the industry as a whole, but the biggest benefits have not gone to the PC makers. They have mostly gone to consumers in the form of lower prices, and to Microsoft and Intel, as software and microprocessors account for an ever greater share of the total cost of a PC.[2]

To understand innovation in the industry, it is important to look at the structure of the innovation network, the innovation processes, the key personal computing products, and interdependencies among innovation processes, products, and the structure of the network.

---

[1]An exception is Apple, which emphasizes attractive design and close integration of hardware and proprietary software in its products. While this has been very successful in its iPod line, Apple's market share in PCs is under 4 percent worldwide, so it is unclear that its innovative PCs have done more than satisfy a small core of Mac users who are willing to pay a premium for its products. By adopting Intel processors for all of its products, Apple has abandoned its proprietary hardware platform in favor of global economies of scale and greater compatibility with Windows PCs.

[2]Even these two face challenges: Intel from AMD and Microsoft from Linux in one product category (servers).

The Innovation Network

The PC industry's innovation network consists of component makers, contract manufacturers (CMs) and original design manufacturers (ODMs), branded PC firms, distributors, and resellers (Figure 1).[3]

The industry can be characterized as horizontally specialized, with the branded firms as the "system integrators" doing design and outsourcing development and production to CMs or ODMs. There are less than a dozen globally competitive PC makers and many smaller local assemblers, supported by another dozen major CMs and ODMs. There are several major suppliers of most key components (e.g., motherboards, hard drives, displays, optical drives, memory, and batteries). Farther upstream in the supply chain, there are several thousand suppliers of less expensive parts and components, most of which are small- and medium-sized firms. Distribution is mostly decentralized and local, although there are a few large distributors who operate internationally such as Ingram Micro, Tech Data, and Arrow Electronics. Our main focus in this chapter is on the branded PC vendors and ODMs who collaborate to bring new products to market using components from upstream suppliers.

Most R&D is done upstream in the industry—by the suppliers of microprocessors, software, peripherals, and components. This innovation is global in the sense that there are major component makers in the United States (microprocessors, graphics, memory, hard drives, networking, software), Japan (liquid crystal displays [LCDs], memory, hard drives, batteries), Korea (LCDs, memory), and Taiwan (LCDs, memory, optical drives, power supply, various peripherals). However, although some companies have set up R&D labs around the world, most R&D is still done in the home country. Some PC makers such as HP, Toshiba, Sony, and Samsung also make components and peripherals, but these are generally done in separate business units who sell to competing PC makers as well as their internal PC units.

The pace of this upstream innovation is a major factor shaping innovation by branded PC vendors who innovate through "systems integration." The PC vendors identify new product markets and design systems that incorporate new technologies to serve those markets. For instance, PC makers identified mobile PC users who want network access without having to plug into a phone line or local area network. This capability was made possible when wireless networking technologies such as WiFi were introduced by component makers. It was then up to PC makers to incorporate the technology into their products. More impor-

---

[3]The terms contract manufacturer and original design manufacturer are used commonly, but not always consistently, in the electronics industry. *Contract manufacturers* provide a range of manufacturing services, including subassembly, final assembly, logistics, and even customer service. *Original design manufacturer* is a term coined in Taiwan when its contract manufacturers began to offer product design and engineering as well as manufacturing of notebooks, motherboards, and other products.

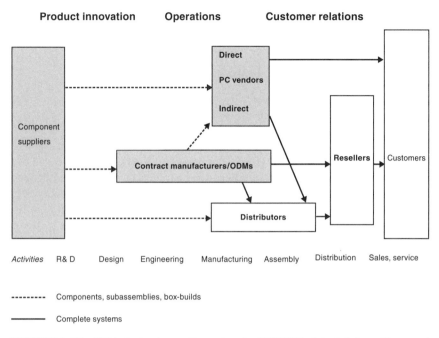

FIGURE 1 The PC industry innovation network. SOURCE: Adapted from Curry and Kenney (1999).

tant, they had to introduce a new technology at a time when the infrastructure to support wireless networking was nearly nonexistent, hoping that this would create the impetus for firms and consumers to invest in wireless networks. Apple initially jumped in by incorporating 802.11 wireless technology in all of its notebooks, and was soon followed by other PC makers. Soon, wireless networks were available in offices, homes, schools, airports, and coffee shops around the world. Apple's early decision was very risky, as there were few networks available, but taking the risk helped to create the market for them.

The creation of new markets by PC makers, in turn, can shape the direction of upstream innovation in components. For wireless notebooks, PC vendors had to decide which networking standard(s) to incorporate as well as find components with low power consumption, longer battery life, and light weight. Available components seldom meet all these needs, so the lead PC vendors each developed their own product roadmaps, which signal to the component suppliers where the firm is headed, the target markets and expected volumes, and the price and performance of components needed to succeed. By doing so, they provided advance knowledge to the upstream suppliers who could respond in terms of feasibility, aggregate demand across PC vendors, plan for the coming changes, and inform their own suppliers. These PC maker roadmaps, which are different from those provided by Intel and Microsoft to the PC makers, are essential to knowledge integration along the supply chain.

Innovation Processes

Product innovation in the industry occurs through two broad processes—R&D and new product development. *R&D* is an ongoing activity that generates knowledge that can be applied to multiple products. *New product development* is a multistage process of design, development, and production that creates physical products for target markets.[4] Although conceptually distinct, there is often a close interaction between the two in practice. New product development integrates knowledge developed by R&D, and R&D is often called on to solve a specific problem in product development. Given that most R&D is done upstream by the component suppliers, the process of knowledge integration occurs between the supplier and the PC maker. The focus is on knowledge needed to integrate a standard component, but occasionally it involves customization or even more intensive joint development. This is especially the case when an entirely new product is being created, such as the wireless notebook that requires integration of communication technologies, or in the case of a new product category such as the Apple iPod.

Products and Innovation Activities

Although new form factors are emerging, desktops and notebooks remain the leading products in the industry, with important differences between them that affect innovation activities. For *desktops*, product innovation mainly centers on conventional systems integration—incorporating new parts, components, and software into a system and ensuring that they work together. The system is largely standardized with respect to components, parts, and interfaces. So innovation involves the selection of components to be included for different target markets (e.g., home, office, game, "value" or "power" user). Most use a standard full tower or midtower chassis with industrial design applied mainly to the bezel (face) to reflect a certain brand image. A few newer models aimed at consumers' living rooms have moved away from the "beige box" to smaller and more stylish designs with unique chassis and industrial designs. PC vendors generally keep concept design and product planning in-house for close control over brand image, user interface, features, cost, and quality. Outsourcing of physical development has occurred in a series of steps since the mid-1990s—first motherboard design, then mechanical design, system test, and finally software build and validation.

---

[4]A detailed discussion of these phases and the activities within each is provided by Dedrick and Kraemer (2006b).

Intel facilitated this trend by providing support and reference designs to ODMs who develop motherboards and full systems.

For *notebooks*, innovation involves high-level system integration with complex mechanical, electrical, and software challenges. Design of such a small form factor presents special challenges with respect to heat dissipation, electromagnetic interference, and power consumption, while the need for portability requires greater ruggedness. Although components such as disk drives and flat panels are mostly standardized, notebooks involve many custom parts. For example, to fit the modular components within the notebook chassis, the motherboard and battery pack may have to be customized for each notebook model. The chassis and other mechanical parts require custom tooling.

PC vendors usually keep notebook design in-house but coordinate physical development jointly with the ODM because there is a strong interdependency between the physical product development and manufacturing. It is critical that product development take manufacturability into account from the beginning; otherwise a product may be developed that cannot be produced at the necessary volume, cost, or quality. Most notebook PCs are designed to be built in a particular assembly plant with specific manufacturing process requirements. As a result, product development and final assembly are almost always handled by one company. In some cases, this means the PC maker keeps both in-house. In most cases it means outsourcing both development and manufacturing of each model to a single ODM.

Thus, the interdependencies of PC form factors and new product development (NPD) activities have led to different organizational arrangements for desktops and notebooks (Figure 2). Because desktops are less complex and more standardized, a complete product specification can be handed off for development and production to ODMs, or a fully developed product can be turned over to a CM for manufacturing. However, because of their greater complexity and customization, notebooks tend to be designed and developed jointly by the PC vendors and ODMs.

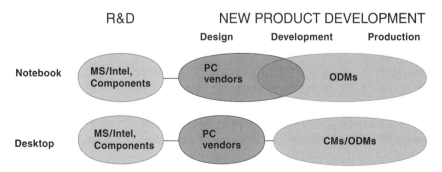

FIGURE 2 Organization of innovation for desktops and notebooks.

As a result of the interdependencies in notebook PC development, leading PC makers HP and Dell have set up design centers in Taiwan to work closely with ODMs, whereas others frequently send staff from the United States. The ODMs may divide product development and manufacturing between Taiwan and China but keep very close interaction between the two locations. For desktops, it is easier to separate development and manufacturing geographically as well as across firm boundaries.

CHANGING INTERNATIONAL STRUCTURE
OF DEMAND AND SUPPLY

Trends in Demand

PC demand has been shifting steadily for over a decade toward smaller, more integrated, and more communications-oriented products. The global demand for PCs is changing in terms of form factor, commercial versus consumer markets, and regional consumption. Portable devices (laptops and notebooks) are the fastest growing form factor, totaling 32 percent of unit demand in 2005 compared to just 10 percent in 1990 (Figure 3), and are expected to exceed desktops in the next 5 years (IDC, 2006b). Other portable devices such as smart phones have seen rapid growth as well. This means that there will be more demand for complex

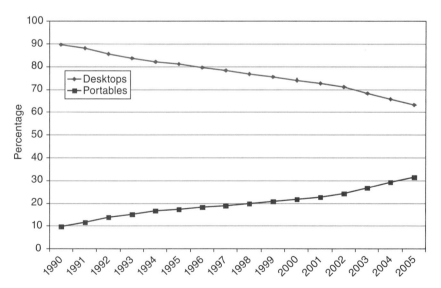

FIGURE 3 Global demand for desktops and portables, 1990-2005 (percent of units sold). SOURCE: Juliussen (2006).

innovation in concept, design, and engineering in the future and that coordination among these stages will have to become closer.

Continued price and performance gains in key components as well as the shift of production to lower-cost locations have driven prices lower, expanding overall demand for PCs. One impact is in consumer markets, whose share of the total market increased from 28 to 38 percent between 1994 and 2005 (Figure 4). Another impact is in emerging country markets where economic growth is providing the income to afford these ever-cheaper PCs. Although North and South America are still the biggest market in the world, followed by Europe, the Middle East, and Africa (EMEA), the Asia-Pacific region is the fastest-growing market (Figure 5). The United States is the single largest market, with 61 million units shipped in 2005, but fast-growing China has surpassed Japan as the second biggest market.

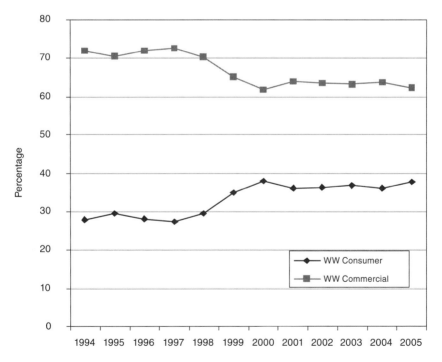

FIGURE 4 Global PC consumption by commercial/consumer markets (percent of units sold). SOURCE: IDC (2006d).

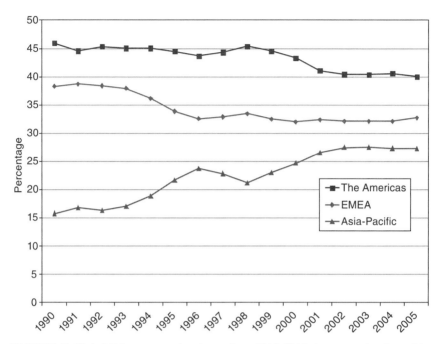

FIGURE 5 Global PC consumption by region, 1990-2005 (percent of units sold). SOURCE: Juliussen (2006).

Geographic Location of Production

With desktop PCs, final assembly by the branded vendors historically was located close to end-user demand because of logistics (they are too heavy to ship affordably by air) and greater customization for national or regional markets. Major PC vendors such as IBM, Compaq, HP, Apple, and Gateway initially had their own production facilities in each world region, but they later outsourced production to CMs such as SCI, Flextronics, Solectron, Mitac, and Foxconn (the registered trade name of Hon Hai Precision Industry Co.), starting in the late 1990s. Dell kept final assembly in-house, but it outsourced base unit production, including chassis with cables, connectors, drive bays, fans, and power supplies. Japanese and Asian vendors generally kept production in-house.

As the branded PC vendors moved offshore and then outsourced, there was a shift in the location of production from the Americas and EMEA to the Asia-Pacific region (Figure 6). Initially, production was spread throughout East Asia in Japan, Malaysia, Singapore, Taiwan, and Korea. Production of desktop base units and various components and subassemblies by Taiwanese companies shifted to the Pearl River Delta in Southern China, but final assembly was usually done regionally: in the United States and Mexico for the Americas, in Ireland

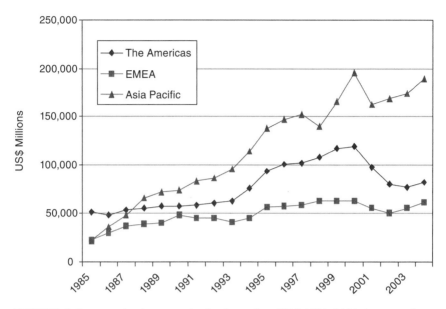

FIGURE 6 Computer hardware production by region, 1985-2004. 2004 data are a forecast. The graph includes parts and subassemblies such as base units that are specifically produced for use in computer equipment. SOURCE: Reed Electronics Research (2005).

and Scotland for EMEA and Malaysia, and in Taiwan and China for the Asia-Pacific region.[5]

Some U.S. companies outsourced notebook production to Japanese, Taiwanese, and Korean manufacturers but eventually shifted mostly to Taiwanese ODMs. In 2001, the Taiwanese government changed investment limitations for Taiwanese firms and the notebook industry moved en masse to the Yangtze River Delta near Shanghai.[6] Japanese firms such as Toshiba moved their own notebook production to the region to take advantage of the supply base, but they also outsourced much of their production. Chinese firms such as Lenovo used these same supply bases for their own production and outsourced some as well.[7]

---

[5]These locations are now changing once again. For example, Dell is moving final assembly and suppliers to Poland for EMEA; both Dell and HP are encouraging their CMs to move to India for the Asia region; and Dell is setting up final assembly in India.

[6]Some notebook ODMs and suppliers moved to the area as early as 1998 so there was already a supply base when most of the industry moved. For example, Asustek had 300 employees in China in 1999 and 45,000 by 2005 (Einhorn, 2005).

[7]This was the case with the IBM PC Company and Lenovo both before and after their integration.

By 2005 China was the single largest producer of PCs and computer equipment in the world. Although the production facilities were located in China, they were mostly owned and managed by Taiwanese firms, such as HonHai/Foxconn and Mitac for desktops, and Quanta, Compal, Wistron, and Inventec for notebooks.[8] The supply chain was also composed largely of Taiwanese firms. Foxconn has a huge facility in Shenzhen that employs over 100,000 workers and produces base units and complete systems for nearly every branded PC vendor, while also assembling products such as game consoles and iPods and making components such as cables, connectors, chassis, and motherboards. Taiwanese ODMs produced 85 percent of all notebooks in the world in 2005 (Table 1), mostly in the Shanghai/Suzhou region of China.

In the past, the location of final assembly was driven by the need for proximity to demand in the United States and Europe but now appears to be driven by growing demand in Asia as well as by the growing capability of firms to exploit lower costs for labor, land, and facilities, the availability of cost-effective skilled labor, and government incentives in China.[9] For instance, low-cost sea shipment of standard (not BTO) desktop PCs from China to the United States, supported by more sophisticated demand forecasting and planning tools, allows PC makers to build a 3-week shipment time into the new product introduction cycle. Notebooks can be economically shipped by air, so even BTO production can be centralized in Asia. Also, with most of the supply chain in Asia, it can be cheaper to assemble there and minimize shipment time for components because the supply base is concentrated there.

GLOBALIZATION OF INNOVATION

The location of NPD activities by the branded PC firms is driven by the product and process interdependencies discussed earlier, the capabilities and relative costs of different locations, and relational factors that tend to "pull" innovation outside the PC vendor or offshore. The relative capabilities and costs of U.S. firms and those in other countries have resulted in a new global division of labor: higher-value architectural design and business management, along with associated "dynamic" and analytical engineering work, is done in the United States, whereas the development and manufacturing of the physical product, along with the more routine, "transactional" product and process engineering, is done in Taiwan and increasingly in China. The result is that both component and system innovation is increasingly global, but U.S. firms continue to play leading roles in both.

---

[8]After IBM sold its PC Division to Lenovo, only Dell (among the U.S. PC companies) had its own final assembly plant in China. Dell's largest assembly site in Asia is still in Penang, Malaysia.

[9]Dell is the only U.S. PC maker who still assembles desktop PCs in the United States; most final assembly of notebooks is centralized in Malaysia. The subassemblies come from the Pearl River Delta (desktops) and the Yangtze River Delta (notebooks) in China. Dell also does final assembly in China and other major markets.

Capabilities and Cost

The *design* of desktops and notebooks involves understanding markets and customer demand, as well as technology trends, anticipating how customer demand and technology trends are converging, and coordinating mixed teams of marketing people and technologists. It requires people with skills and experience in high-level architectural design, with the associated dynamic engineering skills, industrial design, and business and product management.[10] In terms of proximity, it is important to be located in leading markets where new technologies are developed and adopted first.

Development for desktops or notebooks involves more routine, transactional product and process engineering. Therefore, it requires people with mechanical, electrical, and software engineering skills and technical project management experience. In addition, notebook development requires specialized skills in thermal and electromagnetic interference, shock and vibration, power management, materials, radiofrequency, and software. These require a combination of formal training and experience working in a particular engineering specialty, as well as working on the specific product type.

Such knowledge and skill levels vary significantly in different locations due to at least three factors: (1) historical industrial development leading to creation of specialized skills, (2) output of educational systems, and (3) the nature of demand, including market scale and the extent to which the local or regional market may be described as cutting edge, with demanding and innovative customers.

In the United States, there are business skills such as market intelligence and product management that are hard to find elsewhere. There are also leading industrial design firms that specialize in small electronic products such as notebooks and cell phones, and strong software and high-level engineering skills. These skills are taught in universities, invested in by leading domestic firms in the industry, and honed through proximity to leading-edge users.

In Japan, there are industrial designers that are very good at designing for the Japanese market, but who also have experience designing for global markets. Japanese engineering teams have deep skills in design and development, with specialties such as miniaturization that have developed to meet Japanese demand for small, lightweight products. Japan also is very strong in process engineering and manufacturing operations, thanks to its historical and continued emphasis on manufacturing.

In Taiwan, mechanical and electrical engineers are available with strong

---

[10]Gereffi and Wadhwa (2006) distinguish between dynamic and transactional engineers, a classification that we find useful in characterizing the engineering workforces in different countries based on our interviews. Dynamic engineers are capable of abstract thinking and high-level problem solving using scientific knowledge and are able to work in teams and work across international borders. These engineers have at least 4-year degrees in engineering and are leaders in innovation. Transactional engineers have engineering fundamentals but not the skill to apply this knowledge to larger problems. They usually have less than 4-year degrees and are responsible for rote engineering tasks.

practical experience as well as formal training. Taiwan's historical specialization in the PC industry, and with notebooks in particular, has created a pool of engineers with a great depth of knowledge of these products. Taiwan also has strong process and manufacturing skills. These have developed over time as Taiwanese firms have taken on greater responsibilities in PC development and manufacturing. Taiwan mostly lacks marketing skills and industrial design skills that would allow it to take over the concept and product planning stages, because of its focus on original equipment manufacturer/ODM production rather than development of branded products.

China has many well-trained mechanical and electrical engineers, but most lack the hands-on skills that come with experience. Industrial design is weak, and marketing and business skills are very underdeveloped. A large number of engineers are produced each year, but quality varies greatly by university. According to one interviewee, China's engineers "work perfectly at doing what they have been told, but cannot think about what needs to be done; they lack both creativity and motivation. They are good at legacy systems, but not new things; they can't handle 'what if' situations."

In comparing cost across countries, the average salary for electronics engineers in all industries in the United States is about $80,000, compared to $60,000 in Japan, $20,000 in Taiwan, and under $10,000 in China (Dedrick and Kraemer, 2006b). Obviously there are cost advantages to moving engineering to China, but differences in productivity related to education and experience can negate the direct cost differences. Also, it is reported that engineering salaries are rising quickly in China, especially in industry clusters such as the Shanghai/Suzhou area, as multinationals and Taiwanese firms compete with domestic companies for talent. The willingness of multinationals to pay higher salaries gives them access to more experienced engineers and graduates of top universities, but turnover rates are high.

Based on a survey of Taiwanese PC and electronics firms, Lu and Liu (2004) found that the main reason these companies were moving R&D (primarily development) to China was the availability of well-educated and cost-effective local engineers. This finding is supported by our own interviews with Taiwanese companies. As Taiwan's supply of engineers has failed to keep up with demand, the attraction of a large pool of engineers with both linguistic and geographical proximity has been strong. This has enabled Taiwanese engineers to concentrate on more advanced development activities while lower-value activities such as board layout and software testing have moved to China.

The New Global Division of Labor

This confluence of product and process interdependencies with changing capabilities and costs in different locations has led to a new global division of labor (Figure 7). In 1990, the entire NPD process was located in the United States (and

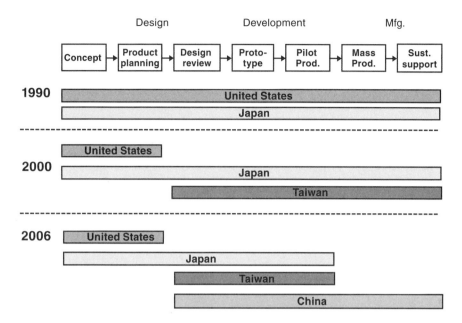

FIGURE 7 New global division of labor in the PC industry.

Japan) in large vertically integrated companies like IBM, HP, Digital Equipment Corporation, and Toshiba, or PC specialists like Apple, Compaq, and Dell, which handled virtually all elements of system-level design and integration. By 2000, only design remained in the United States, while development and manufacturing of notebooks was outsourced mainly to Taiwan and manufacturing of desktops outsourced to major world regions. Japanese PC firms still kept NPD in-house, at least for higher-value products.

In 2006, the U.S. position was unchanged. However, PC vendors like HP and Dell had set up design centers in Taiwan to manage NPD for some products (usually more mature product lines). Locating design in Taiwan allows closer coordination with CMs and ODMs and potentially speeds up NPD, allowing better quality control and problem resolution. They also use these design centers to transfer knowledge to the ODMs and to train locally hired hardware and software engineers to take on more project management and advanced development activities. This division of labor is similar for notebooks and desktops, although some U.S. companies keep desktop development in the United States and then outsource manufacturing to Asia. However, desktop development is being shifted to Taiwanese ODMs in many cases.

The next critical development was the rapid shift of production to mainland China. Encouraged by U.S. PC vendors, Taiwanese manufacturers had moved the production of desktops and many components and subassemblies to the Pearl

River Delta near Hong Kong in the 1990s. Even more dramatic was the shift of notebook production to the Shanghai/Suzhou area after 2000. Many Taiwanese suppliers to the notebook industry had moved to China before 2001. When the Taiwanese government lifted its restrictions on notebook production in China, the ODMs and the rest of their local suppliers moved nearly all of their production to the mainland (Dedrick and Kraemer, 2006a).

In response to U.S. PC makers outsourcing production to Taiwanese ODMs in China, the Japanese PC makers also shifted significant production to China, both through their own subsidiaries and through outsourcing to the Taiwanese ODMs. This further illustrates the compelling economics of the production bases in China as Japanese firms have previously tended to keep production in-house, either in Japan or in Southeast Asia.

China's Expanding Role as a Locus of Innovation

As a result of "production pull" as well as the large pool of lower-cost engineering skills, there is an ongoing shift of product development activities from Taiwan to China. During our interviews with notebook makers in Taiwan and China, one major ODM told us that they did all of their board layout and most packaging design in China, while doing mechanical engineering and software engineering in Taiwan. They were in the process of training people in their electronic engineering methods in China in order to move more development there. As one manager said, "China is a gold mine of human resources, but if you don't get in and train them you won't be able to take advantage of it."

It is expected that more of the NPD process and the associated engineering tests will be conducted in China by many notebook makers (Dedrick and Kraemer, 2006a). These will be relocated from Taiwan and, in some cases, Japan. The shift of product development to China is distinguished not only by which activities have moved or are moving, but also by the type of products that are being developed. Some ODMs are moving product updates to China. However, the development of completely new products and platforms is still done by the ODMs in Taiwan, or by PC makers such as Lenovo (for Thinkpad notebooks) and Toshiba in Japan. More recent interviews with Taiwanese companies suggest that they are hesitant to move these activities to China. This is due in part to the high turnover rate of engineers in China, which makes it hard to create cohesive development teams and also raises the risk of intellectual property loss. Also, unless intellectual property protections are strengthened, China is not likely to become a center for advanced component-level R&D (e.g., in microprocessors, LCDs, or wireless technologies).

A near-term division of labor for product development is likely to be as follows: component-level R&D, concept design, and product planning in the United States and Japan; applied R&D and development of new platforms in Taiwan; and product development for mature products, and nearly all production and sustaining engineering,[11] in China. It is difficult to estimate how long this division of labor will last. A recent study of Taiwanese manufacturers (Li, 2006) shows that the rapid growth of low-margin outsourcing business from foreign multinational corporations has provided Taiwanese firms with the resources and motivation to invest more in R&D to develop greater technology expertise and capture more high-value design work. As the ODMs' expertise grows, multinational corporations have greater incentive to outsource more design activities to further lower costs. Li also shows that Taiwanese firms are attempting to capture value from their innovation efforts by filing for more patents. So the shift from Taiwan to China may be slowing but the shift from the United States to Taiwan could continue.

In addition, Taiwanese manufacturers such as Acer, Asus, BenQ, D-Link, and Lite-on have developed their own brand-name PCs, motherboards, monitors, networking equipment, smart phones, and other products. Acer and Asus brands have captured 14.1 percent of the world market for notebooks (Digitimes, 2006), whereas D-Link has become the top seller of wireless routers for the consumer market. As these companies enhance their R&D, design, and marketing capabilities, U.S. companies may find Taiwan to be a source of competition as well as cooperation.

As China gains experience, it is still possible that the ODMs will shift more of the development process and newer products there, but, unless it becomes a key final market for PCs, it is not likely to capture the market-driven functions of concept design and product planning. As of now, China's PC market is still only about one-third the size of the U.S. market and does not have leading-edge users who are defining what features and standards are developed for the global market. However, as China's PC market continues to grow, and its users become more demanding, it may become the leading market at least for the Asia-Pacific region, and definition and planning of products suitable for the region may be done there. Finally, while Chinese brands remain minor players in the global PC industry for the most part, this may change. Chinese companies such as Lenovo, Huawei, and Haier are already leading brands at home and are expanding to international markets for PCs, network equipment, and other electronics products.

---

[11]Sustaining engineering is the second of two phases in production; the first is mass production. *Mass production* involves the physical manufacturing of a product in large volumes. It requires manufacturing engineers to manage and plan the production process and test facilities and quality engineers to continually improve product and process quality. Over time, these engineers come to know the product extremely well and are best positioned to provide sustaining engineering support that was previously provided by the original product development teams. *Sustaining engineering* deals with changes that occur because of new chips, failing or end-of-life components, or improved components. Each change must be evaluated in terms of its implications for system performance and assembly, and incorporated into the production process. The sustaining engineers also provide the highest level of technical support when problems occur during use during a product's 2- to 3-year warranty period.

Lenovo's acquisition of IBM's PC business has put it directly in competition with HP and Dell around the world, while Huawei uses its relationship with 3Com to access technology and markets and compete with Cisco and others. These companies can use the supply base of Taiwanese and foreign companies in China to match the multinationals on cost, develop products that fit the local market, and then target other emerging markets where innovations developed for the Chinese market are likely to be attractive.

Measurement of the Globalization of Innovation

Measuring the globalization of innovation is more difficult than measuring globalization of manufacturing, which can be captured in national production, trade, and foreign investment accounts. Innovation might be indirectly measured by R&D spending and employees, patents, and new product introductions. While some public data on these measures are available, often the data are not sufficiently disaggregated at the firm level so that they can be tied to a product line such as PCs. This is especially true of multidivision firms such as HP, Fujitsu, Toshiba, Hitachi, Samsung, and Sony. Also, firm-level data do not show the extent to which R&D or other innovative activity is carried out in the home country or other locations.

Given these difficulties, an alternative approach is to measure the innovation effort by the CMs and ODMs who are doing much of the manufacturing in the industry. The share of global notebook shipments produced by Taiwanese ODMs rose from 40 percent in 1998 to 85 percent in 2005 (Table 1). Since manufacturing and development are usually outsourced together, this suggests that the share of offshore product development activity has increased proportionately. This trend is supported by data showing that R&D spending by Taiwanese ODMs and CMs increased significantly from 2000 to 2005 (Table 2), as did the proportion of employees with Ph.D. and master's degrees in these firms. However, most of this R&D spending is on the development side rather than the research side.

Also, reiterating a point made earlier that most innovation is done by upstream component makers, the R&D spending by the ODMs and CMs, as well as nearly all of the PC makers, is minor in comparison to that of upstream suppliers. For example, Table 3 shows that in 2005 some of the lead PC makers[12] spent 1.4 percent of revenues on R&D on average (weighted), the leading ODMs and CMs spent 1.3 percent, and the upstream suppliers, which is where innovation occurs in the PC industry, spent an average of 11.8 percent, or nearly nine times greater than the PC makers, ODMs, and CMs.

---

[12]We could not get public estimates of R&D investment for the PC divisions of large multidivision companies such as HP, Fujitsu, Toshiba, Sony, and NEC, so they are excluded from the table.

TABLE 1 Taiwanese Notebook Industry Share of Global Shipments, 1998-2005

	1998	1999	2000	2001	2002	2003	2004	2005
Shipments volume (thousands)[a]	6,088	9,703	12,708	14,161	18,380	25,238	33,340	50,500
Global market by volume (thousands)	15,610	19,816	24,437	25,747	30,033	37,857	46,110	59,411
Taiwan's share of global market volume	40%	49%	52%	55%	61%	66%	72%	85%

[a]Shipments by Taiwan-based firms, regardless of location of production.
SOURCES: For 1998-2004, MIC (2005); for 2005, Digitimes (2006).

Industry-Level Drivers of Globalization of Innovation

The globalization of innovation in the PC industry has been driven primarily by economic factors and secondarily by relational factors that involve interdependencies of activities, as well as social networks that often influence the choice of suppliers or location. Examples of relational factors include the close interdependence between development and manufacturing of notebook PCs, and the "guanxi" social networks that link Taiwanese firms and managers.

TABLE 2 R&D Investment by Taiwanese ODMs and CMs (million U.S. dollars)

Company Name	2000	2001	2002	2003	2004	2005
Quanta	27.13	38.36	54.55	74.31	92.56	102.36
Compal	24.77		44.69	62.11	70.21	78.78
Wistron			61.12	55.06	68.94	72.49
Asustek Computer	31.97	40.57	53.14	65.87	97.38	128.57
Mitac	24.37	24.70	25.28	32.66	36.90	46.62
Inventec	30.75	25.14	27.38	39.42		48.56
Arima	13.42	12.74	14.85	15.00	19.60	16.71
ECS	3.58	7.20	21.03	14.98	12.74	11.00
First International Computer (FIC)	28.21	10.91	46.72	44.58		
Clevo	8.71	8.10	8.97	9.28	10.28	10.05
Twinhead	7.24	5.31	1.10	0.31	0.43	0.47
Uniwill	7.27	8.20	9.89	11.15	11.55	12.48
Foxconn (HonHai)	32.43	58.14	64.45	66.69	128.78	132.86
Subtotals	239.85	239.37	433.17	491.42	549.37	660.95

NOTE: Blank cells occur where data was not available in annual reports or elsewhere.
SOURCE: Annual reports of the companies.

TABLE 3 R&D Investment as Percent of Firm Revenues, 2005

PC Makers	R&D as % of Revenue	Taiwan ODMs & CMs	R&D as % of Revenue	Component Suppliers	R&D as % of Revenue
Dell	0.9	Quanta	1.1	Microsoft	15.5
Apple	3.8	Compal	1.4	Intel	13.3
Gateway	n.a.	Wistron	1.6	AMD	19.6
Lenovo	1.7	Asustek	1.7	ATI Technology	14.7
Acer	0.1	Mitac	2.0	Seagate (HDD)	8.5
		Inventec	1.4	Western Digital (HDD)	6.6
		Arima[a]	2.8	Maxtor (HDD)	7.5
		ECS[a]	1.6	Chunghwa (Displays)	3.4
		FIC[a]	n.a.	Tatung (Displays)	2.6
		Clevo[a]	4.2	AU Optronics (Displays)	2.2
		Twinhead[a]	0.2	Molex (Cables/ connectors)	5.2
		Uniwill[a]	1.6	Delta (Power supply)	4.8
		HonHai	1.0	Creative (Sound cards)	6.7
Total firm revenues (millions)	$92,535		$76,191		$128,773
R&D (% of revenues) for selected firms (weighted)	1.4		1.3		11.8

NOTE: Large multidivision PC makers like HP, Toshiba, Sony, Fujitsu, and NEC are omitted because R&D investment is not available by division.

[a]Value calculated from data in company annual reports.

SOURCE: Electronic Business Top 300 (2006), unless otherwise indicated.

Regarding economic factors, the manufacturing of desktops was primarily pushed offshore to major world regions to reduce production cost, and secondarily for proximity to markets. Manufacturing was then outsourced to CMs as most PC makers looked to further cut costs and concentrate on product design, branding, sales, and marketing. These CMs are currently moving to new locations within each region (Eastern Europe for EMEA, Mexico for North America, and China for Asia-Pacific)—once again to reduce costs. As noted earlier, for standard build-to-stock desktops, production is increasingly done in China for the U.S. market, because low-cost shipping by sea is viable when fast order turnaround is not necessary.

Cost was also the key factor for notebooks, where both development and manufacturing were outsourced or offshored almost from the beginning—first to Japan, then to Taiwan, and currently to China. Japan's capabilities with development and manufacturing of small form factors provided an initial pull, but lower costs, development of strong indigenous engineering capabilities, and the fact that Taiwanese firms were considered less likely to compete directly with U.S. firms resulted in U.S. PC vendors shifting to Taiwan. In turn, Taiwan has moved manufacturing to China for lower-cost labor, and manufacturing is now pulling some development activities to China as well. Taiwan is trying to expand its role in R&D, design, and other high-value activities, and PC vendors have facilitated this through continued outsourcing and by setting up design centers in Taiwan.

Regarding relational factors in the PC industry, it appears that once production moves to a low-cost location, it will pull some higher-level activities to it. Reinforcing our findings about production pulling knowledge work, Lu and Liu (2004) found that the second major location factor for R&D (after access to low-cost engineers) is proximity to the manufacturing site. This is particularly true for notebook PCs given the importance of design-for-manufacturability. For example, production engineering and sustaining engineering clearly benefit from proximity to manufacturing, because production problems can be addressed immediately on the factory floor and engineering changes in existing products can be tested in production models from the assembly line. It also makes sense to move pilot production to China rather than to maintain an assembly line in Taiwan just for this purpose. Then the question arises whether to move the expensive test equipment from Taiwan to China. If so, then there is more reason to relocate the design review and prototype processes as well.

Beyond proximity considerations in manufacturing, there is a relational "pull" from the ODMs. They often bundle development with manufacturing in order to win contracts. But once the ODM has a contract, the relationship creates incentives for the PC maker to work with the same ODM for future upgrades and enhancements to the product. In addition, there is a great deal of tacit knowledge created in the development process that is known only by the ODM, which creates a further pull. Finally, the close linkage of development activities to manufacturing and the feedback to design from manufacturing has created linkages that favor continuing the ODM relationships.

The concentration of product development and manufacturing in Taiwan and China has reduced cost and accelerated new product innovation, driving down average unit prices, and helping to expand markets. For example, the worldwide average unit price for a PC and monitor has declined markedly over the past 15 years (Figure 8), with desktops and notebooks selling at an average of under $1,100 and $1,400, respectively, in the United States in 2005, and many models

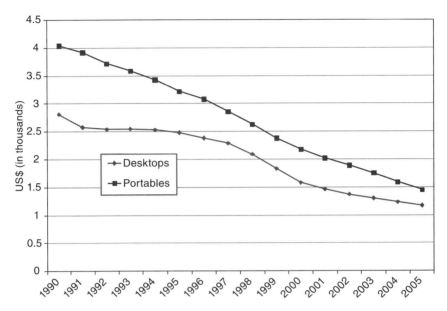

FIGURE 8 Average unit price, desktops and notebooks, 1990-2005. SOURCE: Juliussen (2006).

available for well under $1,000. Of course, when adjusted for quality improvements, the price decline is much more dramatic. Moreover, the price differences between the United States and other regions have declined so that there is now effectively one world price.

Beyond cost reduction, the globalization of innovation also has been driven by a desire to develop a better understanding of the needs of big emerging markets such as China, India, and Brazil to enable the right versioning of existing products. Some PC vendors and ODMs (as well as other suppliers like AMD, Intel, and Microsoft) are seeking new markets in less-developed economies by developing new PCs with much lower price points while also tailoring the technologies to the more extreme environments of these countries. These new product concepts include the One-Laptop-Per-Child design, Intel's Classmate PC, and Asus's eeePC. While previous efforts to develop very-low-cost PCs for developing countries have failed, PC makers and others continue to experiment with new designs.

IMPLICATIONS OF GLOBALIZATION OF INNOVATION

The globalization of innovation has led to a new global division of labor as described earlier. This new international structure of the PC industry has implications for firm competitiveness and strategy, location of innovation, employment, and U.S. policy.

Implications for U.S. Firm Competitiveness

Overall, the changes in the industry appear not to have hurt the competitiveness of U.S. firms. U.S. companies dominate key components such as microprocessors, graphics and other chips, and hard drives, and PC vendors Dell, HP, and Apple hold nearly 40 percent of the world market for PCs. U.S. firms are still unquestioned leaders in operating systems and packaged applications. On the other hand, Asian firms are leaders in displays, memory, power supplies, batteries, motherboards, optical drives, and other components and peripherals. Asia has some leading PC brands such as Lenovo, Toshiba, Acer,[13] and Sony, and Taiwan's CMs and ODMs increasingly compete with U.S. contract manufacturers for outsourced development and manufacturing. On another measure of firm competitiveness, the largest share of industry profits flows to U.S. companies, particularly Microsoft and Intel, but also to Apple, Dell, HP, and to component makers such as Nvidia, TI, and Broadcom. The profitability of most Japanese and Asian companies is generally lower.

Implications for Firm Strategy

For branded PC vendors, the international innovation network described earlier enables faster product cycles with quicker integration of new technologies because the Taiwanese companies are good at fast turnaround and there is a good supply of cost-effective engineers in Taiwan and China to handle more models, changes, and upgrades. It has increased consumer choice, helped grow the market, and for a long time was advantageous for Dell because its direct model gave it an advantage in getting those products to the business customer. But now that most firms are efficient in minimizing inventory and getting new products into the market, the fast product cycles could be seen as an expensive race to the bottom that no PC vendor or component supplier really wins (except Intel and Microsoft).[14] Some PC vendors complain that component innovation is too fast,

---

[13]Acer, which has been a successful Taiwanese branded company, purchased Gateway Computer and Packard Bell in October 2007.

[14]As desktop PCs in particular have become commoditized, business model innovations such as direct sales, BTO, and just-in-time inventory have provided temporary advantage in the industry. They provided an initial advantage to Dell and Gateway, who were the first to adopt direct sales, but Gateway stumbled badly and Dell's efficiency advantage has been reduced as other PC vendors have gone to direct BTO sales. The Dell model also has proved less successful in overseas markets where direct sales are less popular than in the United States. The most important impact of past business model innovation has been a general improvement in the efficiency of the industry as a whole, as most vendors have adopted these practices.

and they feel pressured to introduce too many products for too small markets. For example, one major PC vendor introduces around 1,000 different consumer desktop SKUs (stock-keeping units) in one year globally (Dedrick and Kraemer, 2006b). A question raised by more than one company that we have interviewed is whether the cost of managing so many products might outweigh the benefits of being able to offer products that more closely match the needs of customers.

Beyond desktop and notebook PCs, the growing demand for new products that are smaller, are more mobile, and integrate new functions is bringing new innovation and new players into the personal computing industry. Hit products such as RIM's BlackBerry and Palm's Treo have been developed by firms with no traditional PC business, while Apple's iPod was developed on an entirely different platform from the Macintosh computer line. Such radical or architectural product innovation (Henderson and Clark, 1990; Utterback, 1990) has important differences from the incremental model of development as illustrated in Table 4. The scale and scope of global collaboration is often greater for radical innovation, as existing technologies are adapted to new uses and new technologies are developed. As a result, there is greater need for joint development with partners, while key technologies (particularly software) are developed internally and the entire process is shaped by strong central vision, integration, and control.

An example of the nature of radical innovation is the iPod, which was developed by Apple in collaboration with many external partners in multiple geographic locations. Apple used its internal capabilities to create a closely integrated hardware and software design, while relying on outside partners for both standard and custom components, and for manufacturing. For instance, Apple used a reference design and worked jointly with PortalPlayer to develop the microchip that controlled the iPod's basic functionality. It worked with others for additional chips (e.g., United Kingdom's Wolfson Microelectronics for the digital-to-analog sound chip; New York-based Linear Technology for power management chips; California-based Broadcom for a video decoder chip); with Toshiba for the 1.8-inch hard drive; and with Taiwan's Inventec for manufacturing (Murtha et al., forthcoming).

Apple designed the system architecture that affected critical features such as sound quality and power consumption and developed the distinctive industrial design of the iPod; it developed most of the iPod and iTunes software in-house or adapted others' software. Apple tightly managed the whole process, coordinating closely with outside partners so that it could design the iPod, and its manufacturer and suppliers could concurrently prepare the tooling and supply chain for large-volume manufacturing, and bring it to market in 8 months. As put by the iPod's lead engineer, "Today, there is too much complexity in products for one person or organization to understand. You need a team of internal and external resources working with you to conceive, design, and implement new products" (Murtha et al., forthcoming). The resulting design process is much different from

TABLE 4 Features of Incremental and Radical Innovation

	Design	Development	Production
Radical innovation (iPod, iPhone, Treo)	—Set system architecture, sometimes building on external reference design —Strong central vision and industrial design —Tightly control all aspects of NPD —Develop key software internally —Integrate hardware, software, even services (e.g., iTunes, iTMS) —Design or license complementary assets (SW, content) and distribution system —Collaborate closely with a few key partners for core components	—Collaborate with many partners in multiple geographies —Collaborate with partners of partners —Get partners to adapt existing technologies to proprietary architecture	Outsourced to CM or ODM
Incremental innovation (desktops, notebooks)	—Innovate on Wintel architecture —Control product planning, brand image, marketing, concept design internally —Internal or outsourced industrial design —HW and SW are modular —Leverage existing complementary resources and distribution	—Collaborate with one established ODM in one geography —Outsource detailed physical design, test, and software built within standard architecture	Outsourced to ODM

that in PCs, with more internal development and much closer interaction with key component suppliers.

Finally, for the iPod to be successful in the market, Apple created a new business model that integrated hardware, software, and online content delivery. It developed iTunes software to collect and manage content on a PC or Mac and easily transfer that content to the iPod. It also developed the online iTunes Music Store and tightly integrated that with the iTunes application. Apple licensed content from all the major music labels and subsequently from the audio book, movie, and television industries, and established pricing and digital rights models that were attractive to consumers. The result was a U.S. market share of over 70 percent in both the personal music player and the music download markets.

Given that such design innovation has the potential for creating differentiation in products and gaining competitive advantage, the strategies of at least some branded PC firms are likely to focus more on creating new product platforms. However, examples such as the iPod, Treo, and BlackBerry suggest that radical innovation requires a different process of new product development. As illustrated by our earlier discussion of these innovations, elements of the process include leveraging a firm's unique internal capabilities with those of external partners; working closely with external partners in multiple geographies; engaging in a global search for technologies that can be adapted and integrated into new products; maintaining tight architectural and managerial control over the process; and possibly introducing new business models to provide complementary content and services.

This kind of process is far removed from the incremental innovation within a well-established product architecture and the mature market of the Wintel PC world. As a result, it has been more diversified companies such as Samsung and Sony, wireless specialists such as Nokia, as well as many startups that are trying to innovate with new product platforms that mix communications, entertainment, and computing capabilities in smaller form factors. In these cases, firms have worked with outside partners to exploit external sources of knowledge while keeping their own innovative activities mostly in-house and close to their home base.

Increasingly, hardware-software integration is becoming important as a means of tailoring products to different market requirements such as communications standards, power consumption, language, and customer tastes. Such integration also helps to reduce product costs by enabling standard physical platforms to be produced in large volumes for global sales. More important, it enables greater product differentiation for ever-finer market segments by customizing through changes in software, rather than through costly physical changes in hardware.

Location of Innovation

Innovation at the national level is closely tied to the presence of both technically skilled and entrepreneurial individuals, the quality of infrastructure, and the presence of advanced users who drive firms to innovate. Rapid diffusion of Internet infrastructure in the United States led to ongoing innovation in hardware (e.g., routers, switches), software (e.g., browsers, search engines), and services (e.g., online retailing, banking, stock trading, travel services). The United States has seen strong user-driven innovation (Von Hippel, 1998) such as IT-enabled business process redesign and e-commerce in the corporate world and user-created content in the consumer world. From Cisco and Amazon, to Dell and Wal-Mart to Google and MySpace, innovation on the web has largely occurred in the United States.

By contrast, the relatively slow adoption of broadband and advanced mobile technologies in the United States has left the country falling behind in new areas of innovation. For instance, South Korea is a leader in online computer gaming, thanks in part to its widespread deployment of cheap broadband Internet service. Japan's iMode system for mobile Internet was years ahead of similar services in the United States. High rates of wireless adoption have benefited firms from South Korea, Japan, and Northern Europe, while China's large mobile phone market has attracted firms such as Motorola, Nokia, and Siemens to do product development there. In short, the lack of innovation in industries that are providers of complementary assets (which in turn may reflect the outmoded infrastructure underpinning the large and otherwise highly sophisticated U.S. domestic market) is a major factor hampering innovation in the PC industry. If the United States is to retain its position as a leading market for computing innovation, it cannot afford to remain behind in providing high-quality, low-cost infrastructure to support user-led innovation and drive demand for new personal computing products.

Our field interviews indicate that design innovation, especially concept design and product planning, is likely to remain concentrated in the United States for the major U.S. firms in the personal computing industry. However, there will be increasing use of offshore R&D and design centers in locations that have specialized and cost-effective talent, that lead in particular technical innovations, or that represent important markets in terms of growth potential, special market opportunities (fewer regulatory requirements, government incentives), or challenges (need for cheaper or environmentally friendly PCs), or that may influence technical standards (as China is trying to do in a number of technologies). Private interviews with industry executives indicate that the primary motivation for such offshore outposts is cost reduction, through hiring less costly engineers, programmers, and managers to perform activities previously performed in-house in the United States or in a foreign subsidiary. In time, secondary benefits may also arise as these locations gain capabilities or as local markets develop.

Other product development activities tend to be pulled by production, beginning with manufacturing process engineering, then moving up to prototyping and testing and eventually electrical, mechanical, and software engineering. These are in the process of shifting to China from Taiwan and Japan, although R&D, design, and development of the newest generation of products is still likely to be concentrated in the home countries of the manufacturers (Dedrick and Kraemer, 2006a).

Impacts on Jobs and Employment

With respect to U.S. workers, much of the potential shift of jobs offshore has already taken place with the offshoring and outsourcing of production from 1990 to 2005. There has also been a shift in innovation-related jobs after 2000, as production has pulled development and some design activities to Asia (Dedrick and Kraemer, 2006a). Further movement of jobs offshore is likely to occur in the future to meet competitive pressure for continuous cost reduction. The jobs will be in engineering, software, industrial design, engineering management, and project management at all levels. As one PC industry executive told us in interviews, he has to "push" more physical design and project management jobs overseas in order to keep concept design jobs at home.

The number of jobs directly moved offshore is not large and occurs incrementally. However, another indicator of the impact of offshoring is the number of new jobs that are created offshore rather than in the United States to support the industry's continued growth and proliferation of products. One indicator of this impact is the growth of knowledge jobs in the notebook industry in Taiwan as these firms take on more design and development activities for the United States and other firms. Interviews and company data on the top ODMs in the notebook industry indicate that they hired thousands of new R&D personnel and product engineers in Taiwan between 2000 and 2005, while also hiring thousands more for product and process engineering, testing, and production in China. For example, Quanta, which is the largest notebook ODM, has increased the number of R&D engineers from 750 in 2001 to around 7,000 in 2005 (company annual reports).

As software becomes an increasingly important part of new PC products, there will be a proportionately greater increase in software jobs being moved offshore. In one company we interviewed, 50 percent of the 1,000 employees are engineers and 80 percent of these are software engineers. These jobs are currently in the United States, but the firm is experimenting with offshore teams. While there is broad awareness of the shift of jobs to India and elsewhere by software and IT services companies, there is less awareness of the number of software jobs within the computer hardware industry—jobs that are likewise vulnerable to offshoring.

For the United States, the fact that growth and innovation in the industry are not creating new knowledge jobs (engineering, software, design) in the United States but are creating them in Taiwan and China appears to be a negative. But the number of U.S. engineering jobs in the broader computer industry is fairly stable at about 60,000 between 2002 and 2005 (Dedrick and Kraemer, 2006b), and without globalization there may not be as much growth and innovation. The risks of globalization for the United States are that individuals, firms, or related industries will lose technological advantage and the ability to innovate. A Korn/Ferry International report posed the issue for industry executives as follows:

> North American industrial executives must choose between two fundamental responses to their current competitive environment. One approach is to simply accept that their companies need to focus exclusively on marketing, finance and the design and development functions, while offloading their manufacturing needs and technologies to more accommodating locations, usually overseas. While this strategy can generate short-term profits, it almost inevitably guarantees that a company will lose control of its design and production capabilities.

Eventually, if history is a reliable guide, even home office and corporate functions will cease to exist. (Kotkin and Friedman, 2004)

However, earlier industry innovations as well as recent innovations like the iPod, the Treo, and the Microsoft Xbox were developed mostly in the United States, even though some component innovations came from offshore suppliers and all the manufacturing was done offshore. Moreover, there is little evidence thus far that these firms have "lost control" of the designs or technology for these products. Such innovation is less likely to move offshore and should continue to support engineering and other knowledge jobs in the United States, as long as the United States retains the capabilities needed for such innovation.

Implications for Policy: Sustaining U.S. Innovation Leadership

Although U.S. PC vendors still lead innovation in the industry, they are moving more innovation activities offshore both through setting up design centers and through outsourcing design and development activities to ODMs. The U.S. suppliers of key components such as microprocessors, storage, and software are also setting up R&D and design centers offshore, sometimes in locations with specialized skills such as Israel or Japan, and sometimes in big emerging markets with low-cost engineering talent such as India and China.

The engineering, software development, and management skills associated with these activities are key to the innovation capabilities of the United States and therefore consideration needs to be given to developing people with these skills if such innovation is to remain in the United States (Committee on the Engineer of 2020, National Academy of Engineering, 2005). Our interviews with executives indicate there is a growing need across the PC industry for engineers who are specifically trained to work at the interface between hardware engineering, communications, and computer science. The executives also indicate that many U.S. engineering schools produce specialists in a single engineering discipline, but few schools produce people who can work at the interfaces of these disciplines. There is a need, for example, for hardware engineers who can work with communications standards, and software engineers who can produce embedded software that enables customization of products for markets. When universities fail to develop such talent, firms may rely on on-the-job training, look offshore for experienced people with the needed skills, or develop the skills offshore through on-the-job training of low-cost specialists.

It is also likely that U.S. firms need to make greater efforts to hire rookies and develop them. Several of the companies we interviewed prefer to hire fairly experienced engineers rather than beginners and report no problems in doing so in Silicon Valley or elsewhere. They simply hire people away from other companies, or bring in engineers from foreign countries under immigration policy. However, one highly innovative company we interviewed hired engineers as interns from the best engineering schools in the United States (e.g., Cornell, MIT, UC Berkeley, Carnegie-Mellon) and, if they worked out, made commitments to hire them even before they graduated. Starting as interns, they worked as part of project teams with operational roles and real challenges to overcome. Such on-the-job training can help sustain a career ladder for new engineers as firms offshore more lower-level jobs that would normally be filled by entry-level engineers. An executive for the firm argued that this process benefits the firm as well, by giving it access to the best talent available and the chance to incorporate that talent into product development teams and learn how the company works before the engineers develop bad habits elsewhere.

From a policy perspective, the U.S. government can encourage cross-disciplinary education and more university-industry cooperation through its funding choices, and by documenting and publicizing the need for such changes. While universities are responsive to employer needs, there can be significant inertia in academic departments and university bureaucracies, and external resources and pressure can encourage greater responsiveness and flexibility.

All of the firms we interviewed indicated a need for more H-1B visas, or for reform of the visa process. One issue involves procedures for keeping people who have been educated in the United States and perhaps interned with the firm. Another involves recruiting from abroad for skills for which the U.S. supply of talent is limited, but for which other countries are noted for having people with the needed skills. For example, it appears that the supply of engineers in analog fields in the United States such as radiofrequency is limited, whereas there is a good supply in some European countries. A reported problem with the current immigration process is that the nature of U.S. supply of talent is not considered. From an immigration standpoint, an engineer is an engineer regardless of education level (bachelor, master's, Ph.D.) and there is no way to identify and respond to shortages of very specific skills or levels (e.g., bachelor vs. Ph.D.).

In addition to such human resource issues, another key concern is sustaining the demand for innovation. PC demand, and associated innovation, has been driven in the past decade largely by the Internet and networking in general. With the United States leading in Internet adoption, the PC industry was quick to adopt networking technologies such as Ethernet and wireless networking, and new products such as the BlackBerry and Treo were developed in the United States. However, the United States has fallen behind a number of countries in both wireless and broadband adoption and is not the lead market for products and services such as mobile phones and online gaming. As a result, innovations in new personal computing devices such as smart phones, video game consoles, and other network devices are likely to target foreign markets initially, making it more likely that innovation will occur in those markets rather than in the United States.

While specific policy issues with regard to telecommunications, Internet regulation, content, and pricing are beyond the scope of this chapter, those decisions should be made with an awareness of their potential impact on U.S. innovation in industries such as personal computing. Innovation in PCs can require cooperation by providers of complementary assets, such as content or communication infrastructure. Government policies on telecommunications can influence the speed of diffusion of infrastructure like broadband, 3G, or municipal WiFi networks. Similarly, government policies on copyright can influence the terms under which content can be distributed. While these policy issues are usually debated in terms of impacts on competition, intellectual property rights, or even consumer choice, policy makers also should consider their impact on innovation in high-technology industries.

ACKNOWLEDGMENTS

The research on which this chapter is based has been supported by grants to the Personal Computing Industry Center of the Paul Merage School of Business at UC Irvine from the Alfred P. Sloan Foundation, the U.S. National Science Foundation, the National Academy of Sciences, and the California Institute for Telecommunications and Information Technology. Any opinions, findings, and conclusions or recommendations expressed in these materials are those of the author(s) and do not necessarily reflect the views of these sponsors. The authors acknowledge the very helpful comments of David Mowery, Jeffrey Macher, and anonymous reviewers on drafts of this chapter.

REFERENCES

Committee on the Engineer of 2020, National Academy of Engineering. (2005). *Educating the Engineer of 2020: Adapting Engineering Education to the New Century.* Washington, D.C.: The National Academies Press.

Curry, J., and M. Kenney. (1999). Beating the clock: Corporate responses to rapid change in the PC industry. *California Management Review* 42(1):8-36.

Dedrick, J., and K. L. Kraemer. (1998). *Asia's Computer Challenge: Threat or Opportunity for the United States and the World?* New York: Oxford University Press.

Dedrick, J., and K. L. Kraemer. (2005). The impacts of IT on firm and industry structure: The personal computer industry. *California Management Review* 47(3):122-142.

Dedrick, J., and K. L. Kraemer. (2006a). Is production pulling knowledge work to China: A study of the notebook computer industry. *Computer* 39(7):36-42.

Dedrick, J., and K. L. Kraemer. (2006b). *Impacts of Globalization and Offshoring on Engineering Employment in the Personal Computing Industry. Report for the National Academy of Engineering.* Irvine, CA: CRITO.

Digitimes. (2006). *ICT Report—1Q 2006: Taiwan's Notebooks.* Taipei: Digitimes Research.

Economist. (2006a). The dream of the personal computer: The PC's 25th birthday. 380(8488):17.

Economist. (2006b). Getting personal: The PC's 25th birthday. (July 27):57-58.

Einhorn, B. (2005). Why Taiwan matters. *Business Week.* May 16.

Electronic Business Top 300. (2006). *Electronic Business* August:3-14.

Gereffi, G., and V. Wadhwa. (2006). Framing the Engineering Outsourcing Debate: Placing the United States on a Level Playing Field with China and India. Duke University, Master of Engineering Management Program.

Grove, A. (1999). *Only the Paranoid Survive.* New York: Time Warner.

Henderson, R. M., and K. B. Clark. (1990). Architectural innovation: The reconfiguration of existing product technologies and the failure of established firms. *Administrative Science Quarterly* 35(1):9-30.

IDC (International Data Corporation). (2006a). *Worldwide Black Book, Q2 2006.* Framingham, MA: IDC.

IDC. (2006b). *Worldwide PC Market: 4Q05 and 2005 Review.* Framingham, MA: IDC.

IDC. (2006c). *Worldwide Software 2006-2010 Forecast Summary.* Framingham, Mass.: IDC.

IDC. (2006d). *IDC Worldwide Quarterly PC Tracker.* December 2006.

Juliussen, E. (2006). *Worldwide PC Market: May, 2006 Version.* Arlington Heights, IL: eTForecasts.

Kotkin, J., and D. Friedman. (2004). Executive leadership in the industrial economy. Korn Ferry International. Available at: http://www.kornferry.com/Library/ViewGallery.asp?CID=865&LanguageID=1&RegionID=23.

Kraemer, K. L., J. Dedrick, and S. Yamashiro. (2000). Dell Computer: Refining and extending the business model with information technology. *Information Society* 16:5-21.

Li, W. C.-Y. (2006). Global Sourcing in Innovation: Theory and Evidence from the Information Technology Hardware Industry. Los Angeles: UCLA. Working paper. 46 pp.

Lu, L. Y. Y., and J. S. Liu. (2004). R&D in China: An empirical study of Taiwanese IT companies. *R&D Management* 34(4):453-465.

Mann, C. (2003). Globalization of IT services and white collar jobs: The next wave of productivity growth. *International Economics Policy Briefs* PB03-11:1-13.

MIC (Market Intelligence Center). (2005). The Greater Chinese Notebook PC Industry, 2004 and Beyond. Taipei: Asia IP Report

Murtha, T., F. Giarratani, and T. Sturgeon, eds. (Forthcoming). *Massive Coordination: Creating and Capturing Value in Global Knowledge Networks.*

Porter, M. (1996). What is strategy? *Harvard Business Review* (November-December):1-22.

Reed Electronics Research. (2005). *Yearbook of World Electronics Data.* Vols. 1-4.

Utterback, J. M. (1990). Radical innovation and corporate regeneration. *Journal of Research Technology Management* 37(4):10-21.

Von Hippel, E. (1998). Economics of product development by users: The impact of "sticky" local information. *Management Science* 44(5):629-644.

2

Software

ASHISH ARORA
Carnegie Mellon University
CHRIS FORMAN
Georgia Institute of Technology
JIWOONG YOON
Kyung Hee University, Seoul, South Korea

INTRODUCTION

The global movement of software services activities (defined to include software engineering services and research and development [R&D] as well as the development of software products) to locations outside of the United States is an important and growing phenomenon that has recently attracted widespread attention. Over the period 1995-2002, exports of business services and computer and information services grew at an average annual rate of over 40 percent in India and at a rate of 20 percent in Ireland. These changes have received widespread attention within the United States and have led to concerns of a "hollowing out" of the American information technology (IT) sector and about the potential loss of American technological leadership.

However, despite these changes in the location of production of IT services, there is relatively little evidence of global changes in the location of new software product development. U.S. companies have historically been and continue to be the leading exporters of software products. Moreover, evidence from software patents suggests that inventive activity in software continues to be concentrated in the United States. In the short run, the United States will continue to enjoy a significant lead over other countries in the stock of highly skilled programmers and software designers that provide it with an advantage in the production of new software products. Moreover, proximity to the largest source of IT demand and potential agglomeration economies arising from proximity to competitors and complementors provide software product companies located in the United States with a significant advantage.

DISPERSION OF INVENTIVE ACTIVITY IN SOFTWARE

In this chapter we provide evidence on the geographic distribution of inventive activity in software. Economists have long made a distinction between innovation and invention in the study of technological change. Schumpeter (1934) defined innovations as new, creative combinations that upset the equilibrium state of the economy. Mokyr (2002) defines invention as an increment in the set of technological knowledge in a society. Schumpeter pointed out that invention does not imply innovation, and that it is innovation that provides capitalism with its dynamic elements. Because it is more easily measured, in this chapter we will focus on the geographic dispersion of inventive activity. However, we adopt the position of Mokyr (2002), who argues that in the long run invention is a necessary precursor to innovation.

Unlike some of the other industries studied in this volume, one feature of software development is that it is frequently performed both by suppliers of software packages and services and by users. As a result, software development occurs throughout all industries in the economy, and so to understand the location of inventive activity in software it is insufficient to examine where one or two industries are located.

To understand this point further, it is helpful to gain a better understanding of the types of software development activity. The design, installation, implementation, and use of software consist of several phases. Messerschmitt and Szyperski (2002) identify two distinct value chains in software development. First, there is a *supply value chain* in which software creators develop software artifacts that provide value for the end user. This part of the software value chain consists primarily of design and development activities that can be thought of as software "production." In the past this role had been played primarily by independent users, third-party programmers, or independent software vendors creating custom software, but over the past 20 years this role has passed increasingly to independent software vendors creating software products.

The output of this value chain contains all of what we would traditionally define as software products, such as word processors, operating systems, enterprise software such as enterprise resource planning (ERP) and business intelligence software, as well as middleware software, such as some transaction processing middleware and enterprise application integration. The total value of production in the software product industry was $61,376.9 million in 1997,[1] and 195,200 persons were employed in this industry in the same year.[2] Firms that operate in

---

[1]Data from the U.S. Bureau of Economic Analysis input-output tables. This figure includes the total value of products made in NIPA industry 511200 (Software Publishers); 1997 is the latest benchmark year for the input-output tables. More recent years do not separate software producers from other information publishers.

[2]Data from the Bureau of Labor Statistics (BLS) on the number of employees in the software publishing industry (NAICS 5112), available at http://www.bls.gov/ces/home.htm.

this value chain include all of the well-recognized names that are traditionally regarded as "software" firms, including Microsoft, Adobe, Oracle, and the SAS Institute, as well as smaller firms such as Oblix and Primatech.

This value chain also includes the activity of third-party firms involved in custom programming and software analysis and design. Such firms create custom software products for their customers and include firms like CIBER, Inc., Intergraph Corp., and xwave Solutions. The total value created in custom programming and design services was $115,834.6 million in 1997 while total employment was 675,000 in 1997, indicating that both revenue and employment in this sector are greater than that in the packaged software industry.[3] Moreover, custom programming and design services are also growing faster than is the software publishing industry. Though 1997 is the last year for which we have data on revenues by industry, we can compare employment growth across these two industries. Employment in custom programming and design services has grown from 675,000 in 1997 to 1,025,300 in 2005, for an average annual growth rate of 5.8 percent. In contrast, employment in software publishing has grown from 195,200 in 1997 to 238,700 in 2005, for an average annual growth rate of 2.5 percent.

Second, there is a software *requirements* value chain in which users add functionality to software to meet their own needs. Users engage in co-inventive activity (Bresnahan and Greenstein, 1996) to translate general-purpose software into a specific application. Such co-inventive activity may include modifications to packaged software applications or development of new applications. However, in business software it also involves changes to business processes or organization design.

Activity in this value chain includes both programming by professional programmers and software designers employed by IT-using firms and programming activities performed by users themselves. The activity of both groups is difficult to measure but represents a major share of value created. Scaffidi, Shaw, and Myers (2005) estimate that there were approximately 80 million end-user programmers in 2005,[4] compared to 3 million professional programmers. Moreover, occupation data from the United States indicate that over two-thirds of software professionals do not work for IT firms but rather work for IT-using industries.[5] Neither this software development activity performed by users nor the work performed by software professionals working for IT users is measured in any systematic statistics.

---

[3]These calculations are based on total sales in custom computer programming services (NAICS 541511) and computer systems design services (NAICS 541512). This latter category may include activities outside of programming, such as IT systems design and integration. A conservative estimate of the value and employment of third-party custom programming services uses only NAICS 541511 and yields estimates of $86,326.8 million and 522,300, respectively.

[4]This estimate includes those who create user-developed software that is not sold in markets.

[5]Data from BLS Occupational Employment Statistics.

Though systematic evidence is rare, what we do know suggests that economic activity in this value chain is likely to be far greater than that in the supply value chain. According to Gormely et al. (1998), though the typical cost of implementing an ERP application suite is $20.5 million, only $4 million of this cost is related to hardware and software; the rest is due to the costs of implementing and deploying the software within the business. Using data on sales of software products and services in several Western European countries, Steinmueller (2004) estimates that for every €1 spent on software there is an additional €2.36 spent on IT-related business services. However, this estimate is likely a lower bound, because it includes only software services conducted through market transactions and excludes software development activities within IT-using firms themselves.

The importance of the software requirements value chain has two implications for the measurement of where inventive activity in software takes place. First, a large part of value creation in software takes place outside of firms that reside in what is considered the software product industry. The value of this activity goes largely unmeasured in traditional government statistics, as it often occurs as a labor expense within firms developing or implementing packaged software.

Second, it is very difficult to place a precise definition of what exactly constitutes inventive activity in software. Creation and modification of source code is of course one major component, but so are user modification and business process change. Should these latter activities be included as well?[6] Moreover, how should we treat changes to software code that are embedded in IT hardware? Are these hardware or software inventions? As we will discuss next, given available data, a precise estimate of inventive activity in software is probably not feasible. Instead, we provide a variety of metrics that enable us to estimate broad trends and orders of magnitude in economic and inventive activity in software.

In the section "Trends in the Location of Value Creation" we provide evidence of recent trends in globalization of software services. These data provide evidence on globalization of activity in the software requirements value chain and some inventive activity conducted by services firms in the supply value chain, though they will largely miss changes in cross-country software service activities that are undertaken by firms outside of the software services industry. In the section "Empirical Evidence on the Location of Inventive Activity" we use U.S. software patent data to examine changes in the global dispersion of inventive activity in software product development.

TRENDS IN THE LOCATION OF VALUE CREATION

In this section we investigate broad trends in the location of value creation activities in software. We begin with some statistics describing global variation in

---

[6]It is interesting to note that the U.S. Patent Office has struggled with similar definitional issues, within the context of so-called business method patents (Allison and Tiller, 2003).

the exports and imports of software products and services, followed by a qualitative description of recent trends in countries that have been known to be active producers in the market for software products and services.

Statistical Trends

Software Products

Figure 1 shows the percentage of total 2002 software *product* exports and imports by selected Organisation for Economic Co-operation and Development (OECD) countries. The figure shows that among OECD countries the United States continues to be the leader by a wide margin in the export of software products, accounting for 21.7 percent of total software exports. The next closest country is Ireland, which accounts for 16 percent of software exports. However, as we will discuss in further detail, most of Ireland's software exports arise from U.S. multinational companies that utilize Ireland as a base of operations to localize

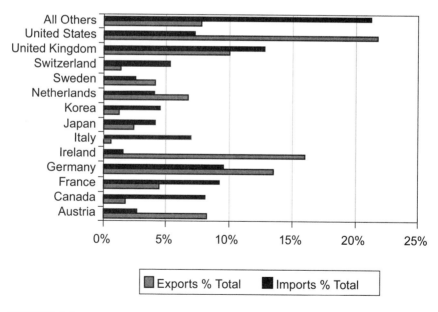

FIGURE 1 Percentage of total 2002 software product exports and imports by OECD country. SOURCE: OECD (2004, Table C.1.8; OECD trade in software goods, 1996-2002). Compiled from International Trade Statistics database.

U.S. software products to be shipped to countries in the European Union.[7] Since the bulk of software product exports from Ireland are due to U.S. multinationals in Ireland—Sands (2005) shows that over 92 percent of Irish software exports are from foreign firms—this suggests that the share of U.S. software exports in global trade flows is probably closer to one-third rather than the one-fifth that the OECD statistics indicate. Following that, the next largest exporters are Germany (due in part to software exports from ERP giant SAP) and the United Kingdom. No other country accounts for more than 10 percent of software exports. Most notably, Japan accounts for only 2.5 percent of total software exports.

Figure 2 presents total packaged software product sales by region. The story here remains the same: North America represents the largest share of packaged software sales, and this percentage has been increasing over time from 47 percent in 1990 to 54 percent in 2001. We explore why other countries have not been more successful in developing software products in further detail in the next section.

Software Services

Figure 3 shows data from the OECD Economic Outlook (2006) and reports the global share of 1995 and 2004 exports in IT services, obtained by summing the categories "computer and information services" and "other business services" from the IMF Balance of Payments data. Though subject to a variety of caveats about measurement and coverage, Figure 3 suggests that the distribution of IT service exports is more evenly distributed across countries than is the distribution of software product exports. Many smaller countries are experiencing rapid growth in their exports of IT services, though some are starting from a very small base.

To explore trends in imports, we use data from the U.S. Bureau of Economic Analysis (BEA) on International Trade in Services. Table 1 provides data on interfirm trade in exports and imports of IT services in 1998 and 2004, calculated by summing the categories "Computer and Information Services" and "Royalties and License Fees."[8] Exports of these services grew from $6,900 million to $10,862 million from 1998 to 2004, while imports grew from $1,992 to $2,591 million from 1998 to 2004.

Cross-border exports to and imports from unaffiliated foreign firms of com-

---

[7]Localization activities include activities such as manual translation or adapting software products to local markets.

[8]The columns labeled "Computer and Information Services" provide data on exports and imports of private services among unaffiliated firms. The columns, "Royalties and License Fees" in the same table include computer-related services that were delivered to foreign markets through cross-border software licensing agreements. These data do not include intrafirm exports of computer services because BEA does not in general release statistics on many of the countries in Table 1. They also do not include wages of U.S. residents who provide computer services to nonresidents.

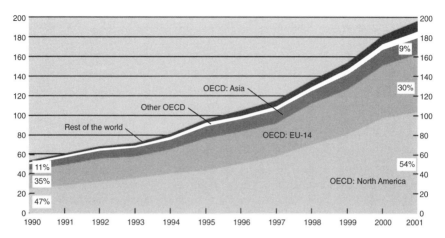

FIGURE 2 Packaged software sales by region, 1990-2001 (U.S. dollars). SOURCE: OECD (2002) using International Data Corporation data. Reported in Thoma and Torrisi (2006).

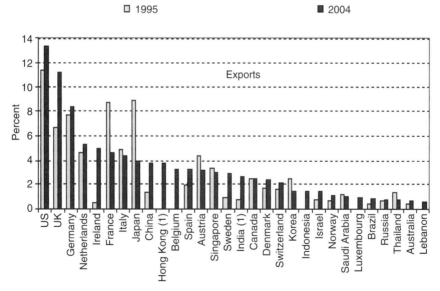

FIGURE 3 Top 30 country shares of reported exports of other business services and computer and information services, 1995 and 2004 (2004 data not yet available for all countries). For Hong Kong (China), India, and the Slovak Republic, data are for 2003. Republished with permission from OECD Economic Outlook (2006). Based on IMF Balance of Payments Database, March 2006.

puter and information services are shown in Table 1. Computer and information services (NAICS 518) include the categories computer and data processing services (NAICS 5181) and database and other information services (NAICS 5182). This table was reorganized based on the tables of Business, Professional, and Technical Services with Unaffiliated Foreigners from BEA. Ireland is included in all other EU and is not identified in BEA's tables. These export and import transactions with unaffiliated foreigners are interfirm transfers, which are traditional trades. Note that "affiliated foreigners" are locally established affiliates of multinational firms. The Asian Tigers consist of Korea, Singapore, Taiwan, and Hong Kong. There are three things to notice about this table. First, at present the numbers are small relative to total U.S. trade in services: exports and imports of software services represent 3.3 and 1.0 percent of total exports and imports of services, respectively. Second, the United States maintains a positive overall balance in trade and services; moreover, over the period 1998-2004 exports of computer services grew at a faster rate than imports (7.86 vs. 4.48 percent average annual growth rate [AAGR]). Third, although imports of computing services from India grew rapidly from 1994 to 2004, overall U.S. imports from India and the other software underdogs are small relative to other estimates.

Data from other sources suggest that the U.S. data may underestimate imports of software services. An OECD estimate indicates that over 90 percent of Indian service exports to OECD countries are not accounted for in the data on service imports published by these countries (OECD, 2004). Other analyses report similar difficulties in tracking Indian software services exports to the United States. A recent General Accounting Office (GAO) report notes that, for 2002, the United States reported $240 million in unaffiliated imports of business, professional, and technical (BPT) services from India, whereas India reported about $6.5 billion in affiliated and unaffiliated exports in similar services categories (GAO, 2005).[9] For 2003, the United States reported $420 million in unaffiliated imports of BPT services from India, whereas India reported approximately $8.7 billion in affiliated and unaffiliated exports of similar services to the United States. The bulk (40-50 percent) of the difference, according to the GAO, is because the United States does not count the earnings of temporary workers resident in the United States in services imports. Other sources include differences in coverage (e.g., embedded software is counted as exports of goods by the United States, or IT-enabled financial services are not classified as IT services by the United States), and because U.S. data do not indicate affiliated imports by country of origin.

As noted earlier, services trade data do not capture intrafirm migration of software activity abroad. The BEA data on U.S. MNCs provide detailed information on the investment and production activities of U.S. companies abroad.

---

[9]Affiliated trade occurs between U.S. parent firms and their foreign affiliates and between foreign-owned firms in the United States and their foreign parent. Unaffiliated trade occurs between U.S. entities and foreign entities that neither own nor are owned by the U.S. entity.

TABLE 1 Computer and Information Services with Unaffiliated Foreigners (million dollars)

	Years										
	1994	1998			2004			AAGR, 1998-2004		
	Computer and Information Services	Computer and Information Services	Royalties and License Fees	Total	Computer and Information Services	Royalties and License Fees	Total	Computer and Information Services	Royalties and License Fees	Total
Exports										
All countries	2,332	3,705	3,195	6,900	6,601	4,261	10,862	10.10	4.92	7.86
Canada	333	430	125	555	1,144	279	1,423	17.71	14.32	16.99
Europe	899	1,767	1,508	3,275	3,281	1,328	4,609	10.87	-2.10	5.86
Japan	177	306	724	1,030	327	1,568	1,895	1.11	13.75	10.70
Asian Tigers	117	200	…	…	163	…	…	-16.34	…	…
Underdogs										
Brazil	48	136	…	…	149	81	230	1.53	…	…
Israel	51	24	32	56	38	13	51	7.96	-13.94	-1.55
China	17	29	46	75	48	51	99	8.76	1.73	4.74
India	9	38	17	55	227	29	256	34.70	9.31	29.21
Imports										
All countries	286	1,494	498	1,992	2,002	589	2,591	5.00	2.84	4.48
Canada	34	589	9	598	1,189	12	1,201	12.42	4.91	12.32
Europe	122	259	449	708	400	562	962	7.51	3.81	5.24
Japan	20	41	26	67	15	1	16	-15.43	-41.90	-21.23
Asian Tigers	6	18	…	…	31	…	…	55.98	…	…
Underdogs										
Brazil	1	1	1	2	1	…	…	0.00	…	…
Israel	0	9	2	11	7	3	10	-4.10	6.99	-1.58
China	2	6	…	…	7	…	…	2.60	…	…
India	7	100	…	…	315	6	321	21.07	…	…

NOTE: Omitted cells include either transactions below $500,000 or data that were omitted to maintain confidentiality. AAGR, average annual growth rate.
SOURCE: BEA Data on U.S. International Trade in Services.

Table 2 shows that growth in employment in IT services and computer design industries has been faster for foreign affiliates of U.S. firms than for their domestic operations (AAGR 5.1 vs. 3.9 percent) due to faster growth among foreign affiliates in computer design and related services.

Financing of Software Products and Services

Table 3 includes data on one of the inputs to software product and service firms: financial capital. It includes data on disclosed rounds of venture capital financing by year and by destination country as reported in the Venture Economics VentureXpert database. As is well known, venture financing exhibits significant yearly variation (e.g., Gompers and Lerner, 2006) and our data may not capture all venture financing rounds. However, some broad trends are suggested. First, similar to our data on inventive outputs (described in further detail later), the United States clearly dominates in inputs of financial capital to emerging software firms. However, based on data from 2002-2005, there is some evidence that rounds of venture financing to the software underdogs declined less from their 2000 peak than did financing to U.S. firms.[10] However, there was an apparent decline in venture financing to these countries in 2005. In short, more years of data are needed to discern whether there is a trend of increasing venture capital financing to the software underdogs.

Regional Trends in Packaged Software and Software Services

In the previous section we showed that the United States represents the majority of world sales in packaged software. However, other regions of the world have a large and increasing percentage of software services. In this section we discuss some regional trends that are partially responsible for the geographic variance in economic activity in packaged software and services.

Software Producers in Europe and Japan

In Western Europe, the software industry has long been dominated by custom software development and software services (Malerba and Torrisi, 1996; Steinmueller, 2004).

Table 4 shows sales of software products and IT services in the EU15 during 2003-2005.[11] IT professional services such as consulting, implementation,

---

[10]The software underdogs consist of India, Ireland, Israel, Brazil, and China.

[11]The EU15 comprised the following 15 countries: Austria, Belgium, Denmark, Finland, France, Germany, Greece, Ireland, Italy, Luxembourg, Netherlands, Portugal, Spain, Sweden, and the United Kingdom.

TABLE 2 Growth in Employment for Foreign Affiliates of U.S. Firms vs. Growth for All U.S. Establishments, Selected Industries, 1999-2002

	1999	2002	AAGR
Information services and data processing services			
Foreign affiliates of U.S. firms	104.5	132.0	8.1
All U.S. establishments	371.9	473.8	8.4
Computer system design and related services			
Foreign affiliates of U.S. firms	157.9	172.9	3.1
All U.S. establishments	997.0	1,061.3	2.1
Total			
Foreign affiliates of U.S. firms	262.4	304.9	5.1
All U.S. establishments	1,368.9	1,535.1	3.9

NOTE: AAGR, average annual growth rate.

SOURCE: Data on foreign affiliates of U.S. firms from table on selected data for majority-owned nonbank foreign affiliates and nonbank U.S. parents in all industries, 2003. From BEA International Economic Accounts, U.S. Direct Investment Abroad: Financial and Operating Data for U.S. Multinational Companies. Data on all U.S. establishments from U.S. County Business Patterns data.

TABLE 3 Disclosed Rounds of Venture Financing by Country, 1988-2005 (thousands of dollars)

	United States	Other G-7	Underdogs	All Other	Total
1988	2,565	660	0	0	3,225
1989	15,000	2,465	0	0	17,465
1990	6,350	464	248	0	7,062
1991	1,100	0	0	0	1,100
1992	1,607	1,418	0	0	3,025
1993	15,247	582	0	0	15,829
1994	7,403	138	0	0	7,541
1995	14,340	0	0	0	14,340
1996	92,784	1,466	0	2,766	97,016
1997	242,873	0	0	7,049	249,922
1998	300,355	9,359	0	6,039	315,753
1999	1,068,310	68,011	28,666	21,102	1,186,089
2000	2,036,591	221,297	73,307	169,636	2,500,830
2001	460,911	83,944	32,256	16,629	593,740
2002	99,836	23,295	6,831	3,815	133,777
2003	173,205	14,607	15,251	167	203,230
2004	151,025	9,492	10,600	1,848	172,965
2005	138,428	2,000	2,000	59	142,487

SOURCE: Venture Economics VentureXpert database, and author's calculations. Software includes rounds of financing from software and e-commerce software firms. Dates are round date of financing.

TABLE 4 Sales of Software Products and IT Services in the EU15

	2003	2004	2005	Average Growth (%)
Software products	59,235	61,707	64,979	4.74
System software	30,944	32,537	34,536	5.64
Application software	28,291	29,169	30,443	3.73
IT services	112,472	116,149	120,913	3.68
Professional services	81,376	84,380	88,147	4.08
Support services	31,096	31,769	32,766	2.65
Total software	171,707	177,856	185,892	
Percent services	52.67%	53.13%	53.74%	

SOURCE: European Information Technology Observatory (2006).

and operations management are larger than the entire software products market. Malerba and Torrisi (1996) identify several reasons for this focus on software services, including a weak local IT hardware industry, first-mover advantages by U.S. software product firms, fragmentation of local demand, and relatively little interaction between European universities and industry. The largest European producer of packaged software is SAP, the producer of enterprise software. SAP is currently the third largest software product company by sales, behind Microsoft and Oracle.

One surprising result in Figures 1 and 2 is that, in contrast to many other technology industries, Japanese firms account for a very small share of the total export market for packaged software. This is not a recent result; Japanese firms have not ever been major players in the world market for packaged software, despite their success in video games and in other IT markets. Japan runs a significant negative trade imbalance in software: In 1997, Japan imported US$3.93 billion of software but exported only US$23.33 million (Asahi Shimbun, reported in Anchordoguy, 2000).

A number of reasons have been provided for the relative weakness of Japanese software producers, including challenges created by the Japanese language, weak venture capital markets, weakness in intellectual property protection, and weak university computer science education (Anchordoguy, 2000; Baba et al., 1996; Cottrell, 1996; Fransman, 1995). Cottrell (1996) argues that weakness in Japanese PC software production was due historically to a fragmented standards environment, while Anchordoguy (2000) argues that the aforementioned proximate reasons were ultimately caused by Japan's economic system of "catch-up capitalism."[12]

---

[12]In particular, she argues that some of the key elements of the Japanese economic system—including state targeting policies, its keiretsu industrial groups, bank-centered financial system, and weak intellectual property system—have been benefited by its development of successful industries in steel, semiconductors, and IT hardware but have hindered the development of its IT software industry.

Other Countries That Are Large Software Producers

Rapid growth in the size of the Indian software industry has recently attracted much attention in the academic and popular press (e.g., Athreye, 2005a; Arora et al., 2001). Data from NASSCOM show that Indian IT services exports grew from $22 million in 1984 to $10 billion in 2005, with an additional $3 billion due to R&D services, engineering services, and software products. As this makes clear, the Indian software industry has largely been built around software services rather than products. Athreye (2005a) estimates that in 2000, revenue per employee among Indian software firms was approximately $35,100, up from only $6,200 in 1993.

Some anecdotal evidence suggests that Indian firms are increasingly performing more R&D-intensive activities. Athreye (2005a) notes the growth of a new innovative sector of small niche companies. Moreover, there is evidence of a deepening of R&D skills and the emergence of informal networks among local firms in India. This is also some evidence of success in certain niche technologies such as wireless and embedded systems (Parthasarathy and Aoyama, 2006; Ilavarasan, 2006); software for mobile phones represents a substantial category. Some Indian firms have also had success in developing software products for the developing countries market: one example is CITIL (now i-flex), a Citibank subsidiary that initially produced software products for developing country markets before eventually moving on to head-to-head competition with the established incumbent producers in developed countries (Arora, 2006; Athreye, 2005b). There are also some data on substantial and growing R&D activities in countries such as India; Arora (2006) reports that total revenues for engineering services and R&D by Indian producers in 2006 were estimated to be US$4.8 billion, a 23.1 percent increase over the prior year. In the next section we attempt to shed some additional light on this issue by examining U.S. patent data.

The Irish software industry consists of two very separate subindustries, each with very different characteristics. First, there is an overseas sector that is dominated by MNCs. These firms primarily are engaged in software logistics (such as media replication and printing and packaging production and distribution), localization (such as translating and adapting software to suit European markets), and development (O'Riain, 1999). Second, there is an indigenous sector that is populated by smaller firms that is engaged in software development and product development activities.

The number of MNCs in Ireland grew rapidly throughout the 1990s, from 74 foreign firms in 1991 to 140 foreign firms in 2000. As Arora, Gambardella, and Torrisi (2004) note, this rapid growth was due to a number of factors, including the liberalization of economic policies that began in 1991, a large and well-educated English-speaking workforce, an advantageous site for localization activities, as well as potential agglomeration economies that were ignited after the Irish software-producing industry reached sufficient scale. MNC subsidiaries are engaged primarily in "low-value-added, low-skill activities such as porting of legacy products on new platforms, disc duplication, assembling/packaging, and localization" (Arora et al., 2004). Revenues and exports in the Irish software industry have always been dominated by these MNCs. Sands (2005) notes that total industry revenues grew from $2.66 billion in 1991 to over $18 billion in 2002, with MNCs continuously accounting for over 90 percent of the total. In contrast, the indigenous sector is more product-based: it accounts for just under half of employment; however, it accounts for only 9 percent of revenues. Indigenous companies are usually young and small, and often produce primarily for niche or vertical (i.e., industry-specific) markets (Sands, 2005).

The software industry in Israel looks considerably different from that in either Ireland or India. Compared to locally owned Indian or Irish firms, Israeli firms are more product-based and are more R&D intensive. Breznitz (2005) notes that revenue per employee for Israeli software firms was US$255,172 in 2000. By his calculations, the similar statistic in 2000 for U.S. software publishers was US$231,621 and for locally owned Irish software producers was US$90,000. Breznitz (2005) examines the reasons for Israel's product-based industry. He provides several reasons: tight links between the R&D activities of Israeli universities and high-tech industries in the country; the presence of a highly successful indigenous hardware industry; the presence of local market demand for new products; the presence of American MNCs locating R&D facilities in Israel; and the ability of the Israeli IT industry to raise capital in U.S. financial markets.

EMPIRICAL EVIDENCE ON THE LOCATION
OF INVENTIVE ACTIVITY

In this section we examine the global geographic distribution of inventive activity in software. The data presented in the preceding section pointed to expanding markets for software services abroad. Those data also show that the market for packaged software continues to be highly concentrated in the United States, and little evidence indicates that this trend is reversing. However, authors such as Athreye (2005a) report increasing inventive activity in Indian firms, and other authors have reported similar trends in Ireland (Sands, 2005) and China (Tschang and Xue, 2005), as well as well-established software product industries in Israel (Breznitz, 2005) and Brazil (Botelho et al., 2005). Software product sales are a lagging indicator of inventive activity in software: Could inventive activity in software be picking up in other areas of the world but not yet reflected in product sales? If so, how significant are these developments in terms of number of inventions and their importance? To answer these questions, one needs a measure of R&D and inventive activity that is comparable across countries.

Patent data have long been used as one measure of inventive activity. Patents have also been found to be correlated, although weakly, with R&D spending, so they provide a weak measure of raw inputs into innovation (Griliches, 1990).

There are, of course, significant limitations to the use of software patents as a measure of inventive activity. As Jaffe and Trajtenberg (2002) note, not all inventions meet the U.S. Patent and Trademark Office (USPTO) criteria for patentability,[13] and inventors must make an explicit decision to patent an invention, as opposed to relying on some other method of intellectual property protection. In particular, there may be incremental inventive activity that is not patented and therefore is not reflected in patent statistics. Moreover, firms may sometimes choose to use trade secrecy rather than patenting to protect groundbreaking inventions because of incomplete enforcement of property rights. To the extent that intellectual property regimes differ across countries, this may make comparison of levels of patents across countries more difficult.

Conversely the high growth rate of patenting that we observe in our sample may be influenced by strategic patenting behavior. As Hall and Ziedonis (2001) document in the semiconductor industry, firms may patent not to protect standalone technological inventions but rather to protect against holdup by external patent holders or to negotiate access to external technologies. Thus, when interpreting our results, readers should be aware of how patent statistics may deviate from the level of inventive activity across countries. However, so long as the propensity to patent does not change significantly over time, these biases should not appreciably affect our interpretation of the time trends of patenting behavior across countries (and their interpretation as a metric of inventive activity).

Historically, inventions in software were not patentable[14] and for a time copyright was the predominant form of formal intellectual property protection in software. However, a series of court decisions widened the scope of software patents. Eventually, this culminated in the Commissioner of Patents issuing guidelines for the patenting of software that allowed inventors to patent any software embodied in physical media (Hall and MacGarvie, 2006). In contrast, over the same period, a series of cases, including several copyright infringement cases brought by Lotus Development, weakened the intellectual property protection offered by copyrights. Graham and Mowery (2003) show that over this period the number of granted software patents has increased dramatically while the propensity of firms to copyright has declined.[15] Recent research has shown that the stock of patents is correlated with firm success in the software industry

---

[13]Note that not all inventions also meet the criteria for patentability for the European Patent Office (EPO) and Japanese Patent Office (JPO).

[14]The following provides a necessarily brief overview of the history of intellectual property protection in software. For a more detailed overview, see Graham and Mowery (2003) and Hall and MacGarvie (2006).

[15]The set of patentable inventions is narrower in Europe than in the United States. To be patentable, the European Patent Convention requires that inventions address a particular technical problem and suggest a technical means to solve this problem (Thoma and Torrisi, 2006). The implication of this requirement is that "inventions having a technical character that are or may be implemented by computer programs may well be patentable" (EPO, 2005).

(Merges, 2006), suggesting that patents may be a potentially useful metric of the inventive output of firms.

A second issue in using software patents to measure inventive activity in software is identifying exactly which patents are software patents.[16] Software patents are not assigned to a particular class or subclass in either the USPTO or International Patent Classification (IPC) schemes. Moreover, there is no unique software classification field for patents. Graham and Mowery (2003) were the first to systematically identify software patents for research purposes. They identified the IPC classes used by the six largest producers of PC software over the period 1984 to 1995. This search resulted in a list of 11 IPC classes, which account for over half (57 percent) of the more than 600 patents assigned to the 100 largest packaged software firms in 1995 (as identified in the trade news publication *Softletter*).

The Graham-Mowery approach of using the patent classification system to identify software patents has been used and revised by others. Graham and Mowery (2005) identify software patents using USPTO classifications. Hall and MacGarvie (2006) identify software patents by finding the USPTO class-subclass combinations in which 15 large software firms patent. To identify their final sample, they intersect the resulting set of patents with another keyword definition used by Bessen and Hunt (2004).

Bessen and Hunt (2004) identify software patents through the use of a Boolean query that searches for keywords in the text of patents. They arrive at a patent sample that is broader than that used by other researchers (Layne-Farrar, 2005). Other researchers have identified a smaller sample of patents by reading them manually. Allison and Tiller (2003) identify Internet business method patents and Allison et al. (2005) identify university software patents.[17]

For this chapter, we use a version of the Graham-Mowery approach based on the IPC system. We began by identifying the top 10 firms by revenue volume in 1995 according to the Corptech Directory of Technology Companies.[18] We then examined the IPC classes in which they patented. Because we found that the Graham-Mowery set of IPC classes covered only 46 percent of the patents of these top 10 firms, we added two additional IPC categories. Our complete list of patent classes covered more than 80 percent of the patents of these top 10 firms. Table 5 provides a list of the included IPC classes and subclasses and their descriptions.

---

[16]This section provides an overview of the issues in identifying software patents. For a more complete discussion, see Layne-Farrar (2005) and Hall and MacGarvie (2006).

[17]Thoma and Torrisi (2005) compare several of these methods in a study of European software patents.

[18]These are Adobe, Autodesk, Cadence, Macromedia Inc., Microsoft, Novell, Oracle, SAP, Sybase, and Symantec Corp. Note that Corptech's coverage of foreign firms is more limited than its coverage of U.S. firms; however, so long as the distribution of patent classes used by software patenting firms does not vary substantially for U.S. and non-U.S. firms, this issue should not appreciably influence our results.

TABLE 5 List of IPC Patent Classes Used in Analyses

Class/Subclass	Description
G06F 3/00	Input arrangements for transferring data to be processed into a form capable of being handled by the computer; output arrangements for transferring data from processing unit to output unit (e.g., interface arrangements)
G06F 5/00	Methods or arrangements for data conversion without changing the order or content of the data handled
G06F 7/00	Methods or arrangements for processing data by operating upon the order or content of the data handled
G06F 9/00	Arrangements for program control (e.g., control unit)
G06F 11/00	Error detection; error correction; monitoring
G06F 12/00	Accessing, addressing, or allocating within memory systems or architectures
G06F 13/00	Interconnection of, or transfer of, information or other signals between memories, input/output devices, or central processing units
G06F 15/00	Digital computers in general
G06F 17/00	Digital computing or data processing equipment or methods, specially adapted for specific functions
G06K 9/00	Methods or arrangements for reading or recognizing printed or written characters or for recognizing patterns (e.g., fingerprints)
G06K 15/00	Arrangements for producing a permanent visual presentation of the output data
G06T 11/00	Two-dimensional image generation (e.g., from a description to a bit-mapped image)
G06T 15/00	Three-dimensional image rendering (e.g., from a model to a bit-mapped image)
G09G 5/00	Control arrangements or circuits for visual indicators common to cathode-ray tube indicators and other visual indicators
H04L 9/00	Arrangements for secret or secure communication

NOTE: Class names are as follows: G06F, Electric Digital Data Processing; G06K, Recognition of Data, Presentation of Data, Record Carriers, Handling Record Carriers; G06T, Image Data Processing or Generation, in General; G09G, Arrangements or Circuits for Control of Indicating Devices Using Static Means to Present Variable Information; H04L, Electric Communication Technique. SOURCE: International Patent Classification System, World Intellectual Property Organization, http://www.wipo.int/classifications/ipc/ipc8/?lang=en.

By using a broader set of IPC classes than Graham and Mowery, we are more likely to include patents that may be assigned to the aforementioned classes but are not software patents. As we will see, software patenting outside of the United States is relatively rare, so we utilize a conservative definition that includes as many such patents as possible in hopes of achieving an "upper bound" on the stock of software patents invented outside of the United States. However, we recognize that, if the rate of patenting in related technologies outside of software is higher than that inside and if the share of inventive activity in these other technologies is higher in the United States than abroad, then our measure may artificially inflate the gap in software patenting between the United States and other nations. To address this possibility, we compare our results using several software patenting definitions, including those of Graham and Mowery (2003, 2005).

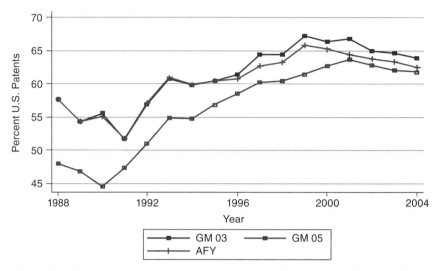

FIGURE 4 Percent of U.S. patents invented in United States under different software definitions. SOURCE: USPTO data and authors' calculations.

As an illustration, we computed the percentage of patents produced by inventors who reside in the United States (regardless of assignee location) under different definitions and then compared them.

Figure 4 presents these results. All three definitions show similar percentages for U.S. patents. Moreover, the three definitions have similar trends: increasing throughout the 1990s before reaching a peak around 2000 before declining slightly. Given the similarity in results across these different definitions, we will continue to focus on our original definition described earlier.

Results of Patent Data Analysis

Figure 5 shows the number of U.S. patents invented in the United States, Japan, other G-7, and all other nations (based on inventor address) by year of patent grant. The steep increase in the number of patents granted after 1995 is consistent with prior work that has shown an increase in the propensity to patent software after increases in the scope of intellectual property rights afforded by software patents (Graham and Mowery, 2003; Hall and MacGarvie, 2006). In 2004, 4,695 software patents were issued to inventors in the United States—a larger number of patents than inventors from all other areas of the world combined (2,811). The average annual growth in software patenting between 1988 and 2004 was also greater in the United States than in all other G-7 nations: patenting by U.S. inven-

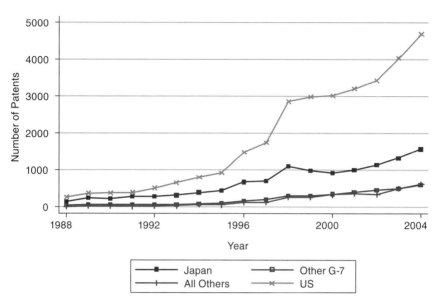

FIGURE 5 U.S. software patents invented in United States and other countries. SOURCE: USPTO data and authors' calculations.

tors grew at an average annual rate of 19.5 percent, compared to 16.1 percent for inventors in Japan and 18.0 percent in other G-7 nations.

These figures may reflect a "home country bias": U.S. firms may be more likely to patent in the U.S. market than foreign firms. Thus, in our data on patenting by location of inventor, the high percentage of U.S. patents may reflect (1) higher rates of U.S. patenting by U.S. firms (compared to firms in other countries) and (2) a higher propensity for U.S. firms to invent in the United States. More broadly, there may be some concern that there are potential differences between the site of inventive activity in U.S.-assigned U.S. patents that have EPO or JPO equivalents and the site of inventive activity in U.S.-assigned U.S. patents that do not have such equivalents. We address this potential concern in two ways. First, we look at the location of inventive activity for patents assigned to firms from outside of the United States. Second, we compare our results to recent work that has examined software patenting behavior in European patents.

We examined the percentage of patents assigned to the home country by country of assignee firm, based on year in which the patent was granted. Figure 6 shows that Japan-assigned U.S. software patents are predominantly invented in Japan, although this share appeared to decline during 2000-2004. Similarly, the location of invention in Israeli- and G-7-assigned patents (excluding the United

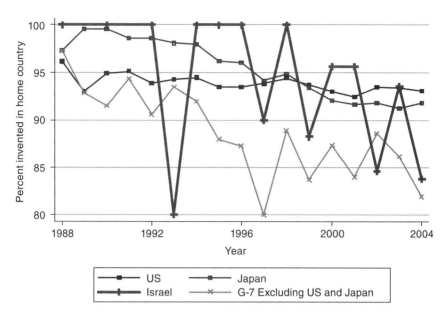

FIGURE 6 Percentage of U.S. software patents invented in home country by country of assignee. SOURCE: USPTO data and authors' calculations.

States and Japan) is predominately sited in those countries and regions. To be clear, comparing the propensity of U.S. software patents assigned to U.S. firms to be invented in the United States with the propensity of U.S. software patents assigned to firms from other countries to be invented in that (home) country is not an "apples to apples" comparison. However, given this important caveat, this figure does not suggest that patents assigned to U.S. firms are significantly more likely to be invented in the home country (United States) than are the patents from other countries. In fact, for several years, the U.S. patents assigned to Japanese and Israeli firms were more likely to be invented in the home country than U.S. patents assigned to U.S. firms. In recent years, however, this "home" percentage has been higher for patents assigned to U.S. firms than for others, though this is largely attributable to a decline in the home invented share for patents assigned to firms from other countries.

Thoma and Torrisi (2006) examine the rate of software patenting in European patents. Figure 7 shows the number of patents granted by country of patent assignee and year of patent application. There are some differences in the way Thoma and Torrisi define software patents and other differences in their sample construction: in particular, Thoma and Torrisi examine the distribution of patent-

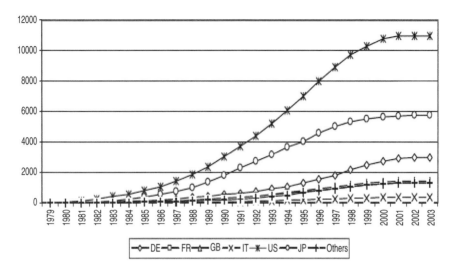

FIGURE 7 European Patent Office software patent grants by country of the assignee and year of application. SOURCE: Thoma and Torrisi (2006).

ing activity by site of assignee rather than inventor.[19] However, the broad trends are very similar to those in Figure 5: U.S. firms are responsible for the majority of software patenting activity, followed by Japanese firms, and then all others. Moreover, Thoma and Torrisi (2006) note that, of the European software patents in their database, 80.3 percent have also been granted by the USPTO and 73.8 percent have also been granted by JPO. If the majority of European software patents assigned to U.S. firms are also invented in the United States (not an unreasonable assumption given evidence presented in the earlier paragraph that the majority of U.S. software patents assigned to U.S. firms are invented in the United States), then the graph suggests that, even using European software patent data, a large share of the inventive activity in software takes place in the United States. Further, we note that while the levels of software patenting expressed in Figures 4 and 5 may be influenced by home country bias, so long as this bias does not change systematically over time, the time trends shown in these figures will not be as influenced by such bias.

---

[19]In particular, Thoma and Torrisi use a variant of Hall and MacGarvie's (2006) method of constructing a software patent sample based on patent classes and Bessen and Hunt's (2004) keyword method. Moreover, this graph shows patenting by assignee country rather than inventing country; however, according to our data 93.4 percent of patents assigned to U.S. firms were also invented in the United States. Last, this figure shows patenting by year of application rather than year of granting; however, the broad trend of greater patenting among U.S. assignees is robust to this difference.

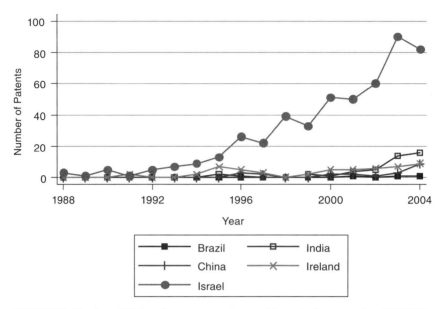

FIGURE 8 Number of U.S. software patents invented in underdog countries. SOURCE: USPTO data and authors' calculations.

Patenting Activity by Region

Figure 8 shows the number of U.S. patents invented in the underdog countries based on inventor location. Israel is the only one among them to have a significant number of U.S. patents. Israeli patenting activity increased from 3 in 1998 to a high of 90 in 2003. No other country has had more than 20 patents in any one year, though the number of patents invented in India has risen slightly in recent years, from an average of 0.5 throughout the 1990s to 16 in 2004.

Figure 9 shows the number of patents invented in the East Asian Tigers based on inventor location.[20] The number of patents invented in these countries is significantly higher than that of the underdogs. However, evidence suggests that many of these patents may be related to electronics.[21] Patenting among these countries is dominated by inventors from indigenous electronics companies in Korea and Taiwan: in 2004, 264 of the 280 patents granted were from this set of assignees.

---

[20]For the purposes of this paper, the Asian Tigers consist of Korea, Taiwan, Singapore, and Hong Kong. This is a separate and distinct set from the software underdogs.

[21]The top patenting firms in these countries include Daewoo Electronics Co. Ltd. (33), Electronics and Telecommunications Research (60), Hyundai Electronics Industries Co. Ltd. (57), Industrial Technology Research Institute (55), Inventec Corporation (25), LG Electronics (102), and Samsung Electronics (463). All of these companies are heavily involved in electronics research.

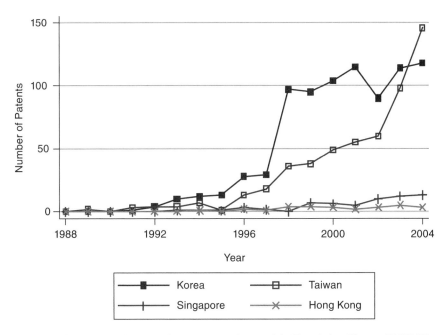

FIGURE 9 Number of U.S. software patents invented in East Asian Tigers. SOURCE: USPTO data and authors' calculations.

Assignee Location for Patents Invented Abroad

As noted earlier, multinational firms have played a major role in the development of software industries in other countries such as India and Ireland and may be driving the patenting activity by overseas inventors. To investigate this question further, we examined the location of U.S. software patent assignees for U.S. software patents invented in different countries. The overwhelming majority of patents invented in the United States were also assigned to U.S. firms. This fraction ranges from 93 to 97 percent over the period 1988-2004. No other region ever exceeded 6 percent in these data.

Figure 10 shows the distribution of assignee country for patents invented in the underdog countries. Here, the fraction of patents assigned to U.S. firms has generally been increasing over time, ranging from 20 percent in 1990 to a high of 65.7 percent in 2002. Excluding Israel (which has a robust software product industry) from the software underdogs, the top assignees in the software underdogs are 3Com (12), IBM (25), and Texas Instruments (12); no other company has more than five patents. The percentage of patents invented in underdog nations that are assigned to underdog firms has similarly been declining over time, from 80 percent in 1990 to 32.7 percent in 2004.

The increasing share of patents invented abroad in one of the software un-

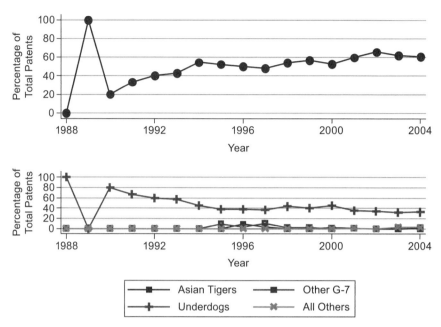

FIGURE 10 Distribution of assignee country for U.S. software patents invented in software underdogs. Top panel, United States; bottom panel, all other countries.

derdogs but assigned to U.S. firms suggests that there may be some shift in the location of inventive activity for U.S. firms to offshore locations. There is some evidence of a shift to more offshore invention for patents assigned to U.S. firms. However, the shift is small and offshore software invention in underdog countries by U.S. firms accounts for a very small share of the total patents assigned to U.S. firms. We also examine the trends in the site of inventive activity for U.S. software patents assigned to U.S. firms; note that these trends, because they only examine the site of inventive activity for patents assigned to U.S. firms, are not subject to concerns of home country bias. The percentage of U.S.-assigned patents invented in the United States fell from 93.5 percent in 1996 to 92.1 percent in 2005.[22] This decrease in the share of U.S. patents is due in large part to the increase in offshore activity in the underdogs: the percent of U.S.-assigned patents invented in the underdogs rose from 1.1 percent in 1996 to 1.8 percent in 2005.

We next examined whether there were any systematic differences in the industrial classification of the patent assignees by region where the patent was in-

---

[22]The share of U.S.-assigned patents invented in the United States was 96.2 percent in 1988 and 93.0 percent in 1989, though the number of software patents in these years was much lower than in 1996 (260 in 1988 and 387 in 1989 compared to 1,519 in 1996) which, as described earlier, was one of the first years in which software patenting began to grow rapidly.

TABLE 6 Assignee Industry for U.S. Patents by Region, 1988-2005

	Industrial Machinery (SIC 35)	Electrical & Electronic Equip. (SIC 36)	Holding Companies (SIC 67)	Software Publishers (SIC 7372)	All Other
United States	9,741	5,291	4,770	2,217	13,434
	(27.48)	(14.92)	(13.45)	(6.25)	(37.89)
Other G-7	1,023	153	12,561	98	3,585
	(5.87)	(0.88)	(72.11)	(0.56)	(20.58)
Asian Tigers	47	51	781	1	818
	(2.77)	(3.00)	(46.00)	(0.06)	(48.17)
Software underdogs	133	136	56	59	354
	(18.02)	(18.43)	(7.59)	(7.99)	(47.97)
All other	85	44	609	15	570
	(6.42)	(3.33)	(46.03)	(1.13)	(43.08)

SOURCE: Author's manipulation of data from USPTO and Corptech Database of Technology Companies. Numbers represent frequencies of row and column combinations. Numbers in parentheses represent the percentage of assignees in an industry conditional on invention in the country in the row. The unit of observation in the table is a patent.

vented. To do this, using assignee (company) name, we merged our U.S. software patent data with the Corptech Database of High Technology Companies.

Table 6 shows the distribution of assignee industry for patents by inventor region. Due to the way the Corptech data are collected, the industries of many Asian and G-7 countries in the Corptech database are classified as holding companies, so we focus our analyses on patents invented in the United States and in the underdog countries. One fact that is immediately apparent across all rows of the table is that patents are assigned to companies that belong to a variety of industries. Outside of the "Holding Company" category, most patents are from the "Other" category. Moreover, most patents are not assigned to firms in the "Software Publishers" industry (SIC 7372)—the SIC industry for packaged software producers. Second, the distributions of industries in the United States and underdog countries are broadly similar, with U.S. firms slightly more likely to be in industrial machinery and equipment and the underdogs more likely to be in electronics.

In Table 7 we provide some descriptive statistics on top patenting firms in major software-producing countries. To construct this table, we identified the five firms with the largest number of U.S. patents in each of nine countries: China, Germany, the United Kingdom, Ireland, Israel, India, Japan, South Korea, and the United States. Two major facts emerge. First, as noted earlier, the top patenting firms in software are usually not packaged software producers. Second, the top patenting firms in the underdog countries are usually large U.S. producers of

TABLE 7 Leading Recipients of U.S. Software Patents, by Country of Inventor, 1988-2005

Name	Year Company was Founded	Industry	Number of Employees	Revenue	Home Country	Number of Patents
CHINA						
Microsoft	1975	Software	71,553	$44 billion	United States	6
IBM	1888	IT hardware, software, services	330,000	$91 billion	United States	5
United Microelectronics Corp.	1980	Electronics	12,000	$39 billion	Taiwan	4
Intel	1968	Electronics	99,900	$8.2 billion	United States	2
Huawei Technologies	1988	Telecommunication	44,000	$8.2 billion	China	1
GERMANY						
Siemens	1847	Conglomerate	472,000	$75 billion	Germany	252
Robert Bosch GmbH	1886	Automotive	251,000	$55 billion	Germany	178
IBM	1888	IT hardware, software, services	330,000	$91 billion	United States	98
Infineon Technologies	1999	Electronics	36,000	$7 billion	Germany	41
Daimler Chrysler AG	1998	Automotive	383,000	$150 billion	Germany	38
UNITED KINGDOM						
IBM	1888	IT hardware, software, services	330,000	$91 billion	United States	140
International Computers Limited	1968	Computers			United Kingdom	40
British Telecommunications PLC	1846	Telecommunications	104,400	$37 billion	United Kingdom	38
Sun Microsystems Inc.	1982	IT hardware	31,000	$11 billion	United States	35
Philips Corporation	1891	Electronics	159,226	$36 billion	The Netherlands	32
IRELAND						
3Com Corporation	1979	Networks	1,925	$800 million	United States	11
Analog Devices Inc.	1965	IT Hardware	8,800	$2.4 billion	United States	3
Richmount Computers Limited						3
Hitachi Ltd.	1920	IT hardware, electronics	323,072	$80.5 billion	Japan	3
IBM	1888	IT hardware, software, services	330,000	$91 billion	United States	3

ISRAEL

Company	Year	Business	Employees	Revenue	Country	
IBM	1888	IT hardware, software, services	330,000	$91 billion	United States	69
Intel	1968	Electronics	99,900	$39 billion	United States	58
Motorola Inc.	1928	Electronics	88,000	$37 billion	United States	32
Scitex Corporation (now Scailex Corporation)[a]		IT hardware, now venture capital		$128.2 million	Israel	12
Applied Materials Inc.	1967	Semiconductor	12,576	$7 billion	United States	11

INDIA

Company	Year	Business	Employees	Revenue	Country	
IBM	1888	IT hardware, software, services	330,000	$91 billion	United States	17
Texas Instruments	1930	Hardware	30,300	$13 billion	United States	12
Honeywell International Inc.	1886	Aerospace	116,000	$26 billion	United States	3
Veritas Operating Corporation (acquired by Symantec)[b]	1989	Software	16,000	$4.1 billion	United States	3
Sun Microsystems Inc.	1982	Hardware	31,000	$11 billion	United States	2

JAPAN

Company	Year	Business	Employees	Revenue	Country	
Hitachi Ltd.	1920	IT hardware, electronics	323,072	$80.5 billion	Japan	1,403
Canon	1937	Imaging	100,000	$35 billion	Japan	1,286
Fujitsu	1935	Hardware	158,000	$40 billion	Japan	1,127
NEC Corporation	1899	Electronics	148,540	$41 billion	Japan	976
Toshiba	1904	Electronics	165,000	$60 billion	Japan	820

SOUTH KOREA

Company	Year	Business	Employees	Revenue	Country	
Samsung	1938	Electronics		$80 billion	South Korea	460
LG Electronics[c]	1958	Electronics	66,614	$23.5 billion	South Korea	100
Electronics and Telecommunications Research Institute	1976				South Korea	55
Hyundai Electronics (now Hynix Semiconductor)		Semiconductors	13,000	$5.6 billion	South Korea	49
Hyundai Motor Company	1967	Automotive	51,000	$57 billion	South Korea	30

continued

TABLE 7 Continued

Name	Year Company was Founded	Industry	Number of Employees	Revenue	Home Country	Number of Patents
UNITED STATES						
IBM	1888	IT hardware, software, services	330,000	$91 billion	United States	4,981
Intel	1968	Electronics	99,900	$39 billion	United States	1,648
Microsoft	1975	Software	71,553	$44 billion	United States	1,136
Sun Microsystems Inc.	1982	Hardware	31,000	$11 billion	United States	1,088
Hewlett-Packard Inc.	1939	Hardware	150,000	$89 billion	United States	682

[a]Data are for Scailex.

[b]Data are for Symantec.

[c]LG Electronics and LG Semicon Co. Ltd. are each part of the LG Group. LG Phillips is a joint venture with the LG Group and Phillips. Data missing are because data on some subsidiaries of the LG Group are not separately available.

SOURCE: The top five firms with the largest number of U.S. patents, identified from our calculations of USPTO data. Company data are from Hoover's Online, company annual reports, company web pages, and Wikipedia. Revenues are in U.S. dollars and for the most current year available. Missing cells represent firms for which we were unable to recover data.

electronics—and to a much lesser extent European and Japanese producers—such as IBM, Intel, Texas Instruments, and Sun Microsystems. One exception is China, where one Taiwanese and one Chinese firm are included among the leading producers. However, as noted earlier, the number of U.S. software patents produced in China is very small.

U.S. MARKET ADVANTAGES FOR INNOVATIVE ACTIVITY

The data in the prior two sections show two very different stories occurring in the globalization of software activity. On one hand, as has been well documented, there has been increasing growth in the production of IT service activity outside the United States. This trend has been going on for some time now and shows no signs of abating. Second, there is evidence that inventive activity in software development (at least as measured by patents) is highly concentrated in the United States and heavily controlled by U.S. firms. Though there is some evidence that inventive activity is picking up outside the United States, at current rates of growth this activity will not catch up with the U.S. software industry any time soon. Moreover, though there is some evidence that some inventive activity by U.S. firms has shifted abroad, at present the shift is small and this remains a small share of U.S. firms' overall inventive activity.

However, these trend rates of growth can change, so it is useful to examine the conditions that are widely thought to be conducive to innovation and inventive activity in new technologies. The literature has long examined some of the factors influencing the variance in innovative activities across countries.[23] These include R&D investments and human capital (e.g., Romer, 1990), supportive public policies (e.g., Nelson and Rosenberg, 1994; Mowery and Rosenberg, 1998), and more localized factors supporting the growth of clusters, including spillovers and user-producer interactions (Porter, 1990). In general, the United States has advantages over other rich and poor countries in all of these dimensions. We focus our attention on one area that we believe has received insufficient attention: the importance of geographic proximity to lead user innovation.

A key factor in the development and growth of a local software industry is the relationship with users. The transition of new inventions to usable economic products is a difficult process. Solving the problems that remain after initial conceptualization requires sustained innovative activity. User innovation and input are often an important part of this process (Rosenberg, 1963), and the willingness and ability of individuals to acquire and use new products and technologies is often as important as the developments of such products and technology themselves (Rosenberg, 1983).

Such user activity is particularly important in software. Business software in particular is often bundled with a set of business rules and assumptions about

---

[23]For a recent overview and review of this literature, see Furman et al. (2002).

business processes that must be integrated with the existing business organization, its activities, and its processes. Recent research indicates that proximity
between software developers and users is particularly important for this activity
to occur.

The software industry has a long history of user innovation and interactions with users leading to path-breaking new products. For example, IBM's
collaboration with American Airlines on the SABRE airline reservation system
in the 1950s and 1960s was an important early use of information technology
in "real-time" applications that would later be used in airline reservations, bank
automation, and retail systems (Campbell-Kelly, 2003; Copeland and McKenney,
1988). The genesis of this project was a serendipitous event: the chance meeting on a flight of R. Blair Smith of IBM's Santa Monica sales office with C. R.
Smith, the president of American Airlines. The eventual outcome of this project
was the SABRE system. Both IBM and American Airlines made extensive investments and contributions to the project: "We tapped almost all types of sources of
programming manpower. The control (executive) program was written by IBM
in accordance with our contract with them. We used some contract programmers
from service organizations; we used our own experienced data processing people;
we tested, trained, and developed programmers from within American Airlines,
and hired experienced programmers on the open market."[24] Similarly, the early
development of ERP software by SAP occurred through a series of incremental
improvements during development of real-time software for clients (Campbell-
Kelly, 2003).

One major challenge to offshoring software product development work will
result from the difficulty of coordinating software development activity across a
globally distributed team. As is well known, partitioning complicated software
development projects across multiple team members is difficult and often substantially increases the costs of software development (Brooks, 1995).

These problems may become still greater when management of such projects
is attempted at a distance. Globally distributed team members do not have access
to the rich communication channels that co-located developers have. Moreover,
differences in language and culture may make it much more difficult to establish
common ground among team members and ensure that miscommunications do
not occur (Armstrong and Cole, 2002; Olson and Olson, 2000). These projects
face other challenges, including an inability to engage in informal communication
as well as the difficulty of managing team members who may believe that such
projects are a prelude to job cuts.

A number of techniques have been proposed for lowering the costs of distributed software development. Going back as far as March and Simon (1958), one
common technique in distributed development is to reduce the interdependencies
among software components. The increasing modularization of software code and

---

[24]Parker (1965), as quoted in Campbell-Kelly (2003).

the use of object-oriented software development techniques has likely reduced some of the costs of distributed development over time. However, schedules and feedback mechanisms are necessary when interdependencies are unavoidable (March and Simon, 1958). The recent successes of large-scale open-source projects such as Linux and Apache have led some to consider whether open-source project management methodologies could be utilized in traditional corporate software development. Globally distributed teams rely heavily on coordination tools such as e-mail, phone, and more recently instant messaging as well as configuration management tools. However, several authors have shown that initial meetings are often necessary both to detail project requirements and for project members to become familiar with one another (e.g., Herbsleb et al., 2005). In general, the literature has demonstrated that, despite the continued development of tools and techniques to manage distributed projects, globally distributed work is difficult and can involve significant coordination costs.

Despite the considerable work that has been done in examining the challenges of software project management in a distributed environment, there has been heretofore relatively little systematic widespread empirical evidence on how distance from software suppliers impacts firm decisions to offshore software development.

Arora and Forman (2007) attempt to gather such systematic evidence by examining which IT services can be effectively performed from a distance or, to put it another way, which IT services are tradable. One way of examining the tradability of IT services is to examine the extent to which they are clustered near local demand. If markets for IT services are local, then we should expect the entry decisions of IT services firms to depend in part on the size of the local market. If markets are not local, then the composition of local demand should matter little; rather, suppliers should locate in low-cost regions. By providing evidence of the geographic reach of markets, this analysis also provides evidence on the tradability of services: Markets for services that are not tradable will be local, whereas those for services that are tradable need not be local.

Arora and Forman examine the clustering of local market supply for two types of IT services: *programming and design* and *hosting.* "Programming and design" refers to programming tasks or planning and designing information systems that involve the integration of computer hardware, software, and communication technologies. These projects require communication of detailed user requirements to the outsourcing firm in order to succeed. Hosting involves management and operation of computer and data-processing services for the client.[25] After an initial setup period, the requirements of such hosting services will be relatively static and will require relatively little coordination between client and

---

[25]While hosting activities do not fit most definitions of "innovation" or "invention" in software per se, they do provide a useful benchmark to compare tradability of services that require complex communication and coordination between supplier and customer and those that do not.

TABLE 8 Average Outsourcing by Size of Metropolitan Statistical Area

	Programming (%)	Programming and Design (%)	Hosting Ex Internet (%)
Rural Area	17.81	24.30	15.91
	(0.38)	(0.43)	(0.37)
Small MSA (< 250,000)	17.87	23.85	15.04
	(0.54)	(0.60)	(0.50)
Medium MSA (250,000 to 1 million)	18.48	26.30	16.41
	(0.35)	(0.40)	(0.34)
Large MSA (> 1 million)	18.54	26.08	15.31
	(0.21)	(0.24)	(0.20)

NOTE: Calculations are for 2002; standard errors in parentheses. The difference between rural/small and medium/large is significant at the 5 percent level for all three types. SOURCE: Arora and Forman (2007).

service provider. Thus, ex ante we would expect that hosting activities may more easily be conducted at a distance than other activities. Using data from U.S. Census County Business Patterns, Arora and Forman (2007) find that the elasticity of local supply to local demand characteristics is higher for programming and design (0.806) than for hosting (0.1899). That is, a 10 percent increase in local market demand will translate into an 8.1 percent increase in the supply of programming and design firms but only a 1.9 percent increase in the supply of hosting firms.

Arora and Forman also examine whether firm decisions to outsource programming, design, and hosting services depend on local market supply. Table 8 shows how 2002 outsourcing varied by the size of geographic area in the United States. Average outsourcing of programming and design is clearly increasing in the size of a location, though the pattern for hosting is less clear. Outsourcing of programming and design increases from an average level of 24.2 percent in small metropolitan statistical areas (MSAs) and rural areas to 26.1 percent in medium and large MSAs, and these levels are significantly different from one another at the 1 percent level. In contrast, outsourcing of hosting declines slightly from an average level of 15.61 percent in rural areas and small MSAs to 15.60 percent in medium and large MSAs; these levels are not statistically different from one other. Since the supply of outsourcing establishments is increasing in location size, these results suggest that the decision to outsource programming and design is increasing in the local supply of outsourcing firms. Controlling for industry differences, establishment size, and other factors yields the same conclusion.

This evidence, combined with that on the costs of distributed software development described earlier, suggests that proximity to users is an important determinant of inventive activity in software. The contrast with other products and industries in this volume is informative. For other products, such as wireless devices or PDAs, lead users have significant concentration in locations outside of the United States such as East Asia. However, the lead users of software are predominantly large organizations, and the leading large organizations in use of software and IT remain in the United States. This is especially true for the large market segment of business applications software, for which software products and services are frequently embedded in business process. User requirements in this setting often involve the transfer of tacit knowledge, and so proximity to lead users is particularly salient. Thus, as long as the United States remains the major market for software products, and the locus of the vast majority of lead users, it is unlikely to lose its technical leadership.

SOME RECENT TRENDS AND PROJECTIONS FOR THE FUTURE

Trends in Computer Science Education

Continued success in any innovative industry like software requires a talented and highly educated workforce. There is widely reported concern about a perceived shortage of domestic-born scientists and engineers in the United States (e.g., Ricadela, 2005).

Figure 11 shows data from the National Center for Education Statistics (NCES) on the number of undergraduate and master's degrees in computer science earned in the United States over the period 1983-2002.[26] The numbers of both undergraduate and master's degrees rose sharply from a combined figure of 35,200 in 1996 to 65,700 in 2002. This increase was influenced by the boom in the IT sector in the late 1990s.

More recent indicators of undergraduate- and master's-level enrollments in computer science are currently unavailable using official U.S. statistics. Figure 12 presents data from an annual survey of incoming freshmen. Mirroring the NCES statistics, these data show intention to major in computer science rising throughout the late 1990s and remaining high until 2001. However, intentions to major in computer science drop sharply thereafter. The Computing Research Association's Taulbee Survey shows similar findings. These data survey Ph.D.-granting institutions in the United States. Aspray et al. (2006) argue that data from the Taulbee Survey closely match trends in the NCES data, and so these data are a good leading indicator of the national educational statistics. Figure 13 shows a sharp decline in newly declared computer science majors after 2000.

Somewhat more recent official data are available for doctoral degrees conferred by U.S. universities. Figure 14 shows the number of doctoral degrees earned in computer science and mathematics during the period 1983-2003. In contrast to bachelor's or master's degrees, the number of doctoral degrees granted has generally been on the decline in the United States over the past decade. The

---

[26]Data from the NCES and other official government statistics in this subsection are from the National Science Foundation publication *Science and Engineering Indicators*.

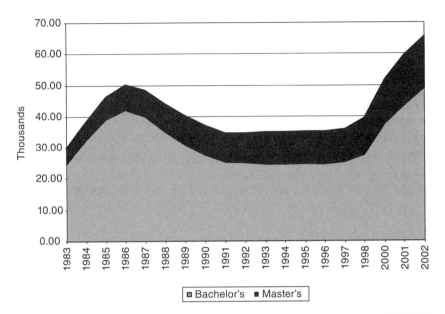

FIGURE 11 Undergraduate and master's degrees earned in computer science. SOURCES: U.S. Department of Education, National Center for Education Statistics, Integrated Post-secondary Education Data System, Completions Survey; and National Science Foundation, Division of Science Resources Statistics, WebCASPAR database, http://webcaspar.nsf. gov. See appendix Table 2-26 from *Science and Engineering Indicators 2006* for further details; 1999 data are not available.

figure shows that the number of computer science Ph.D.s peaked in 1995 at about 1,000 and then has fallen over time. In 2003 the number of such degrees advanced slightly, from 810 to 870. However, due to the very long lag between entry and graduation in doctoral programs, this increase likely reflects enrollment decisions in the middle to late 1990s, when demand for computer scientists was particularly strong.

While the number of students entering computer science programs appears to have fallen recently, there is evidence that such enrollments have been picking up in other countries. Figure 14 also shows the number of doctoral degrees granted in mathematics and computer science in selected countries other than the United States. The number of doctoral degrees in computer science and mathematics has recently been increasing in Asian countries such as China, Korea, and Taiwan.[27]

Unfortunately, similar statistics are not easily available for the production of

---

[27]These statistics, presented in *Science and Engineering Indicators* and collected from a variety of places, are unfortunately available only with some lag, and may not be strictly comparable. Moreover, they do not provide educational statistics on computer science graduates for India.

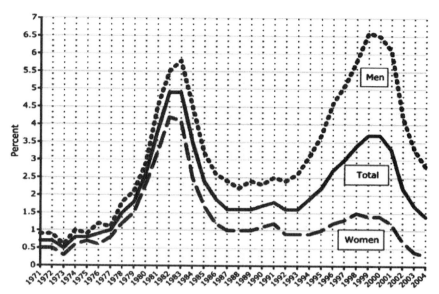

FIGURE 12 Freshman intentions to major in computer science. SOURCE: *Globalization and Offshoring of Software: A Report of the ACM Job Migration Task Force* (2006), eds. Aspray, Mayadas, and Vardi.

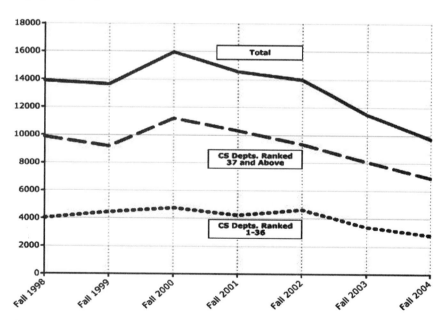

FIGURE 13 Newly declared computer science majors. SOURCE: *Computing Research Association and Globalization and Offshoring of Software: A Report of the ACM Job Migration Task Force* (2006), eds. Aspray, Mayadas, and Vardi.

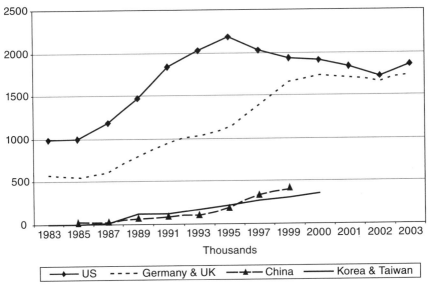

FIGURE 14 Doctoral degrees in mathematics and computer science by region. SOURCES: China—National Research Center for Science and Technology for Development and Educational Yearbook, 2002; Division of Higher Education, special tabulations (2005); South Korea—Organisation for Economic Co-operation and Development, Center for Education Research and Innovation, Education database, http://www1.oecd.org/scripts/cde/members/ EDU_UOEAuthenticate.asp; and Taiwan—Ministry of Education, Educational Statistics of the Republic of China (annual series).

bachelor's and master's degrees. Gereffi and Wadhwa (2005) provide evidence on the number of bachelor's and subbaccalaureate engineering, computer science, and IT degrees for the United States, India, and China in 2004.

Figure 15 shows that the number of degrees awarded in engineering by India and the United States are roughly similar. Although the numbers of engineering graduates in China are much larger than that of either the United States or India, Gereffi and Wadhwa (2005) note that educational statistics on engineers from China include degrees from 2- or 3-year programs that include students graduating from technical training programs that may be qualitatively different from baccalaureate programs in the United States. When normalized by population, the United States continues to lead in the production of bachelor's degrees in engineering, producing 468.3 bachelor's degrees per million compared to 103.7 in India and 271.1 in China.

However, recent work by Arora and Bagde (2006) shows that the number of engineering baccalaureate degrees awarded in India is growing much faster than in the United States. Table 9 shows that, although the number of engineer-

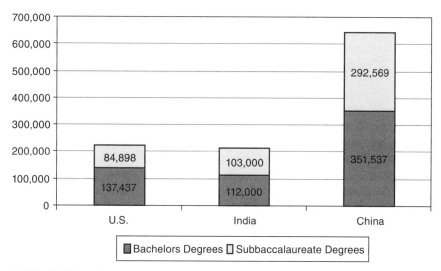

FIGURE 15 Bachelor's and subbaccalaureate degrees in engineering, 2004. SOURCE: Gereffi and Wadhwa (2005).

ing baccalaureate degrees awarded in 2003 is roughly the same as that reported by Gerrifi and Wadhwa, this number has grown steeply over time. From about 42,000 in 1992, the total more than tripled to greater than 128,000 in 2003.[28] Moreover, since the number of baccalaureates produced reflects the capacity added with a 4-year lag, it is important to note that sanctioned engineering baccalaureate capacity in India now exceeds 440,000, although a substantial portion is of dubious quality.

Figure 16 shows that the number of foreign students enrolled in graduate computer science programs in the United States declined in 2003 for the first time since 1995, reflecting visa restrictions imposed after September 11, 2001, the growth in degree-granting programs in other countries, as well as declines in the demand for engineers and computer scientists that took place in the early years of the most recent decade (NSF, 2006).

Overall, the data show that the United States continues to maintain a lead in the production of computer science graduates at all levels. However, recent data suggest that enrollments in computer science may be declining in the United States and picking up in other nations. As we will show in the next section, however, these changes in domestic supply are likely not due to long-term declines in the demand for computer science graduates within the United States.

---

[28]These numbers are based on data reported by 14 states, which include all the major states except Bihar, and probably represent 80-90 percent of the engineering baccalaureates produced in India.

TABLE 9 Output of Engineering Graduates (B.S. and B.E.) in
India, Various Years

Year	Total Number of Engineering Graduates Produced
1990	42,022
1991	44,281
1992	46,762
1993	48,281
1994	52,905
1995	56,181
1996	57,193
1997	61,353
1998	67,548
1999	75,030
2000	79,343
2001	97,942
2002	107,720
2003	128,432

NOTES: These data are based on the figures for the 14 major states (except the
State of Bihar) in India, which account for 80% of the gross domestic product
and likely more than that number of the total production of engineering graduates. These data are based on "Annual Technical Manpower Review" (ATMR) reports published by National Technical Manpower Information System (NTMIS),
India. These reports are prepared by a state-level nodal center of NTMIS and give
details of sanctioned engineering college capacity and outturn for all undergraduate technical institutions in the state. See cited source for more details.
SOURCE: Arora and Bagde (2006).

Labor Market Trends

There is some evidence that growth in the number of computer science degrees awarded over the past 25 years has not been fast enough to keep pace with
demand for workers with computer science training. Figure 17 shows that the
annual growth rate in the production of all mathematics and computer science
degrees averaged 4.2 percent during the period 1980-2000, significantly less than
the average annual growth of 9.3 percent in occupations directly associated with
these fields.[29] In comparison, over the same period, growth of all science and engineering graduates (including math and computer science) averaged 1.5 percent
while growth in all science and engineering occupations averaged 4.2 percent
(Table 10). Thus, the difference between degree growth and employment growth
is larger in mathematics and computer science than it is for science and engineer-

[29]Occupational data from these figures were compiled by the National Science Foundation, Division
of Science Resources Statistics, from U.S. Census data.

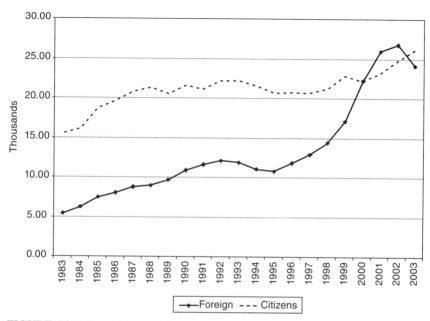

FIGURE 16 U.S. graduate enrollment in computer science by citizenship. SOURCE: *Science and Engineering Indicators 2006.*

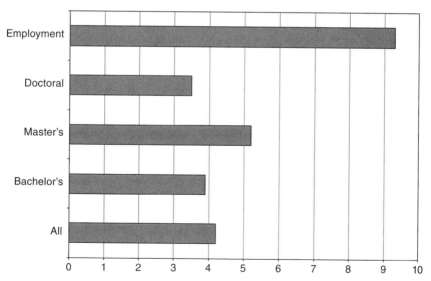

FIGURE 17 Average annual growth of degree production and occupational employment in mathematics and computer science, 1980-2000. SOURCE: *Science and Engineering Indicators 2006.*

TABLE 10 Output of Engineering Graduates (B.S. and B.E.) in India, Various Years

	Degree Growth	Employment in Occupation
All science and engineering	1.5	4.2
Mathematics/computer science	4.2	9.3

NOTES: Degree growth includes undergraduate, master's, and doctoral degrees.
SOURCE: *Science and Engineering Indicators 2006.*

ing overall. These data are now several years old and do not account for students receiving degrees from outside of computer science but moving into computer science professions. However, despite these qualifications, they do suggest that the United States may have relied in part on workers from abroad to make up for the shortfall of native workers with computer and math skills.

Recent data suggest that the inflation-adjusted median salaries for master's graduates in mathematics and computer science rose 54.8 percent between 1993 and 2003, higher than any other broad class of science and engineering graduates and higher than the average across all non-science and engineering graduates.[30] Growth in salaries was similarly competitive for graduates with bachelor's degrees (28.0 percent AAGR, second only to engineering graduates among science and engineering graduates) and those with doctoral degrees (18.6 percent AAGR, second only to graduates in engineering and physical sciences among science and engineering graduates). Furthermore, 2003 median salaries for computer science master's graduates are higher than any other broad category of science and engineering graduates ($80,000), whereas levels for bachelor's ($50,000) and doctoral ($67,000) degree graduates remain similarly competitive. Thus, even when one uses data that include the recent technology downturn, salaries of occupations requiring skills in mathematics and computer science have remained quite competitive when compared to other occupations in science and engineering and compared to the national average.

As noted earlier, there has been a significant shortfall in the rate of computer science degrees conferred relative to the rate of employment growth, and this excess demand for workers with computer science and engineering skills has been partially offset by the immigration of skilled workers from abroad. In fiscal year 2001 there were 191,397 H-1B visa admissions to the United States from computer-related occupations, 57.8 percent of total such admissions and

---

[30]The source for these data is the National Science Foundation, Division of Science Resource Statistics, National Survey of College Graduates.

the largest of any such category.[31] Kapur and McHale (2005a) list the top companies that petitioned for H-1B visas in October 1999 through February 2000, a list that includes some of the leading IT hardware and software firms: Motorola (618 petitions), Oracle (455 petitions), Cisco (398 petitions), Mastech (398), Intel (367), Microsoft (362), Rapidigm (357), Syntel (337), Wipro (327), and Tata Consulting (320).

Changes in immigration represent one mechanism that has the potential to affect the U.S. software industry in the relatively short term, and recent changes in the environment outside the United States can potentially affect immigration flows. The rapid growth in the software industries of countries like India and Ireland has increased the attractiveness of those countries to highly skilled indigenous workers. This has been particularly evident in Ireland, where rapid growth has encouraged an increasing number of highly skilled workers to remain in Ireland or return to Ireland from the United States. Kapur and McHale (2005a) report that emigration of male Irish graduates fell from about 25 percent in 1987 to under 15 percent in 1997, with similar trends for female graduates. Of the 644,444 Irish who had spent one year outside of Ireland in a 2002 census, 42 percent reported taking up residence in Ireland between 1996 and 2002, suggesting that a large fraction are recently returning Irish (Kapur and McHale, 2005a).[32]

With the continuing growth of the software industries in India and Ireland, it is likely that these historically important sources of highly skilled software professionals will retain a growing fraction of their indigenous software workers. Moreover, as noted by Kapur and McHale (2005b), the international market for software professionals is increasingly competitive. Richer countries such as the United States, Canada, Australia, Germany, and the United Kingdom increasingly compete for talent from other countries. In many cases, this competition has manifested itself as a decline in the traditional barriers to short- and long-term migration (Kapur and McHale, 2005b). This competition is likely only to increase with the aging demographics of these countries as well as the increasing requirements for a skilled workforce in software and in other industries.

Federal Government Spending on Software R&D

U.S. federal government investment in computer hardware and software R&D is thought to be one of the contributing success factors to both industries (Flamm, 1988; Langlois and Mowery, 1996). Early government R&D investment in software provided the computer facilities for universities to conduct early software research (Langlois and Mowery, 1996) and federal agencies such as

---

[31]Administrative data from the U.S. Department of Homeland Security, Bureau of Citizenship and Immigration Services.

[32]These data include migration of Irish citizens that have returned after studying in U.S. universities, including those studying for computer science degrees.

National Aeronautics and Space Administration (NASA) and Defense Advanced Research Projects Agency (DARPA) have been long-standing supporters of computer-related research. Federal grants remain a major source of funding for doctoral students in computer science: in 2003, 17.4 percent of full-time computer science graduate students reported that their primary source of funding was from the federal government.[33]

In the 1990s, though funding from the Department of Defense had largely flattened out, R&D spending grew rapidly throughout the decade through expanded funding from agencies such as the Department of Energy and NSF. However, over the period 2001-2003 (the most recent data available), government R&D spending in computer science remained largely flat. Moreover, the percent of total R&D spending on computer science (relative to other fields) declined over the period 2001-2003, from 4.5 to 4.0 percent.[34] We discuss the implications of these spending patterns in the next section.

CONCLUSIONS AND IMPLICATIONS

Public Policy Implications

The trends that we have described in this paper have several public policy implications. First, our results have provided evidence of a sizable export-driven software services sector in countries like India and Ireland, though there is less evidence of substantial inventive activity in software going on outside of the United States. These results suggest that entry- and mid-level programming jobs can be performed away from the point of final demand, though inventive activity that requires proximity with lead users is most effectively done in the United States. However, these entry- and mid-level programming jobs have traditionally provided U.S. IT workers with the skills needed to perform more complicated development activities such as creation of new software programs (Levy and Murname, 2004). In other words, training by U.S. firms has traditionally bestowed a beneficial externality upon entry-level workers by providing them with general human capital that workers appropriate later in their careers. This human capital is not easily provided by traditional publicly funded primary or secondary school education programs (Levy and Murname, 2004). As a result, a declining demand for entry-level programming jobs could hurt U.S. workers' future ability to perform more complex software development activity (e.g., new packaged software development). If this is true, then there are two ways that U.S. workers could obtain the general human capital needed. One would be for U.S. workers

---

[33]National Science Foundation, Division of Science Resource Statistics, Survey of Graduate Students and Postdoctorates in Science and Engineering, WebCASPAR database (*Science and Engineering Indicators*, 2006).

[34]*Science and Engineering Indicators 2006.*

to internalize the externality by accepting jobs for lower salaries. Of course, in the short run, workers may prefer instead to accept jobs in other (relatively higher-paying) fields. Alternatively, the government could attempt to subsidize entry-level employment, for example, by raising the costs of H-1B visas or by direct labor market subsidies. However, if the cost of remote software development remains lower than that in the United States, then clearly implementation of this policy may be problematic.

We have provided evidence of recent declines in computer science enrollments at the graduate and undergraduate levels. In our view, it is too soon to speculate whether these changes are evidence of a new trend or instead reflect temporary student reactions to business cycle fluctuations, particularly the IT downturn that began in the early part of this decade. Still, there is evidence that for some time, U.S. software developers have been using skilled labor from abroad as inputs into their innovation production function, presumably in part to supplement the pool of skilled labor available locally. As noted earlier, there is increasing competition from other industrialized countries for these skilled workers, and there is no sign that this competition will abate in the near future. Decreasing the costs of H-1B visas or lowering the costs of permanent migration is unlikely to be feasible in the short run because of the aforementioned concerns of labor substitution between foreign and indigenous workers. As a result, ensuring an adequate supply of local workers with sufficient basic or enabling skills (Levy and Murname, 2004) in mathematics, computer science, and related fields taught in the nation's school and university system will be important to the long-term success of software producers in the United States.

Another area of public policy concern is in government funding of computer science research. As noted earlier, federal funding of computer science has flattened out in recent years. A continuation of this trend could negatively impact innovative activity in software in the United States in two ways: by decreasing an important source of financial capital for basic research as well as potentially accentuating the negative downturn in enrollments in computer science graduate programs in the United States through a decline in graduate student funding.

Summary and Conclusions

There are currently two very different stories in the globalization of software development. On the one hand, the IT services industries in countries such as India, Ireland, and other countries continue to grow rapidly. The production of IT services is quite dispersed globally, and this dispersion will only increase over time. In contrast, both sales and inventive activity in packaged software are localized in the United States and undertaken primarily by U.S. firms. While differences in the levels of patent activity across countries should be interpreted with some caution because of divergence between the rate of patenting and inventive activity, examination of patent growth rates is less subject to these concerns and they show there is no sign of these trends reversing in the short to medium term.

Recent trends in computer science enrollments have attracted considerable attention in the popular press. We do find evidence of some declines in enrollments in U.S. computer science in recent years.[35] However, of likely equal or greater importance in the short run may be the increasing incentives for skilled foreign workers to remain in their home countries or to depart from the United States immediately or some years after degree conferral. There is already some evidence that improving educational systems and employment opportunities in the underdog countries is causing some skilled software professionals to remain at home or to return.

Nonetheless, there are powerful forces at work that are likely to keep the development of new software products and software innovation concentrated in the United States for some time to come. Despite recent trends, the United States continues to have the best postsecondary educational systems in the world for training computer scientists, and it continues to enjoy substantial albeit declining inward migration that benefits the software (and other) industries. Beyond the education and human capital issues, U.S. software innovators continue to enjoy substantial advantages due to agglomeration economies arising from the preexisting concentration of the industry, as well as a generally favorable business environment. Perhaps the most significant advantage that U.S. software product innovators enjoy is proximity to lead users. U.S. firms have been among the most innovative users of IT in the world, and these users have benefited U.S. software producers in the past and will continue to do so for some time to come.

ACKNOWLEDGMENTS

We thank Jeffrey Macher and David Mowery for their comments and suggestions, and Nicholas Yoder and Kristina Steffenson McElheran for outstanding research assistance.

REFERENCES

Allison, J., and E. Tiller. (2003). Internet business method patents. Pp. 259-257 in *Patents in the Knowledge-Based Economy*, W. M. Cohen and S. A. Merrill, eds. Washington, D.C.: The National Academies Press.

Allison, J., A. Rai, and B. N. Sampat. (2005). University Software Ownership: Trends, Determinants, Issues. Working Paper, Columbia University.

Anchordoguy, M. (2000). Japan's software industry: A failure of institutions? *Research Policy* 29:391-408.

---

[35]In graduate programs these declines appear to be concentrated primarily among immigrants. Among undergraduate degree programs current data are not available to indicate whether these declines are from U.S. nationals or immigrants.

Armstrong, D.J., and P. Cole. (2002). Managing distances and differences in geographically distributed work groups. Pp. 167-186 in *Distributed Work*, P. J. Hinds and S. Kiesler, eds. Cambridge: MIT Press.

Arora, A. (2006). The Indian Software Industry and its Prospects. Working Paper, Heinz School of Public Policy & Management, Carnegie Mellon University.

Arora, A., and S. Bagde. (2006). The Indian Software Industry: The Human Capital Story. Working Paper, Heinz School of Public Policy & Management, Carnegie Mellon University.

Arora, A., and C. Forman. (2007). Proximity and Software Programming: IT Outsourcing and the Local Market. Proceedings of the 40th Annual Hawaii International Conference in System Sciences, Waikoloa, Hawaii. Available at http://www.hicss.hawaii.edu/diglib.htm.

Arora, A., V.S. Arunachalam, V.S. Asundi, and R. Fernandes. (2001). The Indian software services industry. *Research Policy* 30(8):1267-1287.

Arora, A., A. Gambardella, and S. Torrisi. (2004). In the footsteps of Silicon Valley? Indian and Irish software in the international division of labor. Pp. 78-120 in *Building High-Tech Clusters: Silicon Valley and Beyond*, T. Bresnahan and A. Gambardella, eds. Cambridge, UK: Cambridge University Press.

Aspray, W., F. Mayadas, and M. Y. Vardi. (2006). Globalization and Offshoring of Software: A Report of the ACM Job Migration Task Force. Available at http://www.acm.org/globalizationreport/pdf/fullfinal.pdf.

Athreye, S. (2005a). The Indian software industry. Pp. 7-40 in *From Underdogs to Tigers: The Rise and Growth of the Software Industry in Brazil, China, India, Ireland, and Israel*, A. Arora and A. Gambardella, eds. Oxford, UK: Oxford University Press.

Athreye, S. (2005b). The Indian Software industry and its evolving service capability. *Industrial and Corporate Change* 14(3):393-418.

Baba, Y., S. Takai, and Y. Mizuta. (1996). The user-driven evolution of the Japanese software industry: The case of customized software for mainframes. Pp. 104-130 in *The International Computer Software Industry*, D. C. Mowery, ed. Oxford, UK: Oxford University Press.

Bessen, J., and R. M. Hunt. (2004). An Empirical Look at Software Patents. Working Paper 03-17/R, Research on Innovation.

Botelho, A. J. J., G. Stefanuto, and F. Veloso. (2005). The Brazilian software industry. Pp. 99-130 in *From Underdogs to Tigers: The Rise and Growth of the Software Industry in Brazil, China, India, Ireland, and Israel*, A. Arora and A. Gambardella, eds. Oxford, UK: Oxford University Press.

Bresnahan, T., and S. Greenstein. (1996). Technical progress in computing and in the uses of computers. *Brookings Papers on Economic Activity, Microeconomics* 1-78.

Breznitz, D. (2005). The Israeli software industry. Pp. 72-99 in *From Underdogs to Tigers: The Rise and Growth of the Software Industry in Brazil, China, India, Ireland, and Israel*, A. Arora and A. Gambardella, eds. Oxford, UK: Oxford University Press.

Brooks, F. (1995). *The Mythical Man-Month: Essays on Software Engineering, 20th Anniversary Edition*. Reading, MA: Addison-Wesley.

Campbell-Kelly, M. (2003). *From Airline Reservations to Sonic the Hedgehog: A History of the Software Industry*. Cambridge: MIT Press.

Copeland, D. G., and J. L. McKenney. (1988). Airline reservations systems: Lessons from history. *MIS Quarterly* 12(3):353-370.

Cottrell, T. (1996). Standards and the arrested development of Japan's microcomputer software industry. Pp. 131-164 in *The International Computer Software Industry*, D. C. Mowery, ed. Oxford, UK: Oxford University Press.

European Information Technology Observatory. (2006). *European Information Technology Observatory 2006*. Berlin, Germany: European Information Technology Observatory (EITO)—European Economic Interest Grouping (EEIG).

European Patent Office (EPO). (2005). Computer-Implemented Inventions and Patents. Law and Practice at the European Patent Office.

Flamm, K. (1988). *Creating the Computer: Government, Industry, and High Technology.* Washington, D.C.: Brookings Institution Press.

Fransman, M. (1995). *Japan's Computer and Communications Industry: The Evolution of Industrial Giants and Global Competitiveness.* New York: Oxford University Press.

Furman, J. L., M. E. Porter, and S. Stern. (2002). The determinants of national innovative capacity. *Research Policy* 31(6):899-933.

GAO (General Accounting Office). (2005). U.S. and India Data on Offshoring Show Significant Differences. GAO Report GAO-06-116.

Gereffi, G., and V. Wadhwa. (2005). Framing the Engineering Outsourcing Debate: Placing the United States on a Level Playing Field with China and India. Report, Master of Engineering Program, Duke University.

Gompers, P., and J. Lerner. (2006). *The Venture Capital Cycle: Second Edition.* Cambridge: MIT Press.

Gormely, J., W. Blustein, J. Gatloff, and H. Chun. (1998). The runaway costs of packaged applications. *The Forrester Report* 3(5).

Graham, S. J. H., and D. C. Mowery. (2003). Intellectual property protection in the U.S. software industry. Pp. 219-258 in *Patents in the Knowledge-Based Economy*, W. M. Cohen and S. A. Merrill, eds. Washington, D.C.: The National Academies Press.

Graham, S. J. H., and D. C. Mowery. (2005). Software patents: Good news or bad news. Pp. 45-80 in *Intellectual Property Rights in Frontier Industries: Software and Biotechnology*, R. W. Hahn, ed. Washington, D.C.: AEI-Brookings Joint Center for Regulatory Studies.

Griliches, Z. (1990). Patent statistics as economic indicators. *Journal of Economic Literature* 28(4): 1661-1707.

Hall, B. H., and M. MacGarvie. (2006). The Private Value of Software Patents. NBER Working Paper 12195.

Hall, B. H. and R. H. Ziedonis. (2001). The patent paradox revisited: an empirical study of patenting in the U.S. semiconductor industry, 1979-1995. *RAND Journal of Economics* 32(1): 101-128.

Herbsleb, J. D., D. J. Paulish, and M. Bass. (2005). Global software development at Siemens: Experience from nine projects. Pp. 524-533 in *International Conference on Software Engineering.* St. Louis, Mo.

Ilavarasan, P. V. (2006). R&D in Indian software industry. Pp. 134-143 in *Managing Industrial Research Effectively*, R. Varma, ed. ICFAI University Press.

Jaffe, A., and M. Trajtenberg. (2002). *Patents, Citations, and Innovations: A Window on the Knowledge Economy.* Cambridge: MIT Press.

Kapur, D., and J. McHale. (2005a). Sojourns and software: Internationally mobile human capital and high-tech industry development in India, Ireland, and Israel. Pp. 236-274 in *From Underdogs to Tigers: The Rise and Growth of the Software Industry in Brazil, China, India, Ireland, and Israel*, A. Arora and A. Gambardella, eds. Oxford, UK: Oxford University Press.

Kapur, D., and J. McHale. (2005b). *Give Us Your Best and Brightest: The Global Hunt for Talent and Its Impact on the Developing World.* Washington, D.C.: Center for Global Development.

Langlois, R. N., and D. C. Mowery. (1996). The federal government role in the development of the U.S. software industry. Pp. 53-85 in *The International Computer Software Industry*, D. C. Mowery, ed. Oxford, UK: Oxford University Press.

Layne-Farrar, A. (2005). Defining Software Patents: A Research Field Guide. Working Paper 05-14, AEI-Brookings Joint Center for Regulatory Studies.

Levy, F., and R. J. Murname. 2004. *The New Division of Labor: How Computers Are Creating the Next Job Market.* New York: Russell Sage Foundation.

Malerba, F., and S. Torrisi. (1996). The dynamics of market structure and innovation in the Western European software industry. Pp. 165-196 in *The International Computer Software Industry*, D. C. Mowery, ed. Oxford, UK: Oxford University Press.

March, J. G., and H. A. Simon. (1958). *Organizations*. New York: Wiley.

Merges, R. P. (2006). Patents, Entry, and Growth in the Software Industry. Working Paper, University of California, Berkeley.

Messerschmitt, D. G., and C. Szyperski. (2002). *Software Ecosystem: Understanding an Indispensible Technology and Industry*. Cambridge: MIT Press.

Mokyr, J. (2002). *The Lever of Riches*. Oxford, UK: Oxford University Press.

Mowery, D. C., and N. Rosenberg. (1998). *Paths of Innovation*. Cambridge, UK: Cambridge University Press.

Nelson, R., and R. Rosenberg. (1994). American universities and technical advance in industry. *Research Policy* 23(3):323-348.

NSF (National Science Foundation). (2006). *Science and Engineering Indicators 2006*. Washington, D.C.

OECD (Organisation for Economic Co-operation and Development). (2002). *OECD Information Technology Outlook*. Paris: OECD.

OECD. (2004). *OECD Information Technology Outlook*. Paris: OECD.

OECD. (2006). *OECD Information Technology Outlook*. Paris: OECD.

Olson, G. M., and J. S. Olson. (2000). Distance matters. *Human-Computer Interaction* 15(2-3): 139-178.

O'Riain, S. (1999). Remaking the Developmental State: The Irish Software Industry in the Global Economy. Doctoral Dissertation, Department of Sociology, University of California, Berkeley.

Parker, R. W. (1965). The SABRE System. *Datamation* (September):49-52.

Parthasarathy, B., and Y. Aoyama. (2006). From software services to R&D services: Local entrepreneurship in the software industry in Bangalore, India. *Environment and Planning A* 38(7):1269-1285.

Porter, M. E. (1990). *The Competitive Advantage of Nations*. New York: Free Press.

Ricadela, A. (2005). Q&A: Bill Gates on supercomputing, software in science, and more. *Information Week*. November 18.

Romer, P. (1990). Endogenous technological change. *Journal of Political Economy* 98:S71-S102.

Rosenberg, N. (1963). Technological change in the machine tool industry, 1840-1910. *Journal of Economic History* 23:414-443.

Rosenberg, N. (1983). *Inside the Black Box: Technology and Economics*. Cambridge, UK: Cambridge University Press.

Sands, A. (2005). The Irish software industry. Pp. 41-71 in *From Underdogs to Tigers: The Rise and Growth of the Software Industry in Brazil, China, India, Ireland, and Israel*, A. Arora and A. Gambardella, eds. Oxford, UK: Oxford University Press.

Scaffidi, C., M. Shaw, and B. Myers. (2005). Estimating the numbers of end users and end user programmers. Pp. 207-214 in *Proceedings of the 2005 IEEE Symposium on Visual Languages and Human-Centric Computing*. Los Alamitos, Calif.: IEEE Computer Society.

Schumpeter, J. A. (1934). *The Theory of Capitalist Development*. Cambridge, MA: Harvard University Press.

Steinmueller, E. (2004). The European software sectoral system of innovation. Pp. 193-242 in *Sectoral Systems of Innovation: Concepts, Issues, and Analyses of Six Major Sectors in Europe*, F. Malerba, ed. Cambridge, UK: Cambridge University Press.

Thoma, G., and S. Torrisi. (2006). The Evolution of the Software Industry in Europe. Working Paper, CESPRI, Bocconi University.

Tschang, T., and L. Xue. (2005). The Chinese software industry. Pp. 171-206 in *From Underdogs to Tigers: The Rise and Growth of the Software Industry in Brazil, China, India, Ireland, and Israel*, A. Arora and A. Gambardella, eds. Oxford, UK: Oxford University Press.

3

Semiconductors

JEFFREY T. MACHER
Georgetown University
DAVID C. MOWERY
University of California, Berkeley
ALBERTO DI MININ
Scuola Superiore Sant'Anna, Pisa, Italy

INTRODUCTION

The determinants, patterns, and consequences of globalization of the innovative activities of U.S. high-technology firms are the subject of a large empirical literature. Among the central questions addressed within this research is the extent to which international flows of research and development (R&D) investment and the offshore movement of other forms of innovative activity are linked with U.S. firms' foreign investments in manufacturing and related activities. A second important issue concerns measurement of the internationalization of firms' innovative activities—firm-level R&D investment data often do not capture developments within individual technological or industrial fields, and R&D data may provide little information on important aspects of the internationalization of firms' innovation-related activities. Partly because of the imperfections of these data, analyses of the globalization of innovative activity rarely consider developments within individual industries.

This chapter addresses these challenges in an examination of trends in the globalization of innovation-related activities in a single industry—semiconductors. We consider several measures of innovative activity within this industry, including R&D investment, technology-development alliances, and patenting. As is often the case in empirical work, the insights from this approach are obtained at some cost, confining our analysis to a relatively short time period and limiting our discussion of trends in the globalization of non-U.S. semiconductor firms' innovation-related activities. In addition, the data themselves represent imperfect proxies for the actual phenomena that we wish to examine. The different innovation-related indicators also do not aggregate in a straightforward way, which complicates efforts to develop strong conclusions concerning the consequences of these trends. Remarkably, after more than 30 years of intensive study of the internationalization of R&D and other innovation-related activities in the semiconductor industry, the data on these trends remain fragmented and limited in their coverage. Nevertheless, the results of our analysis highlight several distinctive trends in the globalization of innovation-related activities in this industry:

1. The share of industry-funded R&D investment devoted to offshore R&D by U.S. firms in "electronics components" manufacturing (an industry category that includes semiconductors, along with several other electronics product segments) grew only modestly during 1985-2001.

2. The number of technology-development alliances in the global semiconductor industry declined during the 1990s, although alliances among foreign firms appear to have grown more substantially than alliances among U.S. semiconductor firms during this period.

3. Process-technology R&D remains "homebound" in the home countries of U.S. and non-U.S. semiconductor firms, based on trends in the siting of "development" fabrication facilities ("fabs").

4. The patenting activity of large U.S. integrated semiconductor firms (those that both design and manufacture their products) remains predominantly "homebound," with little increase in offshore inventive activity in their patents during the period 1991-2003.

5. Patenting by European, Japanese, and Taiwanese semiconductor firms is similarly dominated by domestic inventive activity and this dominance by "home country" inventive activity appears to have increased slightly during the period 1996-2003.

6. The patenting activity of U.S. "fabless" semiconductor firms, which design and market but do not manufacture their products, indicates modest growth in offshore inventive activity during the period 1991-2003.

7. Although the vast majority of inventive activity undertaken by non-U.S. firms remains homebound, the United States is the predominant location for offshore inventive activity of all but Canadian semiconductor firms.

8. There is little evidence that the changing international structure of U.S. semiconductor firms' innovation-related activities has had negative consequences for engineering employment in the U.S. semiconductor industry, reflecting the limited offshore movement of innovation-related activities documented by these indicators.

Taken as a whole, our findings underscore the importance of a broad view of the array of activities that contribute to innovation in the semiconductor industry. These results also highlight the influence of growing vertical specialization on the globalization of innovation in this industry. Interestingly, the expanded offshore investment by U.S. semiconductor firms in production capacity does not appear to have influenced movement of their R&D activities to non-U.S. locations. Instead, the most important influence on the expanded offshore inventive activities of a subset of U.S. semiconductor firms (the fabless firms) may be the emergence of new segments of market demand that are concentrated in Southeast Asia. But even within the fabless segment of the U.S. semiconductor industry, the contributions of "offshore" innovation-related activities are modest thus far.

STRUCTURAL CHANGE IN THE GLOBAL SEMICONDUCTOR INDUSTRY

The global semiconductor industry experienced significant structural change during the 1990s. The market for semiconductor components shifted from one dominated by personal computers (PCs) to a more diverse array of heterogeneous niches associated with the Internet and wireless communications applications. The integrated device manufacturers (IDMs) that both design and manufacture semiconductor components no longer dominate industry production and innovation; instead, a vertically integrated industry segment coexists with a vertically specialized segment. IDMs compete and often collaborate with firms that specialize in either design and marketing (fabless firms) or manufacturing (foundries). As is the case in other high-technology industries, semiconductor-related market demand and technical expertise are growing in geographic regions that formerly accounted for smaller shares of global demand for semiconductor components (e.g., Malaysia, Taiwan, Singapore, China).

The Decline of the PC and Emergence of New Component Markets

The market for end-use semiconductor components during the late 1990s and early 2000s experienced a gradual shift away from one dominated by computer applications (especially PCs) to a more fragmented market in which wireless communications and other non-PC consumer products are more significant (Linden et al., 2004). Figure 1 depicts the shares of chip consumption accounted for by different end-use markets during the period 1994-2004. Computer applications still represent the predominant end-use market for semiconductor components, but most industry observers agree that non-computer (i.e., communications and consumer product) markets for semiconductor components will grow more rapidly during the next decade.

Differences between PC and non-PC markets for semiconductor components mean that this shift in consumption patterns has important implications for the organization of innovative activities in the semiconductor industry. The PC market is characterized by an entrenched architectural standard (the so-called Wintel standard), with well-defined and stable interfaces among semiconductor components and PC components. This stable architectural standard contrasts with the

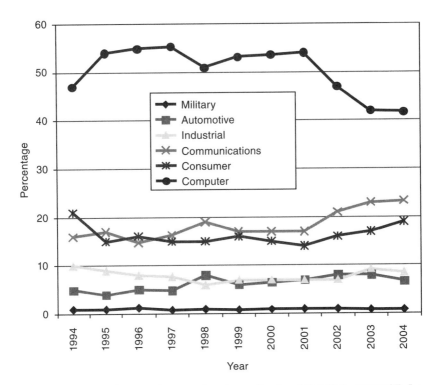

FIGURE 1 Semiconductor end-use markets by application, 1994-2004. SOURCE: Integrated Circuit Engineering (ICE) and IC Insights.

situation in many non-PC markets, where new products require more extensive "design-in" efforts on the part of component suppliers, and the interfaces governing the design and compatibility of components for these products can change significantly through successive product generations. No single product dominates semiconductor end-use demand in these applications—another contrast with PC component markets. As a result, production runs of new component designs are likely to be smaller and the cost savings through production-based learning will decline in significance. Smaller production runs also mean that new semiconductor production capacity, the costs of which continue to rise, must become more flexible and capable of producing a wider variety of component designs.

The relative decline of the PC market for semiconductors has important implications for the geographic location of demand for semiconductor components. The PC market has been dominated by designs developed in the United States and by an architecture that was largely under the control of U.S. firms. But designers

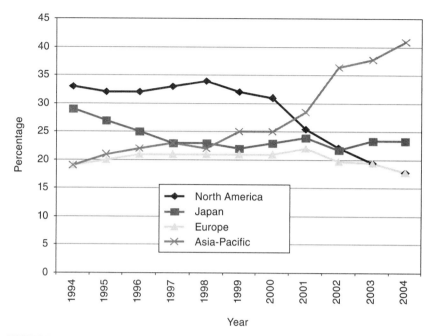

FIGURE 2 Semiconductor end-use markets by geographic region, 1994-2004. SOURCE: Integrated Circuit Engineering (ICE) and IC Insights.

and producers of the systems for which markets are growing more rapidly (e.g., wireless communications and consumer products) are more heavily concentrated in Southeast Asia, especially in Taiwan, Japan, and Singapore. Figure 2 illustrates the shifting geographic structure of demand during the period 1994-2004, highlighting declines in the share of global chip consumption accounted for by Japan and the United States and a corresponding rise in Southeast Asia's share.

Producers of these electronics systems often require that functionality be based on features in the semiconductor components incorporated in the products—so-called system-on-chip designs that are more complex and require more intensive interaction between system and chip designers (Ernst, 2005). Moreover, the number of new applications that use semiconductors has increased dramatically. The needs of an increasing variety of system providers mean that a one-size-fits-all model for semiconductor components is appropriate in only a limited number of cases. As a result, close interaction between designers of components and designers (as well as producers) of these more heterogeneous electronics systems is essential to product development. Proximity to system customers, more and more of whom are located in Southeast Asia, therefore is likely to grow in importance for developers of state-of-the-art semiconductor devices.

GROWTH OF VERTICAL SPECIALIZATION IN THE SEMICONDUCTOR INDUSTRY

For the first two decades of the computer and semiconductor industries, large integrated producers such as AT&T and IBM designed their own solid-state components, manufactured the majority of the capital equipment used in the production of these components, and utilized internally produced components in the manufacture of electronic computer systems that were leased or sold to their customers (Braun and MacDonald, 1978). During the late 1950s, "merchant" manufacturers entered the U.S. semiconductor industry and gained market share at the expense of firms that produced both electronic systems and semiconductor components. Specialized producers of semiconductor manufacturing equipment began to appear by the early 1960s.

Since 1980, the interdependence between product design and process development has weakened in many semiconductor product segments (Macher et al., 1998). This shift has been associated with the entry of new types of firms that specialize in semiconductor component design or production. Hundreds of so-called fabless semiconductor firms that design and market semiconductor components have entered the global semiconductor industry since 1980. These firms rely on contract manufacturers (so-called foundries) for the production of their designs. Contract manufacturers include "pure-play foundries" that specialize in semiconductor manufacturing, as well as the foundry subsidiaries of established integrated device manufacturers (IDMs) seeking to fully utilize excess fabrication capacity. Fabless semiconductor firms serve a variety of fast-growing industries, especially computers and communications, by offering more innovative designs and shorter delivery times than integrated semiconductor firms. Fabless-firm revenues increased from slightly less than 4 percent of global industry revenues in 1994 to more than 15 percent by 2004 (Figure 3). The increasing demand for and variety of communications and consumer products (e.g., GPS systems, game controllers, appliances, automatic lighting) suggests that the demand for special-purpose functionality has also increased. The so-called embedded systems and software market represents a growing and increasingly important segment of the semiconductor industry, with its own vertically specialized market structure. Measuring the growth of this market segment, however, is hampered by a lack of data.

The growth of vertical specialization in the semiconductor industry reflects the influence of developments in markets and technology (Macher and Mowery, 2004). The expansion of markets for semiconductor devices enables vertically specialized semiconductor design and production firms to exploit economies of scale and specialization. Scale economies lower production costs, expanding

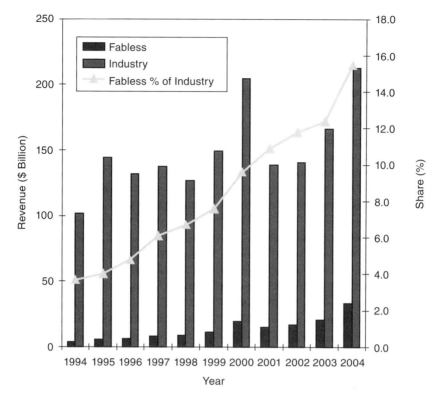

FIGURE 3 Fabless and overall industry revenues, 1994-2004. SOURCE; Fabless Semiconductor Assocation (FSA) and Integrated Circuit Engineering (ICE).

the range of potential end-user applications for semiconductors and creating additional opportunities for entry by vertically specialized firms. The increasing capital requirements of semiconductor manufacturing provide another impetus to vertical specialization, since these higher fixed costs make it necessary to produce large volumes of semiconductor components in order to achieve lower unit costs. The design cycle for new semiconductor products also has become shorter and product life cycles more uncertain. As a result, it is more difficult to predict whether demand for a single product will fully utilize the capacity of a fabrication facility that is devoted exclusively to a particular product, increasing the risks of investing in such "dedicated" capacity. Since foundries tend to produce a wider product mix, they are less exposed to these risks.

At the same time, however, a number of large semiconductor firms (IDMs) still combine semiconductor device design and manufacture. The advantages of integrated management of design and manufacture are greatest in product lines at the leading edge of semiconductor technology, especially in DRAMs (Macher, 2006). The relationship between the specialized foundry producers and the IDMs combines elements of cooperation and competition. For example, U.S. IDMs negotiated license agreements for the supply of product and process technologies to less-advanced semiconductor firms operating in Japan and South Korea during the 1970s, and U.S., Japanese, and European IDMs also supplied product and process technologies to Taiwanese and Singaporean foundry firms during the 1980s and 1990s. Many of these IDMs provided advanced process technologies to foundries in exchange for a guaranteed supply of semiconductor components. The development of a semiconductor intellectual property market also spurred growth in the number and importance of specialized product design firms (Linden and Somaya, 2003). Product and process licensing in the semiconductor industry has facilitated entry by both vertically specialized and integrated firms.

Increased vertical specialization in the semiconductor industry has been associated with the entry of new firms and geographic redistribution in production capacity. Figure 4 shows the regional distribution of fabrication capacity (measured in terms of wafer starts per month[1]) during the period 1995-2003. The North American and Japanese shares of global semiconductor production capacity fell significantly during the period, and the shares attributable to "Asia/ Pacific" countries increased, reflecting capacity growth in China, Taiwan, South Korea, and Singapore. These Southeast Asian countries now collectively account for the largest regional share of global production capacity, and their share will continue to grow in the near future.

Figure 5 reclassifies manufacturing capacity by region of ownership rather than location for the period 1997-2003, revealing a slightly different pattern. The share of global manufacturing capacity owned by firms headquartered in Southeast Asian countries trails that of Japanese and North American producers. North American, Japanese, and (to a lesser extent) European semiconductor firms have shifted much of their production capacity to Southeast Asia since the mid-1990s and have entered joint ventures with Southeast Asian producers. Southeast Asian firms, on the other hand, have invested primarily within their home regions during this period.

The growing concentration of manufacturing capacity in Southeast Asia is attributable in large part to the success of the foundry business model, which is reflected in foundry firms' growing share of semiconductor-industry revenues (Figure 3). The most advanced foundries are located in Singapore and (especially) Taiwan. A few Taiwanese firms have opened foundries in the United

---

[1]There are many possible measures of fab capacity, including the number of wafers processed over a given time period, the total wafer surface area that can be processed, the amount of installed processing equipment, and so on. Leachman and Leachman (2004) measure fabrication capacity as the estimated number of electrical functions that are produced by chip manufacturers, where a function is a memory bit or logic gate.

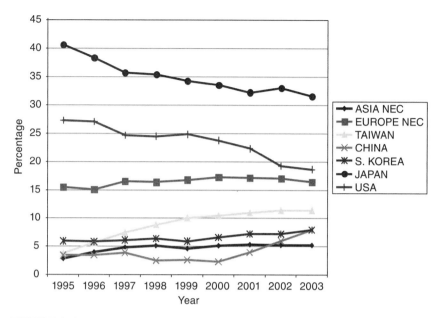

FIGURE 4 Geographic location of semiconductor manufacturing capacity, 1995-2003. SOURCE: Strategic Marketing Associates.

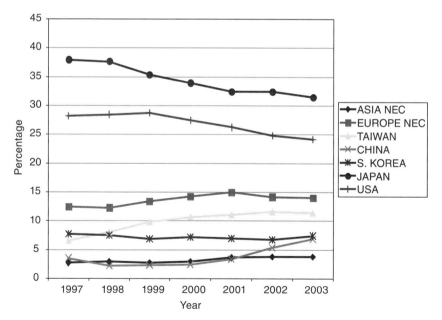

FIGURE 5 Location of ownership of semiconductor manufacturing capacity, 1997-2003. SOURCE: Strategic Marketing Associates.

States, and Taiwan's dominant position in the foundry industry faces competition from lower-cost production sites in other areas of Southeast Asia (particularly Malaysia and China) and elsewhere.

U.S. DOMINANCE IN PRODUCT DESIGN

Although semiconductor manufacturing capacity now is widely distributed among mature and fast-growing regions within the global economy, semiconductor design activities, especially those associated with fabless firms, remain more concentrated. U.S. fabless semiconductor firms accounted for more than 60 percent of the value of orders received by the top four foundry firms (TSMC, UMC, Chartered, and SMIC) during the period 2000-2004.

Nonetheless, the growth of fabless firms in other countries is one indication of the widening geographic distribution of semiconductor-design activity and expertise. Table 1 indicates that several non-U.S. clusters of fabless firms have emerged in Israel, Canada, Taiwan, the United Kingdom, and South Korea. Although hundreds of fabless firms now operate in dozens of other countries, most of these firms are smaller than their U.S. counterparts.[2]

A number of factors have contributed to the success of U.S. firms in semiconductor design. Established regional high-technology clusters in areas such as Silicon Valley, Boston's Route 128, and Austin, Texas, attract large numbers of semiconductor designers. These clusters are located near universities and other research centers that produce new design techniques, design software, and engineering talent. The role of U.S. universities in developing new design software and chip architectures has long outstripped their function as a source of new manufacturing methods, in part because the cost of constantly re-equipping the necessary facilities exceeds the resources of most academic institutions.

Although we lack data to track these trends more systematically, most industry observers suggest that Southeast Asian countries account for a growing share of global semiconductor industry design activities (Brown and Linden, 2006a). As U.S. semiconductor firms, and especially fabless firms, seek to collaborate more closely with the systems firms that are located in Southeast Asia, a regional or local design presence is important. In addition, countries such as Taiwan and South Korea have developed product development expertise in digital consumer electronics and wireless communications, among other areas (Ernst, 2005). Offshore design centers, particularly in China and India, may offer cost savings and comparable productivity in less-sophisticated design activities (Brown and Linden, 2006b). But most observers assess the semiconductor design capabilities of

---

[2]The data in Table 1 that form the basis for this discussion include 640 fabless firms that are members of the Fabless Semiconductor Association (FSA) or nonmembers verified by the FSA. At least 300 other small fabless firms are thought to exist but have not been verified by the FSA.

TABLE 1 Fabless Firms by Country of Location, 2002

Country	Fabless Firms	Non-U.S. City	Fabless Firms
United States	475	Tel Aviv, Israel	14
Canada	30	Ottawa, Canada	13
Israel	29	Hsinchu, Taiwan	13
Taiwan	22	Seoul, South Korea	9
United Kingdom	22	Taipei, Taiwan	8
South Korea	13	Toronto, Canada	8
Germany	8	Cambridge, England	4
France	6		
Japan	5		
Sweden	5		
Switzerland	4		
India	3		
Spain	3		
Others	15		
TOTAL	640		

SOURCE: Arensman (2003).

Chinese and Indian centers as lagging those of Canada, Israel, Taiwan, the United Kingdom, and the United States, as well as other industrial economies.

In summary, the structure of production activities in the global semiconductor industry has shifted from one dominated by vertical integration to a more complex structure that blends vertical specialization and vertical integration. Specialized design and manufacturing firms have entered the industry in large numbers, and the growth of foundry firms has been associated with a substantial shift in production capacity investment to Southeast Asia. Vertical specialization has facilitated the entry of new firms, many of which are located outside of the regions that were homes to established firms. But, thus far, increased vertical specialization in this industry appears to be associated with shifts in the location of production to a much greater extent than shifts in the location of product design and R&D activities.

MEASURING GLOBALIZATION OF INNOVATION-RELATED ACTIVITIES IN SEMICONDUCTORS

Indicators of Offshore Innovation-Related Activities

We use four indicators to examine trends in the offshore R&D activities of U.S. and non-U.S. other firms in the global semiconductor industry: (1) the share of industry-funded R&D expenditures supporting offshore R&D (available only for U.S. firms) during the period 1985-2001; (2) the number and location of development fabs established by U.S. and non-U.S. firms within the global semiconductor industry during the period 1995-2003; (3) the number of international and domestic alliances formed by U.S. and non-U.S. semiconductor manufacturing firms during the period 1990-1999; and (4) the site of inventive activity resulting in U.S. patents issued to U.S. and non-U.S. semiconductor firms during the period 1994-2004. Each measure by itself provides an incomplete portrait of "globalization" of innovation-related activities within the global semiconductor industry, but taken together, they do shed light on the extent of globalization or nonglobalization of innovation within the industry.

Industry-Funded Offshore R&D Investment

Figure 6 displays trends in offshore R&D investment (measured as a share of industry-funded R&D spending) by U.S. manufacturers of electronic components during the period 1985-2001. The data in the figure suggest minimal change in the share of offshore R&D within total industry-funded R&D, which drops to less than 3 percent by 2000 from its 1985 share of more than 4 percent.[3] The sharp increase during 2000-2001 in the offshore share of industry-funded R&D may or may not indicate a significant departure from this flat trend. In addition to the fact that this "reversal" covers only one year of data, the magnitude of the increase in reported offshore R&D during this period (more than doubling, from $327 million in 2000 to $852 million in 2001) suggests that a change in sample composition or other factors may be responsible, rather than a long-term shift in U.S. firms' R&D investment behavior.

The industry-level R&D investment data compiled by the National Science Foundation (NSF) that are the source for Figure 6 have a number of well-known problems. Coverage by the NSF R&D survey of smaller firms (e.g., entrants), particularly for the longer time period depicted in Figure 6, is problematic since the NSF sample frame was not updated frequently during the 1980s and early 1990s. The "electronic components" product line for which these data were compiled by the NSF also includes a number of other products in addition to semiconductor components. Moreover, the definition of this and other product lines for which NSF collects R&D data have undergone some revisions during the period covered in Figure 6.

Even if the reported R&D data accurately summarize the trends in semiconductor-related offshore R&D investment, there is reason to suspect that the R&D investment data reported by semiconductor firms do not capture many of the activities that contribute to innovation in this industry. For example, R&D

---

[3]We examined several sources of R&D data related to the globalization of innovation-related activity in the semiconductor industry. Most of these data, including publicly available information such as company Annual Reports and Form 10-Ks, did not yield time series that were consistent or covered most of the firms in the industry. We therefore narrowed our data sources to indicators that most effectively underscore trends in the offshore R&D activities of U.S. and other firms in the global semiconductor industry.

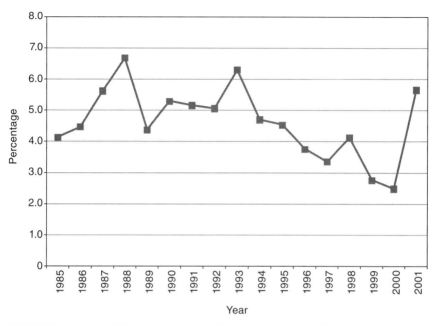

FIGURE 6 Foreign R&D as percentage of industry-funded R&D (electronic components), 1985-2001. SOURCE: National Science Foundation (NSF).

investment data may not include process innovation or the "tweaking" that occurs
within the production facilities of IDMs. As we note later in the chapter, much
of the process innovation within foundries (few of which are operated by U.S.-
headquartered firms) relies heavily on production-facility upgrades that may or
may not be included in the R&D investments reported by firms. Design activities,
especially those carried out by fabless firms, are another important source of innovations that may or may not be reported consistently as R&D investment.

These problems aside, the lack of a strong trend during the period 1985-2001
in reported offshore R&D investment is striking. This widely accepted measure
of "globalization of innovation" indicates that U.S. semiconductor firms do not
appear to be expanding this portion of their offshore innovation-related activities significantly, and the offshore portion of their self-financed R&D investment
remains modest.

Offshore Process-Development Facilities

A second indicator of the offshore movement of semiconductor firms'
innovation-related activities is the location of their process-technology develop-

ment facilities during the period 1995-2003. These facilities are important sites for process innovation in the semiconductor industry, since they are used for the development and "debugging" of new manufacturing technologies (Appleyard et al., 2000; Hatch and Mowery, 1998). But this type of process innovation is irrelevant to innovation by fabless firms, whose designs are manufactured by foundries. Moreover, foundries rarely use development fabs. Rather than developing new manufacturing processes in dedicated R&D facilities and transferring these process technologies to full-volume production sites, foundries instead incrementally upgrade manufacturing processes within their full-volume production facilities. This approach enables the foundry firms to maintain stability and predictability in the evolution of the component designs that they produce for their customers. This indicator of the location of innovation-related activities therefore applies only to one segment of the global semiconductor industry—the IDM and systems firms.

The source for our data on the siting of semiconductor firms' development facilities is Strategic Marketing Associates (SMA), a market research firm that tracks investment in semiconductor manufacturing facilities. The SMA data provide information on the home country for each firm in the dataset; years in operation and facility type (production, R&D, development, etc.) for each production facility; the technological sophistication (e.g., smallest linewidth of components produced), capacity, and size for each facility; and other information. Although the SMA database includes hundreds of public and private semiconductor entities, we exclude government and university-based facilities that are used solely for research and teaching purposes, research labs that possess production facilities for consulting purposes, and similar organizations.

The SMA data are collected through surveys. The development fabs that are included in the following discussion are used for the characterization of new (i.e., state-of-the-art) manufacturing processes and are typically much smaller (less than 5,000 wafer starts per month) than full-volume production facilities. The manufacturing facilities that we categorize as development fabs thus include facilities in which a variety of innovation-related activities are carried out. Nonetheless, these data are the only tabulation of which we are aware that can shed some light on the organization of the process R&D activities of IDMs and systems firms.

Figure 7 displays the number of "domestic" and "foreign" development fabs in operation during the period 1995-2003. Domestic development fabs are defined as those located in the home country of the semiconductor firm and foreign development fabs are located in a different host country. The overwhelming majority of development fabs are located in semiconductor firms' home countries. The "homebound" nature of process-technology development appears to reflect the demanding technical requirements of this activity, the need for close coordination between product design and process-technology development, the need for an iterative approach to the introduction of new manufacturing processes, and the

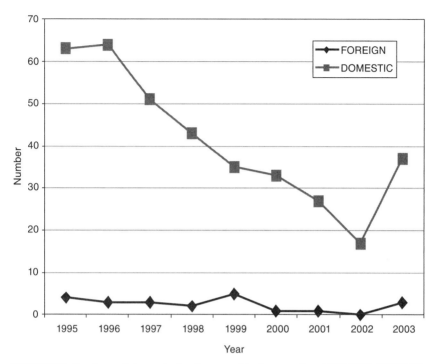

FIGURE 7 Foreign and domestic development fabs in operation, 1995-2003. SOURCE: Strategic Marketing Associates.

importance of close coordination between the process-technology development activities and commercial-scale production operations of the firm. Other empirical work has corroborated the importance of co-location of manufacturing and process-technology development activities (Hatch and Mowery, 1998; Macher and Mowery, 2003).

Figure 7 also reveals a decline in the number of development fabs during the period 1995-2003. This trend is another reflection of the structural change in the semiconductor industry that we discussed earlier. A growing share of global production capacity in the semiconductor industry now is accounted for by foundries, which typically do not use development fabs. Even within the IDM/systems-firm segment of the industry, the high capital costs of new commercial-volume production facilities (now more than $3 billion) have raised the minimum efficient scale of production fabs and appear to have contributed to some shrinkage in the number of new commercial-volume facilities.

Table 2 and Figure 8 provide additional detail on the location of development fabs. Table 2 disaggregates foreign development facilities by country of location and country of ownership. The United Kingdom, Germany, and the United

TABLE 2 Foreign Development Facilities in Operation, 1995-2003

Country of Location	1995	1996	1997	1998	1999	2000	2001	2002	2003
Canada	0	0	0	0	0	0	0	0	0
United Kingdom	2	2	1	0	0	0	0	0	0
France	0	0	0	1	1	0	0	0	0
Germany	0	0	2	1	1	1	1	0	2
Italy	0	0	0	0	1	0	0	0	0
Japan	1	0	0	0	1	0	0	0	0
Netherlands	0	0	0	0	0	0	0	0	0
Taiwan	0	0	0	0	0	0	0	0	1
United States	1	1	0	0	0	0	0	0	0
Country of Ownership	1995	1996	1997	1998	1999	2000	2001	2002	2003
Canada	0	1	0	0	0	0	0	0	0
United Kingdom	0	0	0	0	0	0	0	0	0
France	0	0	0	0	0	0	0	0	0
Germany	0	0	2	1	1	1	1	0	2
Italy	0	0	0	0	1	0	0	0	0
Japan	1	0	0	0	0	0	0	0	1
Netherlands	2	2	1	0	0	0	0	0	0
Taiwan	0	0	0	0	0	0	0	0	0
United States	1	0	0	1	3	0	0	0	0

SOURCE: Strategic Marketing Associates.

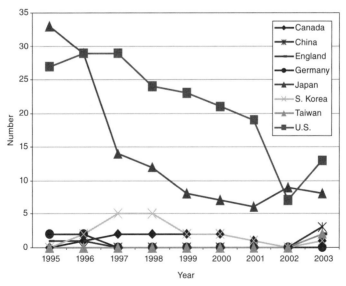

FIGURE 8 Geographic location of domestic development fabs, 1995-2003. SOURCE: Strategic Marketing Associates.

States are the leading locations for foreign-owned development fabs (keeping in mind that these represent a small fraction of total development-fab capacity). Semiconductor firms headquartered in Germany, the Netherlands, and the United States have a greater tendency to invest in foreign development fabs than do semiconductor firms based in other countries. Although interesting, the very small number of observations on offshore development fabs make it difficult to conduct additional analyses of the reasons for these differences.

Figure 8 disaggregates "domestic" development facilities by country of location (as in Table 2, these data refer to the stock of development fabs in operation). Japanese and U.S. semiconductor firms operate the majority of domestic development fabs, and those operated by U.S. semiconductor firms outnumber Japanese "homebound" development fabs during this period. Consistent with Figure 7, the number of domestic development fabs owned by Japanese and U.S. firms fell during the period 1995-2003. Also notable within these data is the lack of Taiwanese-owned domestic production facilities, reflecting the fact that the Taiwanese semiconductor industry is dominated by foundries.

International Technology-Development Alliances

Our third indicator of globalization of innovation-related activities is the innovation-related alliances formed by U.S. and non-U.S. semiconductor firms during the period 1990-1999. The data that we use to track alliances in this industry cover both domestic and international alliances and were obtained from the *Profiles of IC Manufacturers and Suppliers* published by Integrated Circuit Engineering (ICE), a semiconductor industry market research firm.

Alliances in the ICE database include a number of different activities related to innovation and technology development in semiconductors. Some of these alliances consist of production sourcing agreements between fabless and foundry firms, whereas others cover the development and transfer of process technology among firms. These data unfortunately do not include alliances focused on collaborative product development between fabless semiconductor firms and systems firms. The alliance data also lack information on the revenues, investments, or assets associated with individual alliances, which means that we are unable to weight individual alliances by some measure of their economic importance. In spite of these problems, the ICE data have some advantages. They enable us to track the dissolution as well as the formation of alliances, and thus provide information on both the rate of new alliance formation and the "stock" of semiconductor industry alliances in existence during our sample period.

Figure 9 displays the number of newly formed and ongoing alliances in the semiconductor industry during the period 1990-1999. The number of alliances grew during the early 1990s, reached a peak during the middle of the decade, and has gradually declined since 1995 as the rate of alliance formation has decreased.

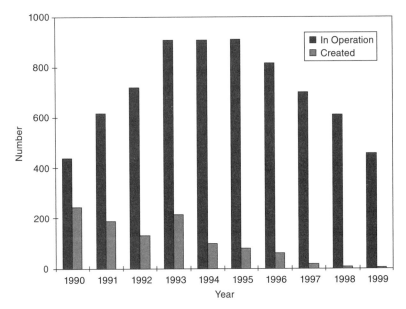

FIGURE 9 Semiconductor industry alliances, 1990-1999. SOURCE: Integrated Circuit Engineering (ICE).

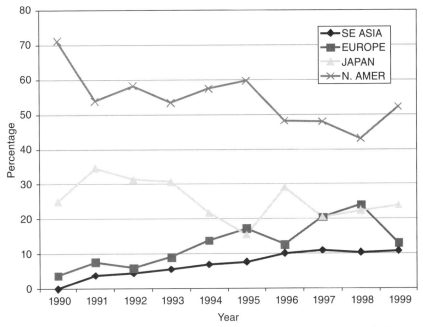

FIGURE 10 Home region of semiconductor alliance participants, newly formed alliances, 1990-1999. SOURCE: Integrated Circuit Engineering (ICE).

The annual "stock" of alliances in operation averaged more than 400 during the period 1990-1999.

The regional composition of alliances also changed substantially during the 1990s. Figure 10 depicts trends in the regional composition of alliance partner firms for newly formed alliances during the period 1990-1999, based on partner-firm home countries. The share of North American firms in newly formed alliances decreased from roughly 70 percent in 1990 to slightly more than 50 percent in 1999, while the share of European and Southeast Asian firms increased.

Figure 11 disaggregates the stock of alliances in Figure 9 by the nationality of the firms participating in them. Domestic alliances are those in which all partner firms are headquartered in the same country, and international alliances are those for which at least one partner firm is headquartered in a different country. U.S. domestic alliances declined from roughly 35 percent of semiconductor industry alliances in 1990 to slightly more than 20 percent in 1999, although the 1999 share represented an increase from its low point of 10 percent in 1996. The share of alliances between U.S. and non-U.S. firms ("U.S. Intl") within total industry alliances was essentially unchanged in 1999 from its 1990 level of 40 percent, although this share grew to more than 60 percent by 1996 before declin-

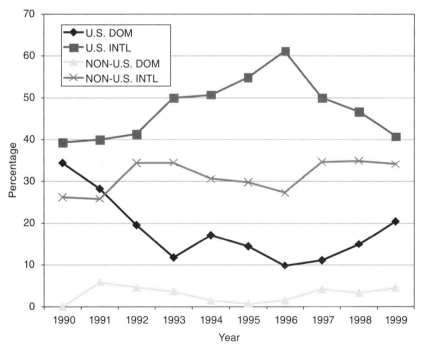

FIGURE 11 Home region of semiconductor industry alliance participants, alliance "stock," 1990-1999. SOURCE: Integrated Circuit Engineering (ICE).

ing. The share of non-U.S. domestic alliances also declined during the period 1990-1999, while non-U.S. international alliances increased from slightly more than 25 percent of total alliances in 1990 to roughly 35 percent by 1999.

Taken together, Figures 9, 10, and 11 suggest a slowdown in the rate of formation of alliances for R&D and technology development by all firms in the semiconductor industry during the 1990s, keeping in mind that we lack information on the size or economic significance of individual alliances. But the behavior of U.S. and non-U.S. semiconductor firms presents some contrasts that are not well understood. U.S. semiconductor firms experienced a period of significant growth during the early 1990s in their international alliance activities, followed by a decline in international alliance formation. These trends contrast with those for non-U.S. firms, which appear to have expanded their share of the shrinking number of newly formed international alliance activities throughout the 1990s. Non-U.S. firms also use alliances to team with firms from other nations or regions. If anything, globalization through alliances therefore appears to have been more persistent and intensive for non-U.S. semiconductor firms during the latter half of the 1990s.

Another analysis of alliance formation during the late 20th century by Hagedoorn (2002) used a different source of data but also concluded that the share of international undertakings declined from the 1980s through the 1990s. Moreover, Hagedoorn found that "information technology" R&D alliances (including semiconductors and a number of other industries) were less likely to involve cross-border relationships than was true of other sectors. The reasons for the apparent decline in the propensity of semiconductor firms to enter international R&D alliances, along with the (apparently) lower rate of formation of international R&D alliances by firms in semiconductors and related industries, remain unclear. Hagedoorn's (2002) examination of R&D alliance data also does not shed light on the relative importance of participant firms from the United States and other industrial economies. Nonetheless, the results of the Hagedoorn analysis are broadly consistent with our findings, suggesting that this vehicle for globalization of innovation-related activities in the semiconductor industry declined somewhat in importance during the 1990s, especially for U.S. firms.

Patenting

Patents, our final indicator of globalization in innovation, are an input to the innovation process rather than a direct measure of innovation. Nevertheless, these data, which are based on the semiconductor technology classification developed by the U.S. Patent and Trademark Office (USPTO), have several useful features, not least in making it possible to locate the site of the inventive activity that resulted in the patent. We are also able to weight our patent observations by various citation-based measures of the significance of individual patents, to control somewhat for the skewed distribution of patents by economic or technological importance.[4]

The semiconductor industry is a high-technology sector for which patents for many years were viewed as unimportant, reflecting the fact that (among other things) much of the most important innovative activity related to process innovation was protected through secrecy or inimitability, and cross-licensing of patents was widespread. As Hall and Ziedonis (2001) point out, however, patenting by both IDMs and fabless semiconductor firms grew rapidly during the 1980s, following changes in U.S. patent policy and some large financial awards in patent-infringement cases. We therefore believe that the period covered by our patent analysis is one during which semiconductor firms patented extensively, and patents therefore should serve as a reasonable proxy for inventive activity.

Our empirical analysis utilizes patents assigned to 217 U.S. and non-U.S. semiconductor firms during the period 1991-2003 (based on year of application). This dataset includes almost 114,000 patents from more than 80 patent classes, as identified by the USPTO's Office of Technology Assessment and Forecast and Hall and Ziedonis. (The Appendix in this chapter lists the patent classes included in our dataset.) We also collected data on citations to these patents in subsequently issued patents in all patent classes.

We use information on the location of the inventors listed on each patent as an indicator of the site of the inventive activity that produced the patent. Based on a comparison of the reported site of the invention with the headquarters location for each company in our dataset, each patent is assigned to one of the following mutually exclusive categories.

- **Domestic Patents:** patents whose inventors are all located in the home country of the controlling company;
- **Foreign Patents:** patents whose inventors are all located in countries other than the home country of the controlling company; and
- **International Collaboration Patents:** patents that have at least one inventor located in the home country and at least one inventor located in another country.[5]

We corrected patent data for self-citations (excluding citations to other patents filed by the assignee), and created forward citation windows of 2, 3, and 4

---

[4]Patent citation data were collected with the help of the Metrics Group Division of UTEK-EKMS, an IP strategy company based in Boston.

[5]Bergek and Bruzelius (2005) argue that the "inventor location" data often yield misleading information, because of firm-specific differences in the attribution of patents to inventors or inventor mobility between the time of application and the time of issue of the patent. It is plausible that such "noise" may affect inferences from cross-sectional analyses of patent data. We use the "inventor location" information for longitudinal analysis, however, and there is little reason to believe that the problems identified by Bergek and Bruzelius have become significantly more severe during the time period of this analysis.

years after publication for each patent, yielding a total of 402,865 forward citations that omit self-citations. In our citation analysis that follows, we use a 3-year
"window" for patent citations (i.e., we limit the count of citations to those in the
first 3 years after the year of issue of the relevant patent).

Any comparisons of patents issued to U.S. and non-U.S. firms must be interpreted carefully. For most non-U.S. firms, the higher costs of seeking a patent
in the United States in addition to a home-country patent mean that U.S. patent
protection will be obtained for only the more valuable patents in these firms'
portfolios. A simple comparison of U.S. patents assigned to U.S. firms with those
assigned to non-U.S. firms thus may yield misleading results because of the different underlying "quality" of the two patent groups.

Accordingly, our comparisons of U.S. patents assigned to U.S. and non-U.S.
firms use only U.S.-assigned patents with "equivalents" in either the Japanese
Patent Office (JPO) or the European Patent Office (EPO). In other words, our
discussion of U.S. and non-U.S. semiconductor firms' patenting that follows (but
only that discussion) includes only the semiconductor patents assigned to U.S.
firms for which a patent on the same or a very similar invention also has been
issued in one of these other major markets. We used the Delphion international
patent database to identify U.S.-assigned patents for which an equivalent patent
has been issued by either the JPO or the EPO.

Figure 12 depicts trends during the period 1991-2003 in the share of all semiconductor patents assigned to U.S. firms that were created by domestic inventors
(either an individual located in the United States or an "all-domestic" inventor
team) and the share of semiconductor patents assigned to U.S. firms that involved

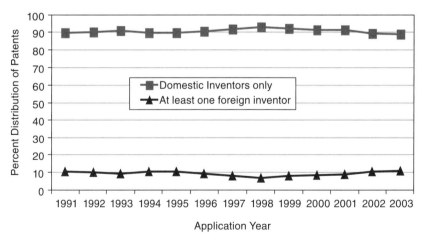

FIGURE 12 Domestic and offshore U.S.-assigned semiconductor patents, 1991-2003
SOURCE: Thomson Delphion Consulting.

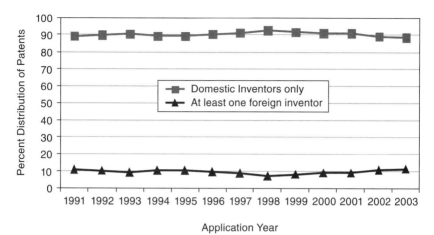

FIGURE 13 Domestic and offshore U.S.-assigned semiconductor patents, IDM and systems firms, 1991-2003. SOURCE: Thomson Delphion Consulting.

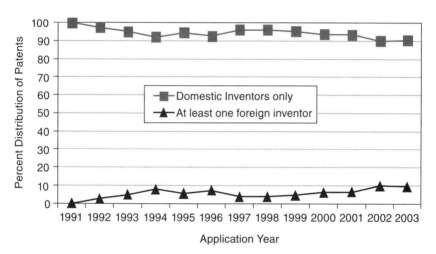

FIGURE 14 Domestic and offshore U.S.-assigned semiconductor patents, fabless firms, 1991-2003. SOURCE: Thomson Delphion Consulting.

at least one offshore inventor (combining foreign and international collaborative patents). The absence of any strong trend during the decade in this measure of offshore inventive activity is striking—the domestic share remains stable at roughly 90 percent throughout the period. A similar lack of growth in offshore inventive activity also is apparent in Figure 13, which shows the same "site of invention" trends for the IDM and systems firms in our sample. Although the number of patents for fabless firms in our sample is considerably smaller, Figure 14 reveals

TABLE 3 Firm HQ and Location of Inventive Activity, U.S. Patents

		One Inventor Located in								
	1994-2003	USA	Europe	Japan	Taiwan	Korea	Israel	Canada	Singapore	Others
Firm HQ in	USA	86.5%	6.5%	3.8%	0.2%	0.2%	0.9%	0.4%	0.3%	0.7%
	Europe	28.4%	60.0%	1.5%	7.1%	1.3%	0.2%	0.3%	0.5%	0.4%
	Japan	4.2%	0.7%	94.8%	0.0%	0.0%	0.0%	0.0%	0.1%	0.1%
	Taiwan	7.6%	0.1%	1.0%	90.7%	0.1%	0.0%	0.0%	0.1%	0.5%
	Korea	2.5%	0.1%	0.5%	0.0%	96.1%	0.0%	0.1%	0.1%	0.8%
	Israel	5.3%	0.0%	0.0%	0.0%	0.0%	91.6%	3.1%	0.0%	0.0%
	Canada	13.3%	19.1%	1.2%	0.0%	0.0%	0.0%	66.5%	0.0%	0.0%
	Singapore	18.5%	1.0%	0.0%	0.0%	0.0%	0.0%	0.0%	80.4%	0.3%

		One Inventor Located in								
	1996-1999	USA	Europe	Japan	Taiwan	Korea	Israel	Canada	Singapore	Others
Firm HQ in	USA	86.9%	6.4%	4.1%	0.1%	0.2%	0.7%	0.4%	0.4%	0.5%
	Europe	30.6%	57.2%	1.4%	9.1%	0.6%	0.3%	0.2%	0.3%	0.3%
	Japan	4.2%	0.7%	94.8%	0.0%	0.0%	0.0%	0.0%	0.1%	0.1%
	Taiwan	9.2%	0.1%	1.5%	88.7%	0.0%	0.0%	0.0%	0.0%	0.4%
	Korea	2.4%	0.0%	0.5%	0.0%	96.1%	0.0%	0.0%	0.0%	0.9%
	Israel	6.7%	0.0%	0.0%	0.0%	0.0%	88.9%	4.4%	0.0%	0.0%
	Canada	13.1%	20.4%	1.5%	0.0%	0.0%	0.0%	65.0%	0.0%	0.0%
	Singapore	25.0%	0.8%	0.0%	0.0%	0.0%	0.0%	0.0%	73.9%	0.8%

		One Inventor Located in								
	2000-2003	USA	Europe	Japan	Taiwan	Korea	Israel	Canada	Singapore	Others
Firm HQ in	USA	85.1%	8.1%	3.8%	0.2%	0.2%	0.3%	0.3%	0.2%	0.9%
	Europe	23.8%	62.6%	2.0%	6.4%	2.7%	0.2%	0.5%	0.7%	0.7%
	Japan	4.0%	0.8%	95.0%	0.0%	0.0%	0.0%	0.0%	0.1%	0.0%
	Taiwan	5.7%	0.1%	0.5%	92.8%	0.2%	0.0%	0.0%	0.2%	0.3%
	Korea	8.7%	0.0%	0.0%	0.0%	82.6%	0.0%	0.0%	4.3%	4.3%
	Israel	3.8%	0.0%	0.0%	0.0%	0.0%	94.9%	1.3%	0.0%	0.0%
	Canada	16.0%	15.4%	0.6%	0.0%	0.0%	0.0%	67.9%	0.0%	0.0%
	Singapore	14.2%	1.4%	0.0%	0.0%	0.0%	0.0%	0.0%	84.5%	0.0%

SOURCE: Thomson Delphion Consulting.

a modest shift toward greater offshore inventive activity among fabless firms.[6] Even for this group of firms, however, patenting remains dominated by home-country inventive activity.

Table 3 includes patents assigned to U.S. (for which an international "equiva-

---

[6]We also analyzed the share of forward citations accounted for by the "home-invented" and "off-shore participant" subsamples in our patent database. Interestingly, citations are proportionate to the patent shares (i.e., there is no evidence that home-invented patents are cited much more intensively than those for which offshore inventors are involved).

lent" patent, as defined earlier, was found) and non-U.S. firms, disaggregating the data by home country of patent assignee and site of inventive activity for 1994-2003, and further splitting the data into 1996-1999 and 2000-2003 subperiods. Japanese firms' inventive activity is dominated by home-country inventors to a greater extent than is true of either U.S. or European semiconductor firms for the 1994-2003 period. Japanese inventors are listed on almost 95 percent of Japanese firms' U.S. patents, whereas U.S. inventors are listed on almost 87 percent of U.S. patents and European inventors appear on 60 percent of European firms' U.S. patents. Looking at the off-diagonal portions of Table 3, European inventors account for almost twice as large a share of U.S. semiconductor firms' patents as Japanese inventors. U.S. inventors appear on nearly 30 percent of European firms' U.S. patents, and Japanese inventors on less than 2 percent. A comparison of the two subperiods does not reveal significant differences, although there is some indication that the "homeboundedness" of the inventive activities of European, Japanese, and Taiwanese semiconductor firms is increasing slightly. For all but Canadian semiconductor firms, the single most important offshore site for inventive activity is the United States, which on the basis of this evidence remains the dominant site for the offshore inventive activities of most non-U.S. semiconductor firms. For U.S. semiconductor firms, the leading offshore site for inventive activity in both the 1996-1999 and the 2000-2003 subperiods is Europe, followed closely by Japan.

Figure 15 depicts the leading locations of offshore inventors in the semiconductor patents assigned to U.S. firms for the 1996-1999 and 2000-2003 subperiods. The figure highlights considerable change in these locations over time and differences in the location of offshore inventive activity between fabless and other firms in the U.S. semiconductor industry. Canadian inventors play a more prominent role in the offshore patenting of U.S. fabless firms than is true of IDM and systems firms, accounting for less than 4 percent of the offshore inventive activity of IDMs and system firms versus more than 20 percent for fabless firms, in both periods. The Japanese share of U.S. fabless firms' offshore patenting also declines between the two subperiods, perhaps reflecting the growth in systems design in non-Japanese Asia (notably Taiwan, South Korea, and Singapore). European inventors are of comparable importance for both U.S. IDMs and U.S. fabless firms in both time periods. A comparison of the two subperiods for both groups of firms also highlights the shift in the importance of non-Japanese Asian inventors. The share of "other Asia" inventors (particularly Singapore) for fabless firms increases more than sixfold (albeit from a very modest initial level), and the "other Asia" share for IDMs and systems firms nearly doubles.

Figures 16 and 17 compare the invention locations for U.S. semiconductor patents assigned to U.S. and non-U.S. firms, using only the U.S.-assigned patents for which an international "equivalent" patent exists. The data in the figures are based on a random sample of 5,000 patents from the semiconductor patent portfolios of Asian, European, and U.S. firms. The random-sampling procedure, which

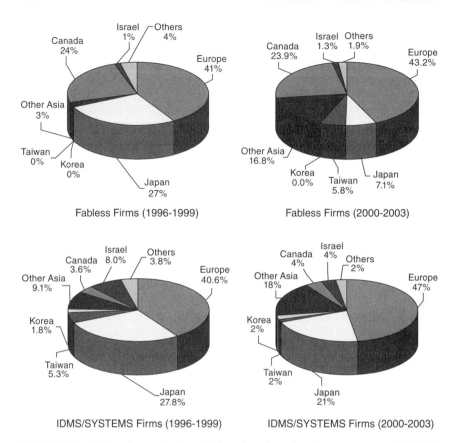

FIGURE 15 Offshore invention sites, U.S.-assigned semiconductor patents, 1996-2003
SOURCE: Thomson Delphion Consulting.

is stratified by year and patent class within our overall semiconductor-patent "family," was adopted to provide a patent sample that was not affected by the different propensities of U.S., Asian, and European firms to obtain foreign as well as domestic patents. Figure 16 displays data on the reported site of inventions for the 1996-1999 and 2000-2003 subperiods, and Figure 17 displays data on the shares of citations within the first 3 years after issue for these patents. Although the United States is the largest single site of inventive activity resulting in patents throughout the 1996-2003 period, its dominance declines modestly between the 1996-1999 and 2000-2003 subperiods. The share of patents attributable to Japanese or Taiwanese sites is essentially unchanged throughout the 1993-2003 period, whereas the European share of inventive activity increases between the 1996-1999 and 2000-2003 subperiods.

Comparing the shares of patents with the shares of citations in Figure 17

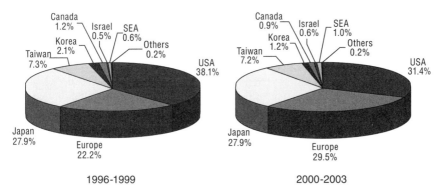

FIGURE 16 Distribution of invention sites, all USPTO semiconductor patents, 1996-1999.

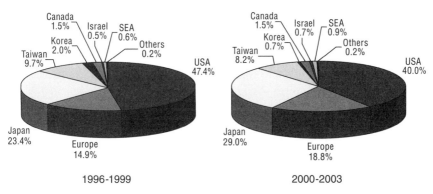

FIGURE 17 Distribution of citations by invention site, all USPTO semiconductor patents, 1991-2003. SOURCE: Thomson Delphion Consulting and UTEK-EKMS.

reveals that patents from Japanese and European invention sites tend to be "undercited" (their share of citations is smaller than their share of overall patents), whereas U.S. patents are "overcited" relative to their share of overall patenting. Patents with Taiwanese invention sites are cited slightly more intensively relative to their share during the 2000-2003 subperiod.

Overall, the results of our analysis of the site of inventive activity resulting in U.S. patents support the original findings of the analysis of patenting by firms from a broader sample of industries by Patel and Pavitt (1991), who found that large multinational firms' patenting was dominated by home-based inventive activity. The evidence on post-issue citations suggests that the most "important" semiconductor patents are slightly more likely to result from domestic inventive activity, but this conclusion should be qualified by a recognition of the small size of the sample on which it is based. The trends in patenting for fabless firms suggest that the demanding requirements for close collaboration between semiconductor component designers and systems firms may be causing some shift within this segment of the semiconductor industry toward greater reliance on foreign inventive activity in patenting; but any such trend is very modest. Although additional analysis is required, this finding concerning fabless firms' offshore patenting is consistent with the "market demand-exploitation" motive for locating R&D offshore discussed by Gerybadze and Reger (1999)—an important factor in the location of firms' R&D is their need to be near their innovative customers.

IMPLICATIONS FOR ENGINEERING EMPLOYMENT

The evidence presented in this chapter suggests that globalization and structural change in the semiconductor industry have resulted in significant growth in offshore manufacturing capacity, much of which remains owned by U.S. producers. But this growth in offshore production capacity has had far less significant effects on the location of innovation-related activities within the semiconductor industry, including the innovation-related activities of U.S. semiconductor firms. Indeed, the growth of offshore foundry production capacity in Southeast Asia has helped sustain the growth of employment of engineers and designers in U.S.-based fabless semiconductor firms.

Innovation-related activities in this industry include a number of different activities, such as semiconductor chip design, process-technology development, and product-technology development. The limited evidence on these three activities discussed earlier in this chapter suggests that semiconductor chip design is the least "homebound" of the three. Process-technology development, as measured by the siting of development fabs, does not appear to have moved offshore to any extent, while patenting (which includes product- and process-technology development) also displays little evidence of significant offshore relocation.

A recent analysis of engineering-employment trends in the U.S. semiconductor industry (Brown and Linden, 2006b) suggests that the employment effects of shifts in the location of chip design for U.S. engineers have been modest thus far. Brown and Linden (2006b) found that employment and earnings growth were higher during the 2000-2005 period for engineers employed in the U.S. semiconductor industry than in other U.S. industries. Employment growth during this period was strongest for electrical, computer hardware, and electronic engineers, but was negative for industrial engineers within the semiconductor industry. Brown and Linden (2006b) report some evidence that median earnings for "mature" engineers (50 years and older) employed in the semiconductor industry are lower than for younger engineers, but this tendency appears throughout the time period (2000-2004) that they analyze, rather than becoming more pronounced in the most recent year. Moreover, interpreting these trends is difficult, since they reflect some tendency for engineers to move into management positions as they mature. The individuals classified as "engineers" among this older cohort thus have remained out of management, which may affect their reported earnings.

Similarly to many other dimensions of globalization of innovation-related activities in the semiconductor industry, the available data are not well suited to analysis of the employment effects of any movement of different categories of innovation-related activities to non-U.S. locations. The most detailed analysis of employment trends in the U.S. semiconductor industry during the 2000-2005 period does not reveal significant erosion in engineering employment or earnings during the period (Brown and Linden, 2006b). The available evidence, imperfect as it is, thus does not support grave concern about the employment consequences of recent shifts in the location of innovation-related activities in the semiconductor industry.

The U.S. employment data are largely "backward-looking" indicators of the employment consequences of globalization of innovation-related activities in the semiconductor industry. The implications for future engineering employment trends in the U.S. semiconductor industry associated with the growing numbers of design engineers in China, India, and Taiwan are unclear (Brown and Linden, 2006b). Despite government sponsorship and local access to systems firms, China's chip design capabilities do not yet represent a viable product design outsourcing alternative. India offers certain product design advantages given the predominant use of the English language and its thriving software sector, but zero government involvement, limited manufacturing capacity, and no major fabless firm presence have hampered success. Taiwan appears best poised to become a viable offshoring and outsourcing alternative, as well as a significant competitive threat, to U.S. semiconductor firms in the near term, given its focused government programs, locally owned fabless design segment, and close proximity to systems houses and foundries. These factors suggest that U.S. semiconductor firms' dominance in product design could be challenged in the future, potentially reducing U.S. employment in this innovation-related activity. But here too, even a partial "hollowing out" of the U.S. product design segment seems unlikely.

IMPLICATIONS FOR FIRM STRATEGY AND PUBLIC POLICY

An earlier study of the U.S. semiconductor industry's innovative and competitive performance (Macher et al., 1999) noted that U.S. firms' development of new products (including microprocessors and digital-signal processing chips) and new business models (notably the growth of fabless firms) had enabled them to overcome a significant competitiveness crisis during the late 1980s and early 1990s. U.S. semiconductor firms remain globally competitive in the face of rapid innovation by non-U.S. semiconductor firms, but structural change in the global semiconductor industry has resulted in considerable change in the structure of the innovation process.

Fabless firms in particular seek to develop closer collaborative relationships with major systems firms based outside of the United States. Shorter product life cycles and the increased variety of individually smaller applications that utilize semiconductor components mean that the IDMs face greater risks from swings in demand as the costs of their production facilities continue to rise. The design and (especially) the manufacturing capabilities of foreign regions also have improved significantly since the 1980s, creating new opportunities for U.S. firms to exploit a global division of labor in semiconductor design and manufacturing. Among other things, this emergent division of labor has supported the rapid growth of fabless semiconductor firms in the United States.

In contrast to the industry challenges of the 1980s that threatened the viability and very survival of the U.S. semiconductor industry, the challenges of the early 21st century stem from the need to manage this global division of labor effectively and strategically while maintaining leadership in innovation. These challenges are hardly new, as lower-productivity, labor-cost-sensitive functions in many U.S. manufacturing industries (and a growing array of U.S. nonmanufacturing industries, such as software and financial services—see Chapters 2 and 10, respectively) have moved to lower-cost areas of the global economy. Many of these regions have developed strong educational and economic infrastructures that can support the creation of productive labor forces and contribute to such innovation-related activities as product design and process engineering.

The offshoring and outsourcing of various activities by U.S. firms has a long history, but so too do the innovative responses of successful U.S. firms. Even as the more cost-sensitive, lower-value-added activities have been shifted to offshore locations, U.S. firms have maintained their global competitiveness by developing and introducing innovative new products (e.g., PCs and communications) and business models (e.g., fabless semiconductor firms). In the semiconductor industry, product innovation will remain central, and manufacturing-process innovation is likely to focus on a narrower range of products in which U.S. IDMs remain dominant. In other words, the strategic management of innovation becomes even more important for the competitive performance of semiconductor firms that seek to exploit the emerging global division of labor in product design and manufacturing while maintaining strength in product and process innovation.

It seems likely, for example, that the remaining U.S. IDMs will continue to exploit offshore sites for manufacturing while relying on foundries to serve a larger portion of their production requirements for products that are slightly behind the "bleeding edge" of technology—the "fab-lite" model of production. The continuing growth of semiconductor foundries will provide further opportunities for expansion by U.S. fabless firms, although these firms also are likely to shift at least some of their design-related activities to offshore locations because of the presence of major customers in these areas.

In spite of the powerful forces that are shifting some design, manufacturing, and other functions to offshore locations, the bulk of U.S. semiconductor firms' "inventive activity" did not shift during the 1990s. As measured imperfectly by the reported residence of inventors listed on U.S. patents, the inventive activities of U.S. semiconductor firms remain concentrated in the United States. Moreover, the inventive activities of non-U.S. semiconductor firms, as measured by similar information for their U.S. patents, also appear to be concentrated in their home countries. This tendency for inventive activity to remain homebound was first pointed out in an analysis by Patel and Pavitt (1991) of patenting by multinational firms. The "nonglobalization" of patenting activity seems to reflect the strong dependence of inventive activity on domestic sources of fundamental research and skilled researchers. Despite remarkable advances in the codification and global transmission of scientific research, access to such research results for purposes of inventive activity remains surprisingly national in scope. And the apparent importance of the national science and engineering base for domestic inventive activity reinforces the importance of another key governmental function—funding the scientific and engineering research and education that support this domestic knowledge pool.

The most important implications of this study for U.S. public policy thus relate to (1) the importance of continued (and arguably renewed) federal funding for R&D in the engineering and physical sciences in industry and universities and (2) the importance of public support (which may be financial or regulatory) for more rapid development of the "information infrastructure" (e.g., broadband communications) that can support the growth of a large domestic market of demanding and sophisticated consumers that will in turn spawn innovations in information and electronics technologies.

Much of the remarkable record of innovation in the U.S. semiconductor and related IT industries that spans the 1945-2006 period rested on substantial investments of public funds in R&D that supported industrial research and innovation, as well as the training of generations of engineers and scientists. Much of this federal R&D investment was linked to national-security goals, and the end of the Cold War and associated defense "build-down" led to significant reductions in growth and in some cases the level of federal funding for R&D in the physical and engineering sciences, especially in academic institutions. Growth in federal R&D investment since the late 1980s has been dominated by growth in biomedical R&D funding. Although a portion of this biomedical R&D investment has supported education and training in the physical sciences and engineering, the imbalance in investment trends, if not reversed, could have detrimental consequences for the continued innovative vitality of the U.S. semiconductor industry and related industries.

The importance of market demand in the locational structure of innovation in the semiconductor industry and other high-technology industries (see Chapters 8, 9, and 10, which illustrate the importance of local demand in service industries as well) is difficult to overstate. We have noted that the declining share of semiconductor consumption accounted for by the PC has been associated with the growth of new markets for semiconductor components (e.g., wireless communications devices) that involve major non-U.S. systems firms. Moreover, many of the most innovative, demanding, and sophisticated users of such devices now are located in non-U.S. markets (e.g., South Korea for wireless devices, or Finland for broadband-based applications). Historically, U.S. semiconductor firms have derived enormous competitive advantages from their ability to serve (and learn from) a large domestic market populated by sophisticated and demanding users—in some cases, these demanding users were major institutions, such as the military.

One important reason for the rapid development of browser-based applications and new business models in the early days of the World Wide Web, which relied on innovations developed in Europe, was the broad diffusion and low cost of PCs within the United States, as well as the low costs of accessing computer networks (Mowery and Simcoe, 2002). Government policy can play a significant role in the creation or support of markets for advanced technologies (recall that the Internet was aided by substantial federal as well as private funding) by supporting investment in the infrastructure that proved so fruitful in the early days of computer networking and developing regulatory policies that create incentives for the large private investments in the communications infrastructures needed for the emergence of new applications, products, and services. R&D and related investments from around the globe are likely to flow to markets in which users demand the most advanced technologies and where these users have access to an array of options for developing new applications of these technologies. Such markets are likely to rely in part on a sophisticated wireless and high-speed broadband communications infrastructure.

CONCLUSION

The U.S. and global semiconductor industries have experienced significant structural change since 1980 with the growth of specialized design and manufacturing firms. The growth of new products that use semiconductor components and the entry of firms from Southeast Asia also have contributed to growth in offshore manufacturing capacity within the industry, much of which remains under the control of U.S. semiconductor firms. Nevertheless, there is surprisingly little evidence that the innovation-related activities of U.S. semiconductor firms have moved offshore to a comparable extent. Overall, the results of this descriptive examination of an array of measures of the "globalization" of innovation-related activities in the semiconductor industry support the findings of Patel and Pavitt (1991) from more than a decade ago. The innovation-related activities of otherwise global firms in this industry remain remarkably nonglobalized, even in the face of expanded international flows of capital and technology, far-reaching change in the structure of semiconductor manufacturing, and significant shifts in the structure of demand.

How can one explain these findings? The homebound nature of process innovation investments is perhaps the least surprising, given the complexity of process technology within the semiconductor industry and the demanding requirements for coordination between product and process innovation. Moreover, the emergence of vertically specialized foundries that do not rely on development facilities to the same extent as IDMs means that our data on the location of development fabs exclude process innovation in a segment of semiconductor manufacturing that has grown considerably and gives every indication of continued growth. It is also important to highlight the retrospective nature of these indicators, which (especially in the case of patents) reflect R&D and related investments made years before their effects appear in these data. Trends in patenting in the late 1990s thus reflect actions or strategies that were put in place in the early 1990s and, like most other scholars, we have almost no forward-looking indicators.

Some of our other indicators, such as the NSF R&D investment data, exclude non-U.S. firms, and the data themselves may well omit significant innovation-related activities. It is plausible, for example, that much of the design work of U.S.-based fabless firms is not captured by the NSF R&D surveys. U.S.-based semiconductor firms also benefit from the strength of their home-based innovation system, especially in the product design area. "Home-base augmentation" (Kuemmerle, 1999) thus may be a relatively minor factor for U.S. firms' R&D investment strategies and a significant motive for non-U.S. firms' R&D investments in the United States and elsewhere. Moreover, the exploitation by U.S. semiconductor firms of these "home-base" advantages may not require significant offshore R&D investment to complement offshore production investment. Indeed, one hypothesized motivation for offshore R&D that receives the most support from our analysis is the "market-demand exploitation" hypothesis of Gerybadze and Reger (1999), which may be particularly relevant to the patenting activities of U.S.-based fabless firms.

The trends highlighted in our discussion of technology-development and R&D alliances in this industry also raise interesting questions. The declining rate of formation of domestic and international alliances by semiconductor firms throughout the industry is surprising but may reflect some exhaustion of the pool of potential alliance partners or projects. Nontariff barriers to U.S. firms' access to foreign markets resulting from government procurement restrictions or other policies have been reduced during the past decade in several industries (OECD, 2005), and it is possible that these reductions in market-access barriers have reduced U.S. firms' incentives to pursue collaborative ventures with non-U.S. firms. Our alliance data also do not fully capture the types of alliances that are important to the fabless firm segment of the U.S. semiconductor industry, and thereby understate the significance of international alliances within the overall industry. These data nevertheless suggest some growth in the participation by non-U.S. firms in domestic and foreign alliances, especially among non-Japanese Asian firms. Some portion of this alliance activity may be motivated by access to the Chinese mainland market, where nontariff barriers remain significant.

The results of our analysis of patents provide the strongest support for the original findings of Patel and Pavitt (1991), but these results also must be treated with caution. As we pointed out earlier, patents omit many of the innovation-related activities that are most important to the creation or maintenance of competitive advantage for IDMs and fabless firms alike, and our findings for these indicators accordingly must be qualified.

Does the nonglobalized character of U.S. semiconductor firms' innovation-related activities differ significantly from that of semiconductor firms based in other nations? The data presented in Table 3 indicate that the homebound character of U.S. semiconductor firms' patenting is similar to that of semiconductor firms headquartered in other nations, as is the homebound character of the process-technology development facilities that U.S. and non-U.S. IDMs and systems firms operate.

Are the trends discussed in this paper for the semiconductor industry representative of other high-technology industries, or is this industry unique? Comparing the extent of "nonglobalization" in the semiconductor industry with that of other knowledge-intensive industries is difficult, since few detailed studies of these trends have been undertaken for other industries. Offshore R&D investment by U.S. firms in electronic components accounted for a smaller share of industry-funded R&D during the 1990s than is true of U.S. firms in pharmaceuticals, where more than 14 percent of industry-funded R&D was performed offshore in 2001 (National Science Board, 2006). Like semiconductors, the pharmaceuticals industry underwent considerable structural change and vertical specialization during the 1990s (see Chapter 6), particularly through the entry of biotechnology firms that often specialize in drug discovery and contract research organizations that specialize in drug development (i.e., clinical trials). Unlike semiconductors, however, the structure of market demand in pharmaceuticals has undergone little significant change—the U.S. remains the most profitable single national market, thanks to the peculiar structure of its health care delivery system.

Why do we observe such contrasts between these two industries in the (apparent) share of offshore sites in innovation-related activities? One hypothesis appeals to the more diverse structure of products in the pharmaceuticals industry, combined with substantial scientific research capabilities (in many cases, based on public funding) in many non-U.S. industrial economies. The science underlying product innovation in various therapeutic classes, to say nothing of the growing variety of delivery mechanisms (topical, inhaled, subdermal, as well as oral), arguably spans a wider variety of fields and has become much more diverse during the past 20 years than is true of product innovation in semiconductors. These factors have supported the growth of significant clusters of scientific expertise in specific therapies or diseases that attract the R&D investments of U.S. pharmaceuticals firms. A large part of the "D" in pharmaceutical R&D also represents costs associated with conducting and administering clinical trials to diverse patient populations. The situation in semiconductors arguably is quite different—the structure of products and markets remain less complex, and government R&D programs have not created comparably accessible clusters of scientific and technological expertise.

As this speculative discussion suggests, the dynamics of globalization and nonglobalization in innovation are complex and reflect contrasting paths of evolution (at the industry level) within different national innovation systems, as well as the interplay among these national innovation systems, trade policy, and other influences. Still another important influence on the globalization of at least some types of pharmaceuticals R&D is regulatory policy in the offshore as well as domestic markets in which all global pharmaceuticals firms operate. Maintaining a significant R&D presence in their offshore markets may facilitate the management of clinical trials for new products that U.S. firms seek to introduce into these markets.

Even in the pharmaceuticals industry, however, Narin et al. (1997) have pointed out that the patents filed in the United States by non-U.S. (as well as U.S.) inventors tend to rely disproportionately on "home-country" science, as measured by the citations to scientific publications in their patent applications. The links between science and technology that contribute to much of the inventive activity that is embodied in patenting retain a considerable homebound element, rather than operating seamlessly and frictionlessly across national boundaries.

Overall, this discussion of the globalization of innovation-related activities in the U.S. semiconductor industry does not indicate an imminent policy-related "crisis" in the innovative capabilities of U.S. firms. The implications of our discussion for the employment of engineers in innovation-related activities in the U.S. semiconductor industry also are reasonably positive. As we have noted repeatedly, U.S. firms have reacted to the growth of offshore innovative and productive capabilities by developing novel business models that have enabled them to compete successfully in an array of new markets. The success of these innovative strategies has sustained innovation and employment in the U.S. semiconductor industry. Nonetheless, it seems clear that much of the innovative performance of U.S. semiconductor firms relies on the health of a complex domestic R&D infrastructure that has benefited from large investments of public funds during the past six decades. A second important historical contributor to the innovative performance of U.S. firms is the large domestic market of innovative users that these firms face. Sustaining both of these factors that have contributed to the innovative performance of U.S. semiconductors in an intensely competitive global industry will require innovations in policy by both government and industry for decades to come.

ACKNOWLEDGMENTS

Research for this chapter was supported by the Alfred P. Sloan Foundation, the Ewing M. Kauffman Foundation, and the Andrew W. Mellon Foundation.

A portion of David Mowery's research was supported by the National Science Foundation (SES-0531184). A portion of Alberto Di Minin's research was supported by the In-Sat Laboratory, Scuola Superiore Sant'Anna, Pisa, Italy. Deepak Hedge provided valuable comments on an earlier draft of this chapter.

REFERENCES

Appleyard, M. M., N. W. Hatch, and D.C. Mowery. (2000). Managing the development and transfer of process technologies in the semiconductor manufacturing industry. Pp. 183-207 in *The Nature and Dynamics of Organizational Capabilities*, G. Dosi, R. R. Nelson, and S. G. Winter, eds. London: Oxford University Press.

Arensman, R. (2003). Fabless goes global. *Electronic Business.* March 21.

Bergek, A., and M. Bruzelius. (2005). Patents with inventors from different countries: Exploring some methodological issues through a case study. Paper presented at the DRUID conference, June 27-29, Copenhagen, Denmark.

Braun, E., and S. MacDonald. (1978). *Revolution in Miniature: The History and Impact of Semiconductor Electronics.* Cambridge: Cambridge University Press.

Brown, C., and G. Linden. (2006a). Offshoring in the semiconductor industry: A historical perspective. In *Brookings Trade Forum 2005: Offshoring White-Collar Work*, S. M. Collins and L. Brainard, eds. Washington, D.C.: Brookings Institution Press.

Brown, C., and G. Linden. (2006b). *Semiconductor Engineers in a Global Economy. National Academy of Engineering Workshop on the Offshoring of Engineering: Facts, Myths, Unknowns and Implications.* Washington, D.C.: The National Academies Press.

Ernst, D. (2005). Complexity and internationalisation of innovation: Why is chip design moving to Asia? *International Journal of Innovation Management* 9(1):47-73.

Gerybadze, A., and G. Reger. (1999). Globalization of R&D: Recent changes in the management of innovation in transnational corporations. *Research Policy* 28(2-3):251-275.

Hagedoorn, J. (2002). Inter-firm R&D partnerships: An overview of major trends and patterns since 1960. *Research Policy* 31:477-492.

Hall, B. H., and R. H. Ziedonis. (2001). The patent paradox revisited: An empirical study of patenting in the U.S. semiconductor industry, 1979-1995. *RAND Journal of Economics* 32(1):101-128.

Hatch, N. W., and D. C. Mowery. (1998). Process innovation and learning by doing in semiconductor manufacturing. *Management Science* 44(1):1461-1477.

Integrated Circuit Engineering (1990-1999). Profiles: A worldwide survey of IC manufacturers and suppliers. Scottsdale, AZ: Integrated Circuit Engineering Corporation.

Kuemmerle, W. (1999). The drivers of foreign direct investment into research and development; An empirical investigation. *Journal of International Business Studies* 30(1):1-24.

Leachman, R. C., and C. H. Leachman. (2004). Globalization of semiconductors: Do real men have fabs, or virtual fabs? In *Locating Global Advantage: Industry Dynamics in the International Economy*, M. Kenney and R. Florida, eds. Palo Alto, Calif.: Stanford University Press.

Linden, G., and D. Somaya. (2003). System-on-a-chip integration in the semiconductor industry: Industry structure and firm strategies. *Industrial and Corporate Change* 12(3):545-576.

Linden, G., C. Brown, and M. Appleyard. (2004). The net world order's influence on global leadership in the semiconductor industry. Pp. 232-257 in *Locating Global Advantage: Industry Dynamics in the International Economy*, M. Kenney and R. Florida, eds. Palo Alto, Calif.: Stanford University Press.

Macher, J. T. (2006). Technological development and the boundaries of the firm: A knowledge-based examination in semiconductor manufacturing. *Management Science* 52(6):826-843.

Macher, J. T., and D. C. Mowery. (2003). Managing learning by doing: An empirical study in semiconductor manufacturing. *Journal of Product Innovation Management* 20(5):391-410.

Macher, J. T., and D. C. Mowery. (2004). Vertical specialization and industry structure in high technology industries. Pp. 317-356 in *Business Strategy over the Industry Lifecycle—Advances in Strategic Management* (Vol. 21), J. A. C. Baum and A. M. McGahan, eds. New York: Elsevier.

Macher, J. T., D. C. Mowery, and D. A. Hodges. (1998). Reversal of fortune? The recovery of the U.S. semiconductor industry. *California Management Review* 41(1):107-136.

Macher, J. T., D. C. Mowery, and D. A. Hodges. (1999). Semiconductors. Pp. 245-286 in *U.S. Industry in 2000: Studies in Competitive Performance*, D. C. Mowery, ed. Washington, D.C.: National Academy Press.

Mowery, D. C., and T. S. Simcoe. (2002). Is the Internet a U.S. invention? An economic and technological history of computer networking. *Research Policy* 31:1369-1387.

Narin, F., K. S. Hamilton, and D. Olivastro. (1997). The increasing linkage between U.S. technology and public science. *Research Policy* 26(3):317-330.

National Science Board (2006). *Science and Engineering Indicators 2006*. Arlington, VA: National Science Foundation.

OECD (Organisation for Economic Co-operation and Development). (2005). *Looking Beyond Tariffs: The Role of Non-Tariff Barriers in World Trade*. Paris: OECD Trade Policy Studies.

Patel, P., and K. Pavitt. (1991). Large firms in the production of the world's technology: An important case of "nonglobalisation". *Journal of International Business Studies* 22(1):1-20.

APPENDIX: SEMICONDUCTOR PATENT CLASSES

- 029. Subclasses: 116.1; 592; 602.1; 613; 729; 740; 827; 830; 832; 835; 840; 841; 854; 855; 025.01; 025.02; 025.03
- 065. Subclass: 152
- 073. Subclasses: 514.16; 721; 727; 754; 777; 862; 031.06
- 084. Subclasses: 676; 679
- 102. Subclasses: 202
- 117 Subclasses: all
- 118: Subclasses: 407; 408; 409; 410; 411; 412; 413; 414; 415; 669; 715; 716; 717; 718; 719; 720; 721; 722; 723; 724; 725; 726; 727; 728; 729; 730; 731; 732; 733; 900
- 134. Subclasses: 902; 001.2; 001.3
- 136. Subclasses: 243; 244; 245; 246; 247; 248; 249; 250; 251; 252; 253; 254; 255; 256; 257; 258; 259; 260; 261; 262; 263; 264; 265
- 148. Subclasses: 239; 033
- 156. Subclasses: 345; 625.1; 626; 627; 628; 636; 643; 644; 645; 646; 647; 648; 649; 650; 651; 652; 653; 654; 655; 656; 657; 658; 659; 660; 661; 662
- 164. Subclass: 091
- 174. Subclasses: 102; 261; 015.1; 016.3; 052.4; 052.5
- 194. Subclass: 216
- 204. Subclasses: 192; 206
- 205. Subclasses: 123; 157; 656; 915
- 206. Subclasses: 334; 710; 711; 832; 833
- 216. Subclasses: 002; 014; 016; 017; 023; 079; 099
- 219. Subclasses: 121.61; 385; 500; 501; 505; 638
- 228. Subclasses: 122.2; 123.1; 179.1; 903
- 250. Subclasses: 200; 208; 338.4; 339.03; 341.4; 370; 371; 390; 492; 552; 559
- 252. Subclasses: 950; 062.3
- 257. Subclasses: all
- 264. Subclasses: 272.17
- 307. Subclasses: 201; 270; 272; 291; 296; 355; 443; 446; 454; 455; 456; 463; 465; 468; 473; 475; 530; 651
- 310. Subclass: 303
- 313. Subclasses: 366; 367; 498; 499; 500; 523
- 315. Subclass: 408
- 323. Subclasses: 217; 223; 235; 237; 263; 265; 268; 300; 311; 313; 314; 315; 316; 319; 320; 350; 902; 907
- 324. Subclasses: 207.21;235; 252; 719; 722; 763; 765; 767; 768; 769; 158F; 158R
- 326. Subclasses: all

- 327. Subclasses: 109; 112; 127; 170; 186; 188; 189; 192; 193; 194; 195; 196; 203; 204; 206; 207; 208; 209; 210; 211; 212; 213; 214; 223; 224; 258; 262; 281; 288; 306; 324; 327; 328; 334; 366; 367; 368; 369; 370; 371; 372; 373; 389; 390; 391; 404; 405; 409; 410; 411; 412; 413; 416; 417; 419; 420; 421; 422; 423; 424; 425; 426; 427; 428; 429; 430; 431; 432; 433; 434; 435; 436; 437; 438; 439; 440; 441; 442; 443; 444; 445; 446; 447; 448; 449; 450; 451; 452; 453; 454; 455; 456; 457; 458; 459; 460; 461; 462; 463; 464; 465; 466; 467; 468; 469; 470; 471; 472; 473; 474; 475; 476; 477; 478; 479; 480; 481; 482; 483; 484; 485; 486; 487; 488; 489; 490; 491; 492; 493; 494; 495; 496; 497; 498; 499; 500; 501; 502; 503; 504; 505; 510; 511; 513; 527; 528; 529; 530; 536; 537; 538; 539; 541; 542; 543; 546; 562; 563; 564; 565; 566; 568; 569; 570; 571; 574; 575; 576; 577; 578; 579; 580; 581; 582; 583; 584; 585; 586; 587; 051; 065; 081; 408; 409; 410; 411; 412; 413
- 329. Subclasses: 301; 305; 314; 326; 342; 362; 364; 365; 369; 370
- 330. Subclasses: 114; 116; 117; 118; 124; 127; 128; 129; 140; 141; 142; 143; 144; 146; 147; 148; 149; 150; 151; 152; 153; 154; 155; 156; 157; 168; 172; 181; 182; 183; 185; 186; 192; 193; 199; 200; 202; 250; 252; 253; 254; 255; 260; 263; 264; 267; 269; 270; 272; 275; 277; 282; 285; 290; 292; 296; 297; 004.9; 007; 009; 044; 051; 056; 059; 061; 069; 070; 075; 076; 087; 299; 3
- 331. Subclasses: 107; 111; 008; 052; 1A
- 332. Subclasses: 102; 105; 110; 113; 116; 130; 135; 136; 146; 152; 164; 168; 177; 178
- 333. Subclasses: 103; 247
- 334. Subclasses: 015; 047
- 338. Subclass: 195
- 340. Subclasses: 146; 598; 634; 814; 815.45; 825
- 341. Subclasses: 118; 136; 143; 145; 156
- 346. Subclass: 150.1
- 347. Subclasses: 130; 238; 001; 059
- 348. Subclasses: 126; 294; 390; 391; 420; 801; 087
- 349. Subclasses: 140, 202, 041, 042, 047, 053
- 355. Subclass: 053
- 356. Subclass: 030
- 358. Subclasses: 261; 426; 482; 483; 513; 514; 037
- 359. Subclasses: 109; 248; 332; 342; 343; 344; 359; 360; 006
- 360. Subclasses: 051
- 361. Subclasses: 100; 196; 197; 198; 277; 519; 523; 525; 527; 537; 600; 697; 703; 717; 718; 723; 737; 763; 764; 783; 813; 820; 001; 056; 091
- 362. Subclass: 800
- 363. Subclasses: 108; 109; 114; 123; 125; 126; 127; 128; 131; 135; 159; 160; 163; 010; 027; 037; 041; 048; 049; 053; 054; 056; 057; 060; 070; 077
- 364. Subclasses: 232; 249; 468; 477; 488; 489; 490; 491; 578; 579; 715; 716; 748; 750.5; 754; 760; 787; 862; 927; 954; 514R

- 365. Subclasses: 103; 104; 105; 106; 114; 145; 156; 174; 175; 176; 177; 178; 179; 180; 181; 182; 183; 184; 185; 186; 187; 188; 189; 200; 201; 203; 205; 207; 208; 210, 212; 218; 221; 222; 225; 226; 227; 230; 233; 015; 049; 053; 096
- 368. Subclasses: 239; 241; 083
- 369. Subclasses: 121; 044; 047
- 370. Subclasses: 013; 060; 062; 085; 094.1
- 371. Subclasses: 005; 010; 011; 021; 022; 037; 040; 047
- 372. Subclasses: 043; 044; 045; 046; 047; 048; 049; 050; 075; 081
- 374. Subclasses: 163; 178
- 375. Subclasses: 118; 224; 351; 356
- 376. Subclass: 183
- 377. Subclasses: 127; 057; 069
- 379. Subclasses: 253; 287; 292; 294; 361; 405
- 381. Subclasses: 175; 015
- 382. Subclasses: 144; 145; 151
- 385. Subclasses: 131; 014; 049; 088
- 395. Subclasses: 182.03; 200; 241; 250; 275; 280; 290; 296; 309; 325; 375; 400; 403; 425; 430; 445; 500; 519; 550; 575; 650; 700; 725; 750; 800
- 396. Subclasses: 211; 236; 321; 081; 099
- 414. Subclass: 935
- 417. Subclass: 413
- 422. Subclass: 245
- 427. Subclasses: 457; 074; 080; 098; 099; 523; 524; 525; 526; 527; 528; 529; 530; 531; 532; 533, 534; 535; 536; 537; 538; 539; 540; 541; 542; 543; 544; 545; 546; 547; 548; 549; 550; 551; 552; 553; 554; 555; 556; 557; 558; 559, 560; 561; 562; 563; 564; 565; 566; 567; 568; 569; 570; 571; 572; 573; 574; 575; 576; 577; 578; 579; 580; 581; 582; 583; 584; 585; 586; 587; 588, 589; 590; 591; 592; 593; 594; 595; 596; 597; 598; 599; 600; 601
- 428. Subclasses: 209; 450; 457; 620; 641; 650; 680; 938
- 430. Subclasses: 311; 312; 313; 314; 315; 316; 317; 318; 319; 005
- 436. Subclasses: 147; 149; 151; 004
- 437. Subclasses: all
- 445. Subclasses: 001
- 455. Subclasses: 169; 180.4; 191.2; 193.3; 252.1; 331; 333
- 505. Subclasses: 190; 191; 220; 235; 329; 330; 703; 917; 923
- 510. Subclasses: 175
- 524. Subclass: 403
- 902. Subclass: 026

4

Flat Panel Displays

JEFFREY A. HART
Indiana University

INTRODUCTION

This chapter focuses on the flat panel display (FPD) industry, an industry that manufactures display components for various types of electronic systems from cell phones to high-definition televisions. This is a relatively young but highly competitive and dynamic industry that got its technological start in the United States in the 1960s but quickly migrated to Japan, then to Korea and Taiwan. Despite the fact that FPDs are now manufactured almost entirely in East Asia, a number of U.S. firms (such as IBM, Corning, Applied Technologies, and Photon Dynamics) are central participants in the industry. This chapter examines changes in the structure and geographic location of the industry's innovation process since 1990 and discusses the effects of these changes on U.S. firms and workers.

One way to address these issues is to examine whether innovative activity has followed the movement of investment in FPD manufacturing. Since investment in manufacturing has been primarily in East Asia since 1990—first in Japan and then later in Korea and Taiwan—one might also expect most innovative activity to be located there. In actuality, some important innovative activity is still located outside East Asia, primarily in supplier firms in the United States and Western Europe. U.S. and European firms remain important in the industry's innovative processes, but it will be difficult for them to remain so unless they work closely with the manufacturers in East Asia. Several U.S. firms have done this and have remained, as a result, at the center of the innovation process. A major implication is that public policy should not punish U.S. firms for their efforts to follow the action in globalizing industries like this one.

BACKGROUND INFORMATION ON THE INDUSTRY

In 2005, the total value of FPD[1] sales worldwide was \$65.25 billion (see Figure 1). Liquid crystal displays (LCDs) accounted for over 95 percent of FPD sales by value; thin-film transistor (TFT) LCDs accounted for over 90 percent of LCD sales; and *large-sized* TFT LCDs accounted for about 75 percent of the value of TFT LCD sales.[2] The unit volume of large-sized TFT LCD panels in 2004 was 138.5 million displays. The unit volume of small- and medium-sized LCDs in that year was around 650 million (Young, 2005). The average annual growth rate from 1990 to 2005 in the real value of FPD sales was 23 percent. Real growth rates for TFT LCD sales are likely to be somewhat higher than those for FPDs.

Demand for TFT LCDs is a function of the demand for a wide variety of products, including, among others, televisions, personal computers, PDAs, camcorders, cell phones, and digital cameras (see Figure 2). The market for TFT LCDs and other FPDs became larger and increasingly diversified as the consumer electronics market moved toward digital and high-definition televisions and portable digital devices and as the size and quality of TFT LCDs increased.

Innovations in process technology along with vigorous competition permitted consumers to benefit from steadily declining prices over time. For example, prices of TFT LCDs declined with each successive generation of production equipment. Every time the glass substrate size for processing displays increased, a new generation of production equipment was created to match that size. With the entry of Korean and Taiwanese firms into the market, the demand for TFT LCDs increased in all markets where thinness and low power consumption were valued by consumers.

The potential market for FPDs is enormous. About 780 million cell phones were sold globally in 2005; 176 million TV sets; 145 million desktop personal computers; 62 million notebook computers; 9 million Personal Digital Assistants (PDAs); 10 million camcorders; 50 million MP3 players; and 85 million digital cameras.[3] And yet, while TFT LCDs accounted for almost all notebook computer displays, camcorder viewfinders, PDA displays, and handheld TVs in 2005, they accounted for only around 60 percent of computer monitors and 10 percent of televisions. Until recently, most cell phones used Super Twisted Nematic (STN)

---

[1]The term flat panel display encompasses a variety of display technologies, including LCDs, plasma displays, organic light-emitting diodes, and electroluminescent displays. Many of the statistics collected about the industry focus on the largest segment of the flat panel display market—LCDs. This chapter focuses mainly on LCDs.

[2]An LCD is a thin, flat display device made up of any number of pixels arrayed in front of a light source or reflector. See http://en.wikipedia.org/wiki/Liquid_crystal_display for details. A color filter is required for color displays and, since the mid 1990s, most LCDs sold use a multiplexed active-matrix method of addressing the pixels that depends on the deposition of very small TFTs on the bottom glass panel of the device. See http://en.wikipedia.org/wiki/TFT_LCD. A large-sized panel is 10 inches or more, measured diagonally. Small- and medium-sized panels are less than 10 inches.

[3]Various business press sources. The estimate for PDAs is for 2004.

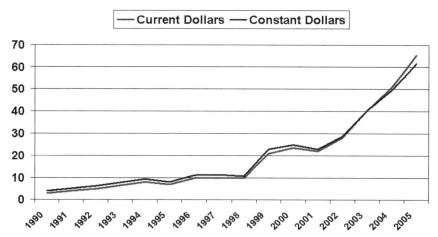

FIGURE 1 Global flat panel display revenues, 1990-2005 (current and constant 2003 dollars).
NOTE: The statistics in the figure include revenues for a variety of FPD products including LCDs, plasma displays, electroluminescent displays, and organic light-emitting diodes.
SOURCE: DisplaySearch.

LCDs because of their lower price. In 2005, however, 47 percent of cell phones had TFT LCD displays, up from 30 percent the previous year (*Softpedia News*, 2005). Even in those display markets where TFT LCDs competed with alternative technologies, growth rates were impressive. For example, in 2005, sales of LCD

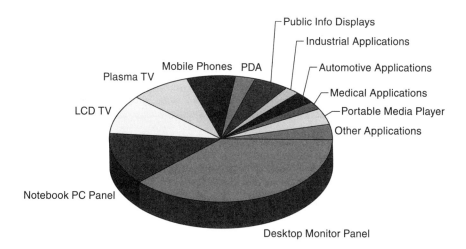

FIGURE 2 Global flat panel display sales by application, 2005. SOURCE: Frost and Sullivan, http://www.frost.com/prod/servlet/cio/FA1F-01-02-01-01/chart2.1.gif.

TVs grew to over 17 million units, up from 181,000 in 2000 (Cantrell, 2005). Figure 2 shows global sales of FPDs in 2005 by application.

Display size was initially a major constraint on demand for TFT LCDs. In the early 1980s, when the maximum size of TFT LCDs was 2 to 3 inches (measured diagonally), sales were limited to handheld TVs and camcorder viewfinders. In the late 1980s, when the maximum size was 13 inches and prices were still relatively high, computer monitor sales were limited primarily to displays for expensive notebook computers. Most notebook computers had passive-matrix[4] STN LCD displays until the price of TFT LCDs came down sufficiently to attract buyers. By the late 1990s, high-quality TFT LCD monitors for computers were being produced in high volume and prices had declined to the point where they were competitive in the marketplace with cathode ray tube (CRT) monitors.

By 2005, when the maximum size of TFT LCDs that could be produced in high-volume factories was over 40 inches, the main constraint on sales was price and quality relative to alternative similarly sized computer and TV displays, including plasma display panels (PDPs)[5] included in the FPD revenues discussed earlier. By 2005, TFT LCDs were competing successfully in television markets with 42-inch or smaller CRT-based televisions and PDPs. Given the previous price declines in smaller TFT LCDs, however, it was clear that TFT LCDs would soon be competing successfully in the larger screen sizes as well.

PATTERNS OF INVESTMENT IN MANUFACTURING

In 1996, over 95 percent of all TFT LCDs produced globally were made in Japan. In 2005, less than 11 percent were made there; the top two production locations were Korea and Taiwan (see Figure 3), each producing roughly 40 percent of global supply. The change over time in the location of TFT LCD production was a result of a sequence of investment decisions on the part of major firms in the three countries. Japan was the dominant production site until 2001, when Korean firms took the lead. Taiwanese production accomplished the same in 2002 but remained a bit below Korean production in 2002 and 2003. By 2004, the Koreans and Taiwanese were running neck and neck. Whereas the Koreans

---

[4]A passive-matrix display is one in which each pixel must retain its state between screen refreshes without the benefit of a steady electrical charge. Pixels in passive-matrix displays are addressed via row and column drivers. TFT LCDs are active-matrix displays because a transistor associated with each pixel holds the steady charge that is lacking in an STN LCD. A key advantage of active-matrix displays over passive-matrix displays is that it is not necessary to address each pixel via row and column drivers during each screen refresh. Only those pixels that need to change are addressed during a refresh. This generally permits active-matrix displays to have faster response times than passive-matrix displays.

[5]A PDP is an emissive FPD in which visible light is generated by phosphors excited by a plasma discharge between two panels of glass.

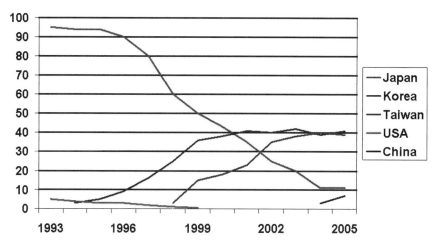

FIGURE 3 Production shares of FTF LCDs by location, 1993-2004 (percentages). SOURCE: Murtha et al. (2001).

began to produce TFT LCDs in high volume around 1995, the Taiwanese did not begin to do so until 1998.

The main reasons for this shift in production within East Asia were lower engineering and labor costs in Korea and Taiwan and the ability of first Korean and then Taiwanese firms to raise the large amounts of capital needed for investing in state-of-the-art fabrication facilities. It is important to note that most Japanese firms were not able to do this after the beginning of the bubble economy in 1991.[6] This provided a window of opportunity for the entry of Korean firms in the mid-1990s. Similarly, a window of opportunity was created for Taiwanese firms in the wake of the Asia Crisis of 1997-1998, as Korean firms temporarily experienced difficulties in financing new plants.

The ownership of production was similar to the location of production with some notable exceptions. Some of the production (less than one-fourth) located in Japan in the mid-1990s was owned by IBM through its joint venture with Toshiba (Display Technologies, Inc.). Some of the production (about one-fourth) located in Korea in the late 1990s was owned by Philips (a European firm) in its joint venture with LG (LG Philips Displays). After 2000, Sony also owned some of the

---

[6]After the collapse of the Japanese stock market in 1990 and a major decline in the value of real estate, Japanese banks suffered from a shortage of capital. Since many loans to small businesses were backed by property and small business loans constituted more than a majority of total loans, the entire banking system began to look shaky after 1990. Financial regulators failed to force the banks to write off their bad loans, so bank depositors began to look elsewhere for places to invest their capital and corporate borrowers began to look to overseas capital markets for loans (see Hutchison, 1998; Wood, 1992).

TABLE 1 Major TFT LCD Manufacturers by Location, 2005

Japan	Korea	Taiwan	U.S.	China
Sharp	• Samsung • LG Philips Display • Sony-Samsung LCD	• AU Optronics • Chi Mei Optoelectronics • HannStar • Quanta Display • Chunghwa Picture Tubes	None	• Beijing Orient Electronics • SVA-NEC

production in Korea through its joint venture with Samsung (S-LCD). Japanese firms provided some of the capital and technology for new entrants in Taiwan and China. They did this in order to have access to dependable supply sources of flat panels so that they would be able to compete with low-cost producers in Japan (such as Sharp) and Korea in end-user markets for computers and televisions. Taiwanese firms supplied assembled displays to Japanese firms on an original equipment manufacturer basis. Table 1 provides a list of the largest producers of TFT LCDs in 2005.

TFT-LCD Manufacturing

TFT LCD manufacturing is technically challenging, expensive, and risky. The seventh generation of TFT LCD fabrication plants required an investment of between $1.5 billion and $2 billion per plant. Because of learning-curve economies (dynamic economies of scale) in TFT LCD production, the price for any given size of display declined over time, just as it did for integrated circuits (ICs). But the competition among display firms was so intense that it was not always possible to enjoy the profits that are sometimes connected with learning-curve economies, hence the risk of making large investments with limited payoffs.

The technology for manufacturing TFT LCDs is quite complex, bearing many similarities to that for ICs. Both TFT LCD and IC production require advanced clean rooms, advanced lithography equipment, chemical or physical vapor deposition equipment, specialized testing equipment, and robotic handling equipment. TFT LCDs, like ICs, require many process steps; an error at any step may produce a faulty device. TFT LCD and IC production is highly capital-intensive, and extensive training is required for clean-room production workers and, even after a factory is producing at full capacity, a large team of production engineers must be on hand to diagnose and fix production problems. The proportion of engineers to production workers increases as one moves from generation to generation (as does the necessity to employ automated handling and conveyance technologies) because of the increasing difficulty of maintaining high yields and throughput.

As the glass substrate size increased in TFT LCD production (see Figure 4), some physical properties changed, thus creating new challenges for equipment manufacturers. For example, large glass substrates required special handling technologies because of the tendency of glass to sag when transported horizontally. Photolithography equipment had to be updated to be capable of transferring designs onto the larger and larger substrates. Filling the LCD cells with liquid crystal materials became more challenging as the cell and module size increased.

One important difference between TFT LCDs and ICs is that ICs are always shrinking in size in order to achieve a greater number of chips per silicon wafer and to speed the performance of the chip itself. In contrast, a significant portion of the global market for TFT LCDs tends to shift toward larger-sized displays (e.g., for computer monitors and TVs) while the demand for smaller displays (e.g., for cell phones) also has tended to grow rapidly, so the only way to reduce average unit costs is by increasing yields and by reducing defects on larger and larger glass substrates—the large sheets of glass on which multiple display panels are processed (see Figure 4).

This episodically requires a shift to the next generation of production technology, new tools and handling equipment geared to the larger substrates, and manufacturers to have flexible strategies with regard to the production of a variety of display sizes. Figure 4 shows that there have been six generational shifts between 1991 and 2005, so the average time between shifts has been 2.3 years. While the IC industry also has gone through transitions to larger wafer sizes (the latest being from 200- to 300-mm-diameter wafers), these transitions are

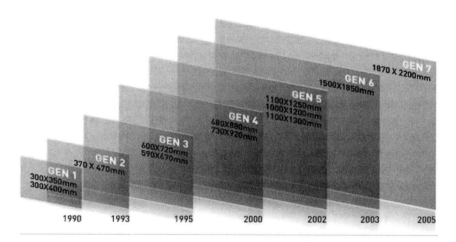

FIGURE 4 Glass substrates, first through seventh generation. SOURCE: Samsung Corning Precision, http://www.scp.samsung.com/content/en/product/generation.asp.

less frequent and ICs of a given type generally do not increase in size over time (rather they tend to shrink).

Firm Strategies

The strategy for manufacturers of TFT LCDs consists mainly of deciding if and when to invest in the construction of a new fabrication facility. Along with the decision to invest in a new line comes the even more important decision of whether to move to the next generation of substrates (see Figure 4). It is a risky decision for a number of reasons: (1) there is always uncertainty about the future demand for TFT LCDs of various sizes, (2) there is uncertainty about how many competitors will match the investment and when (hence uncertainty about future supply), (3) there are uncertainties about both product and process technologies, and (4) there are uncertainties about how well the firm itself will be able to execute its selected strategy.

Product technologies are uncertain because of the potential competition from alternative display technologies. For example, it was not clear that LCD televisions would be able to compete with CRT-based televisions and PDPs in the market for digital televisions until fifth-generation TFT LCD plants were built. Not only was there the question of relative price, there was also a question of relative quality of displays and the premium that consumers would be willing to pay for higher quality.

Process technologies are uncertain because of the problems connected with scaling up equipment and altering handling systems for each generation of substrates. In the move from second- to third-generation substrates, for example, it was necessary to move to new types of conveyor systems and automated guided vehicles to transfer partially processed glass substrates from one machine to another on the factory floor. This was done to reduce both breakage and particle contamination rates.

Besides having to deal with technological and other uncertainties, manufacturers have to decide which suppliers to work with. This can be crucial because of the need to ramp up production quickly in order to exploit whatever temporary advantages might accrue to early investors. Suppliers can fail manufacturers in a number of ways: Materials suppliers might not be able to provide key inputs at the right time; equipment suppliers might not be able to deliver or maintain equipment that is crucial to raising yield and throughput. Because of the extreme time pressures in this industry, most manufacturers work with established suppliers who have extensive experience with high-volume TFT LCD manufacturing. Only if a new and inexperienced supplier has a very important technological edge will a manufacturer be willing to work with them.

Consider the strategies selected by Japan's pioneer TFT LCD manufacturer, Sharp, in the early 1980s. Sharp invested earlier than other Japanese firms in the first generation of TFT LCD plants mainly because management believed that not having an internal source for CRTs had hurt Sharp's ability to compete with other Japanese firms in the television business. As an early producer of handheld TVs and calculators, Sharp saw a bright future for other devices that required information displays. Sharp management accepted the risk of investing in an unproven display technology because they felt they had no other choice.

Sharp initially worked with only a few external suppliers and tried to develop its own manufacturing equipment. When the IBM-Toshiba joint venture, DTI, built a third-generation plant in Japan that outperformed Sharp's earlier-generation plants, it did so by relying more than Sharp had done on external suppliers. As production moved from Japan to Korea to Taiwan, firms in the other East Asian countries generally became increasingly dependent on external suppliers, mainly because they lacked the ability to develop quickly all the necessary capabilities in-house. In addition, they moved some of the more labor-intensive processes, such as module assembly, to lower-wage locations, including China.

Third-generation plants were considerably more automated than earlier-generation plants, so external equipment firms that could work with manufacturers to perfect their automation systems had an opportunity to become key participants in the industry. (We will later see how this approach worked for two U.S. firms—Applied Materials and Photon Dynamics.)

During the bubble economy period in Japan, most Japanese firms were unable to invest in new plants. Instead they retrofitted their old plants to produce higher value-added products like low-temperature polysilicon TFT LCDs, which were mainly used for small displays such as those used in cell phones. Later several of these firms moved their display operations to Taiwan and China through foreign investment and technology transfers.

The decision of the Taiwanese firms to invest in fifth-generation plants when Korean firms were investing in sixth- and seventh-generation plants requires some explanation. The logic of entry via the latest generation may not have held for Taiwanese entry because of the ability of the Taiwanese to find other ways to become cost-effective manufacturers. Being the first mover to a new production technology can be quite expensive, especially if the rest of the industry is not ready to make the jump. Nevertheless, since their entry in the late 1990s, Taiwanese firms have invested in latest-generation plants as soon as they were able.

The development and introduction of each successive production generation occurred in a variety of locations, but importantly the successful integration of each generation of production equipment depended on investment in high-volume production. This meant that developers of equipment had to work with whatever firms had decided to be early adopters of the latest generation in order to remain competitive. Similarly, firms that supplied key inputs, like glass substrates and color filters, also had to do this.

To be more specific, the development of lithographic equipment occurred primarily in Japan, the United States, and Western Europe, even though installation and testing of that equipment was primarily in East Asia from the 1980s onward. Similarly, liquid crystal materials were developed and fabricated primarily in Western Europe and then sold to East Asian producers. Chemical vapor deposition (CVD) equipment was developed primarily in the United States and Western Europe. Testing equipment was developed mainly in Japan and the United States. In short, new materials and equipment did not have to be developed in the same countries that invested in manufacturing, but firms that supplied the manufacturers had to work closely with them to remain competitive.

Unlike the semiconductor industry, where design and manufacturing became less integrated over time (in fabless firms and foundries), no such vertical disintegration of that part of the value chain occurred in the FPD industry. However, in other parts of the value chain it was possible for some disintegration to occur, particularly between the glass-processing phase and the final assembly phase of production.

Materials and equipment suppliers became more important over time because of the time pressures created by the rapid changeover from one generation of production technology to the next. Korean and Taiwanese firms were generally unable to emulate Japanese leaders, mainly Sharp and NEC, in building their own production equipment; instead, they had to rely on external suppliers to a greater extent.

THE STRUCTURE OF INNOVATION

Innovation in this industry (as in other manufacturing industries) occurs in both the design of new products and the refining of manufacturing processes. For example, as TFT LCDs started penetrating the market for televisions, panels had to be improved to meet the need for wider viewing angles than is necessary for displays in notebook and desktop computers. A key innovation was "in-plane switching" technology because it increased viewing angles along with contrast ratios and brightness of displays. The response times required for real-time video in video games and television also required innovations in product technology. In 2001, NEC developed its "feed forward" technology to speed up response times for televisions. Various types of "overdrive" or "response time compensation" technology were developed by Samsung, LG Philips, CMO, BenQ, and ViewSonic for their displays.[7] An example of a major recent innovation in process technology was the invention of "one drop fill"—a new way of inserting liquid crystal material between the two processed glass plates of a TFT LCD (Kamiya et al., 2001). With every increase in the size of substrates came a demand for new processing and handling machines.

While manufacturers must innovate both process and process technologies, they are often helped by suppliers. When suppliers who are not manufacturers themselves provide new materials or processing equipment, they must work

---

[7]"Advanced Technology," *TFT Central*, http://www.tftcentral.co.uk/advanced.htm.

closely with manufacturers to ensure that the materials and equipment will meet the needs of their customers. The intense competition in TFT LCD end-user markets from other manufacturers and from alternative technologies means that manufacturers must bring new plants online as quickly and cheaply as possible, and to do this they increasingly turn to external suppliers.

Patenting activity can be used as a crude indicator of innovation. Location of patenting activity in the TFT LCD industry tends to follow investment in manufacturing with a lag (see Figure 5). For example, the five largest holders of U.S. LCD patents as of 2005 were Sharp, LG Philips, Canon, Hitachi, and Seiko-Epson. LG Philips's U.S. patenting activity began in 2000 just as Sharp's patenting activity declined. Samsung's patenting activity was markedly lower than that of LG Philips but it also took a turn upward from 1995 onward. Between 2000 and 2005, the four major Taiwanese firms (AU Optronics, Chungwha Picture Tubes, Chi Mei Optoelectronics, and HannStar Display) successfully filed for LCD patents in the United States but the total patents granted were considerably fewer in number than those held by Japanese and Korean firms (see Figure 6).

U.S. firms accounted for a decreasing percentage of total patents between 1969 and 2005 (see Figure 6). This figure was generated by counting the annual number of U.S. patents for which patent holders were either U.S.-owned firms or U.S.-located laboratories and comparing that number to the total. Even before IBM decided to exit the TFT LCD market as a manufacturer, U.S. firms were not keeping up with the increased patenting activity of foreign firms.

Because of rapid technological change, the role of tacit knowledge, and the importance of proximity to physical production, much of the innovative

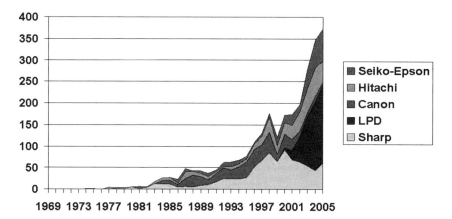

FIGURE 5 U.S. LCD patents granted annually to the top five patent holders, 1969-2005. SOURCE: U.S. Patent Office, http://www/uspto.gov/go/taf/tecasga/349_tor.htm.

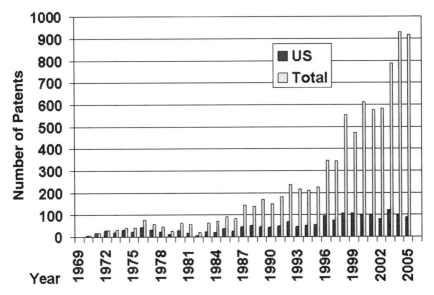

FIGURE 6 Number of LCD patents issued to U.S. firms or laboratories compared with total number of patents, 1969-2005. SOURCE: U.S. Patent Office, http://www/uspto. gov/go/taf/tecasga/349_tor.htm.

activity takes place within the manufacturing firms themselves and particularly within their associated laboratories. In East Asia, most scientists and engineers employed by domestic firms are nationals of the home country. There is very little research and development (R&D) done by these firms outside their home countries. Other than Sharp's laboratory in Camas, Washington, no major display research laboratory was established by an Asian firm in the United States. This contrasts markedly with the nationality and location of scientists and engineers employed by U.S. and European firms. IBM Japan operated an important display laboratory in Yamato; Philips acquired the laboratories of Hosiden in Japan and then worked in collaboration with LG in Korea after the joint venture was established. Many members of the top management of Korean and Taiwanese firms were previously employed by U.S. or European firms and received graduate training in U.S. and European universities. For example, the head of Samsung's TFT LCD operations was Jun Souk, who had previously worked for IBM.

Some important innovations occur in government and university laboratories or in supplier firms and in collaborations between suppliers and manufacturing firms. To demonstrate this, I turn to a discussion of the historical importance of U.S. firms as both suppliers and manufacturers in the TFT LCD industry.

U.S. Participants in the Industry

The story of TFT LCD manufacturing in the United States began in the late 1960s with a number of important successes in the research efforts of major firms like RCA, Westinghouse, Exxon, Xerox, AT&T, and IBM. However, none of these firms (except IBM) decided to invest in volume manufacturing of TFT LCDs. IBM decided in 1986 to invest in high-volume production only in Japan and only in a joint venture with Toshiba. During the 1980s, in the wake of the high dollar and Japanese successes in semiconductors, U.S. firms (other than IBM) decided not to invest in TFT LCD manufacturing. In contrast, all the major Japanese electronics firms had invested in high-volume TFT LCD manufacturing by the late 1980s.

Nevertheless, a number of U.S. firms decided to participate in the industry, most notably IBM, Corning, Applied Materials, and Photon Dynamics (to name the four important firms). These firms remained key players in the market thanks to their ability and willingness to acquire knowledge by working collaboratively with manufacturers outside the United States.

All U.S. manufacturers of TFT LCDs other than IBM were relatively small, niche producers.[8] These firms engaged in a variety of efforts to catch up with the Japanese leaders, some of which involved help from the U.S. government, particularly the Advanced Research Projects Agency of the Department of Defense. They failed mostly because U.S. governmental policies made it difficult for firms receiving government funds to work closely with manufacturers in Asia. But government policy was not solely to blame for this. Most U.S. firms had not grasped the essence of the problem: To succeed in entering the industry at this stage, they had to work with partners experienced in high-volume production.[9] Firms that understood this—IBM, Corning, Applied Materials, and Photon Dynamics—were successful, as discussed next.

IBM

In 1986, IBM and Toshiba entered into 2 years of joint research on TFT LCD manufacturing. The research would be conducted jointly by researchers at IBM's Yorktown Heights laboratories, IBM Japan, and Toshiba. Each company would host the project for 1 year in its respective facilities in Japan, starting at Toshiba, where a rudimentary R&D line was to be erected as soon as possible. At the end of the joint research project, each company would be free to pursue its own manufacturing plans or to walk away. On the strength of these discussions, Toshiba engineers apparently went immediately to work designing the line

---

[8]My collaborators and I have written about these small manufacturers elsewhere (see, e.g., Lenway et al., 2000).

[9]For details, see Lenway et al. (2000) and Office of Technology Assessment of the U.S. Congress (1990).

and ordering equipment. The contract was officially signed and work began on August 1, 1986. One month later the R&D line was up and running.

By July 1988 both Sharp (Japan's pioneer and still-dominant TFT LCD manufacturer) and IBM-Toshiba had developed 14-inch TFT LCD prototypes, demonstrating a potential for flat video reproduction that had seemed remote only 5 years earlier. Sharp publicly announced its achievement, as is customary for Japanese companies. IBM-Toshiba did not at first announce their achievement, which was consistent with IBM company policy.[10] Toshiba later prevailed, and an announcement was made in the wake of the Sharp press conference. Both companies claimed the laurels for largest size and best resolution.

Nearly a year after the IBM-Toshiba prototype announcement, on August 30, 1989, the two companies announced their agreement to form a manufacturing alliance called Display Technologies, Inc. (DTI). The alliance was to be structured as a 50/50 joint venture between Toshiba and IBM Japan. The partners initially capitalized DTI at about $140 million (*Los Angeles Times*, 1989), of which $105 million was earmarked for a high-volume TFT LCD fabrication facility. DTI's headquarters and first fab would be located in Himeji City, next to one of Toshiba's STN LCD fabs. DTI officially started up on November 1, 1989. R&D for DTI was conducted in three laboratories, one in the United States and two in Japan: IBM's Thomas J. Watson Laboratory in Yorktown Heights, New York; IBM Japan's laboratory in Yamato; and Toshiba's laboratory in Himeji.

Thus, IBM was the only large U.S. firm to invest in high-volume TFT LCD manufacturing. It is important to note that IBM decided to do this in Japan with a large Japanese partner, Toshiba, partly because it believed this was the only way to become globally competitive. Japan was where the TFT LCD action was and where learning about the new industry could be maximized. The most important customer for IBM FPDs was the IBM PC Division in Florida, so locating production in Japan had little to do with servicing customers. IBM divested itself of its stake in DTI in 2001 when it no longer saw a need to have an internal supplier of TFT LCD panels. By then, there was plenty of competition in the global TFT LCD market and no difficulty finding the high-quality displays needed for IBM end products. IBM was not interested in LCD television sales (although perhaps it should have been). Like all the large Japanese electronics firms, except Sharp, IBM turned its attention to higher-value-added businesses, including very-high-definition FPDs, and to advanced services where potential profits and revenue growth were higher.[11]

---

[10]Out of sensitivity to U.S. antitrust law, IBM has remained reluctant since the 1950s to announce technology breakthroughs prior to the availability of products in markets.

[11]To date there has been no high-volume production of IBM's very-high-resolution displays. See IBM Research, Roentgen Project Page, Roentgen Introduction, http://www.research.ibm.com/roentgen/.

Corning

In fall 1986, a group of top executives in Corning held a meeting to decide on entering the market for LCD substrates. Corning research in New York and the marketing organization in Japan had followed FPD developments since the early 1980s through several major turning points. Around the time of Matsushita's 1986 pocket television introduction, a number of senior managers in Corning came to envision TFT glass substrates as a major business opportunity for Corning.

Corning gained experience selling glass in Japan for STN LCDs over many years, beginning in the early 1970s with sales to makers of watch and calculator displays. Corning researchers made an effort to develop extremely thin sheet glass for these applications, using a product the company was selling for use as microscope slide covers for medical laboratories. Corning had developed its proprietary fusion glassmaking technology as a method of fabricating extremely thin, optical-defect-free glass without the need for grinding or polishing.

Early LCD technologies, however, did not require the advanced properties of fusion glass. Most of Corning's sales for these applications continued to consist of glass manufactured using more conventional methods. But in the early 1980s, managers in Corning Japan noted with some surprise that the laboratories of several major electronics groups were placing regular, gradually increasing orders for a more advanced product, Corning's 7059 fusion-formed borosilicate glass.

Corning's proprietary fusion glass technology seemed uniquely matched to the apparent technological trajectory of TFT LCDs. Corning Japan's managers had worked diligently to nurture relationships with Japanese manufacturers. Even a technologically well-matched Japanese competitor would have faced difficulties building the same network and familiarity with market needs. For a competitor from outside Japan, these barriers would be insurmountable.

Corning's main research unit, located in Sullivan Park in Corning, New York, developed new products and manufacturing processes for LCD glass substrates. The first fusion glass machine was built there in the late 1970s. In 1982, Sullivan Park developed an ultrathin glass—Corning's 7059 fusion-formed borosilicate glass—that was used in TFT LCD laboratories around the world. Sullivan Park also developed the fusion glass process that was used in Harrodsburg in 1984. The collaboration between Sullivan Park and Harrodsburg continues to the present. The two locations recently co-developed a new glass called Eagle XG, which provides all the desirable properties required for TFT LCD substrates but does not contain arsenic, antimony, or barium. Eagle XG is therefore a greener product than its predecessor, Eagle^{2000}, which was introduced in 2000 and contained the three heavy metals. Corning tries out all new TFT LCD glass products and processes at Harrodsburg before transferring knowledge to its other facilities.

Corning opened a new glass melting and finishing facility for TFT substrates and a new TFT research center in Shizuoka Prefecture in 1989. Its growing TFT customer base in Japan would be served from both Shizuoka and Harrodsburg.

As the industry grew, Corning had to expand production rapidly. Demand for thinner substrates resulted in volume production of 0.7-mm glass in 1990 and 0.5-mm glass in 1998.

Samsung Corning Precision (a 50/50 joint venture established in 1995 between the two firms) began to produce substrates for Korean producers in Kumi in 1996. A major expansion in production capacity in all three locations—the United States, Japan, and Korea—occurred in 2000. A second Korean plant was opened in Cheonan in 2002 and reached volume production in 2003.[12]

In 2004, the firm began to produce TFT substrates in Tainan, Taiwan. It soon opened another plant in Taichung to service the rapidly growing demand for substrates in Taiwan. The governments of both Korea and Taiwan were concerned about the dependence of domestic TFT LCD manufacturers on foreign suppliers and the impact of that dependence on the balance of trade. Accordingly, they encouraged the establishment of domestic suppliers wherever possible but also urged foreign firms to establish local operations as soon as possible.

Corning broke ground for a substrate finishing plant in China in November 2006. Apparently the firm had started its manufacturing operations in Korea and Taiwan with finishing plants before establishing melting operations. Corning was waiting to see how rapidly the Chinese TFT LCD firms would be ramping up production before committing to a melting plant there.

Corning followed the shift in manufacturing of TFT LCDs as it moved successively from Japan to Korea to Taiwan. Corning had to move manufacturing and some research to East Asia, but most of the research on fusion glass remained in the United States (Guan, 2004). By 2004, roughly a third of Corning's total revenues ($3.8 billion) depended on the sales of TFT LCD substrates (*Newsday*, 2006). Corning's ability to remain a key participant in the TFT LCD industry depended on its proprietary fusion glass technology.

Applied Materials/AKT

Applied Materials, the U.S. semiconductor manufacturing equipment maker, started a display arm called Applied Display Technology in 1991. In 1993, Applied Materials initiated a strategic alliance with Komatsu, the Japanese heavy equipment maker, called Applied Komatsu Technology (AKT). AKT developed and manufactured TFT LCD manufacturing equipment in the United States using globally sourced components. The company maintained principal R&D and engineering facilities in Santa Clara, California; funded basic research in outside institutions such as universities; and also relied on the specialized R&D and basic research capabilities of its global supply network. In 1993, AKT established its

---

[12]Samsung Corning Precision Glass, "Company: General Information: History," http://www.scp.samsung.com.

headquarters in Kobe, Japan, and set up a technology center there but most R&D remained in California.[13]

The motivation for forming AKT did not revolve around the conventional joint venture criteria of market, technology, or capital seeking. The R&D was complete, the funds invested, and Applied Materials already enjoyed a prestigious position in Japan's semiconductor industry. Rather, Applied Materials actively sought an alliance partner to address the personnel requirements of sustaining and growing the new business. The industry would grow rapidly (no one at Applied Materials knew how quickly at the time). To keep pace, the new venture would need to expand rapidly in its abilities to conduct site installations and testing as well as continuing servicing of its machines at the customers' premises. Applied Materials offered to ally with Komatsu after a rigorous search for a partner that shared its beliefs about the necessary marriage of cost effectiveness, quality, and technological advancement. Komatsu invested $35 million initially in the new venture (*Electronic News*, 1993).

In October 1993, AKT announced that its commercial CVD tool, the AKT-1600, was ready for sale, for delivery 6-8 months after an order was placed. Beginning with the startup of the first Generation 2 manufacturing lines in mid-1994, all TFT LCD producers had the opportunity to benefit from the innovations incorporated in the new CVD tool. The AKT-1600 sold for about $5 million for a four-chamber production system, which could process about 40 substrates per hour. High-volume Generation 2 fabs needed about four of them.[14] By year's end, AKT had captured CVD market leadership by a wide margin.

Shortly thereafter, Hitachi approached AKT for a design to process a 400 × 500 substrate, out of which they could make four 11.3-inch displays. Hitachi ended up with a CVD tool that could process 370 × 470 substrates, a modification of the DTI design. By the end of the life cycle for Generation 2, AKT had modified the 1600 to accommodate 400 × 500 mm substrates for a Generation 2.5 line that Sharp started up in July 1995. Even the largest manufacturers remained indecisive regarding the best display size to manufacture and the best substrate size on which to manufacture it, but by supplying these firms with tools that matched their diverse specifications, AKT acquired knowledge that would allow it to be the dominant supplier of CVD equipment for years to come. Along with Corning, it would become a key participant in the global effort to establish an industry consensus on standards for next-generation equipment. Despite the movement of manufacturing from Japan to Korea and Taiwan, development of

---

[13]Applied Materials announced the joint venture's creation on June 17, 1993. According to Applied's 1999 Annual Report, the venture ended in 1998. AKT then reorganized as a wholly owned subsidiary of Applied Materials. Since the reorganization, Tetsuo Iwasaki has served as AKT's chairman, in addition to his position as chairman and CEO of Applied Materials Japan.

[14]Material for this paragraph was taken from *DisplaySearch Equipment and Materials Analysis and Forecast* (Austin, Tex.: DisplaySearch, 1999).

new equipment was accomplished primarily in Northern California even though machines had to be built and tested on factory floors in East Asia.

Photon Dynamics

Founded in 1986 by Francois Henley and headquartered in Milpitas, California, Photon Dynamics initially produced inspection and testing tools for semiconductor manufacturing and did not enter the FPD industry until 1991. The firm developed test, inspection, and repair systems for FPD manufacturing that were used to increase yield, reduce materials loss, get new designs from R&D into production, and assist in the rapid startup of new plants.

For TFT LCDs, materials costs (for glass substrates, color filters, and polarizers, for example) represent at least 40 percent of the cost of production. As a result, test and inspection of substrates is a key part of improving manufacturing efficiency. It is critical to identify defects and repair them as soon as possible prior to further processing to avoid wasting materials. When defects cannot be repaired, the substrate needs to be scrapped. The same goes for cells and assembled modules.

Photon Dynamics was able to sell to high-volume manufacturers in Japan, Korea, and Taiwan on the basis of being able to offer products and services competitive with those of its main competitors: Micronics Japan, AKT, and Shimadzu in array testing and NEC, NTN, and Hoya in cell and module testing. Two proprietary technologies played a key role in the early success of Photon Dynamics: voltage imaging and N-aliasing image processing.[15] The firm held over 20 patents in its intellectual property portfolio.

Unlike Corning and AKT, Photon Dynamics had no overseas research facilities. R&D was done mainly at its headquarters in San Jose, California. In 2005, the firm maintained sales and customer support offices in China (Beijing), Korea (Seoul, Daejeon, Kumi, and Cheonan), Taiwan (Hsinchu and Taichung), and Japan (Tokyo and Tsu). Some repair equipment was to be manufactured in Korea in 2007, but all other manufacturing was done in California.

Competition in Two TFT LCD Supplier Industries

Three U.S. suppliers were competitive in two industries that supplied important inputs to TFT LCD manufacturing: Corning in glass substrates, and AKT and Photon Dynamics in manufacturing and testing equipment. Corning's main

---

[15]Voltage imaging produces a two-dimensional image of the voltage distribution across the surface of a TFT array. It greatly reduces the amount of time needed to test an array prior to assembly into a TFT cell, which can greatly increase the yield and throughput of TFT LCD production facilities. N-aliasing image processing refers to special software algorithms used to detect defects or anomalies in the images produced by visual imaging equipment.

competitors were Asahi Glass and Nippon Electric Glass (NEG), both Japanese firms. Neither Asahi Glass nor NEG had developed fusion glass technology; therefore, they were at a disadvantage as the industry turned more and more to fusion glass as substrates grew larger. Corning's careful husbanding of its intellectual property rights in fusion glass technology was crucial to maintaining its competitive advantage.

AKT's main competitors in CVD equipment in 2005 were Unaxis-Balzers of Western Europe and Jusung of Taiwan. Photon Dynamics' main competitors were Micronics Japan, AKT, and Shimadzu in array testing and NEC, NTN, and Hoya in cell and module testing. Both were vigilant in protecting the intellectual property rights associated with their equipment and occasionally engaged in patent infringement suits to protect those rights.

All U.S. suppliers needed to locate warehouses and service facilities in Japan, Korea, and Taiwan close to major customers. As the industry matured, suppliers also felt pressure to locate manufacturing and R&D facilities in East Asia. These pressures arose because of governmental concerns about technological dependency and the impact of technological imports on the balance of trade. Asian governments wanted key technologies to be developed domestically. If domestic firms were unable to do this, then foreign firms would be encouraged to locate their development efforts in the country. Such pressures were generally resisted because the supplier firms wanted to maintain the core of scientific and engineering expertise closer to home. Corning experimented with a joint venture with Samsung that proved successful but the joint venture licensed fusion glass technology from Corning and was not permitted to compete with the parent firm in other markets. Corning located melting facilities in Japan, Korea, and Taiwan and was pressured to locate melting facilities in China, but so far had declined to do so.

PUBLIC POLICY ISSUES

In the background of this limited but important participation by U.S. firms in the TFT LCD industry is the slow and steady relative decline of innovative activity in LCDs in the United States (see Figure 6). At the very beginning of the FPD industry, RCA's Sarnoff Lab was a key location for cutting-edge research. The Sarnoff Lab developed the dynamic scattering mode display that was used in the first calculator with an LCD. Sarnoff licensed the patent to Sharp in 1973, the same year that it decided to end its LCD research program (Castellano, 2005; Johnstone, 1999). The Westinghouse laboratory's LCD R&D program, led by Peter Brody, was terminated in 1978 (Brody, 1996). Xerox's display efforts lasted considerably longer, culminating in the formation of a spinoff firm named dpiX in 1998, but dpiX never attempted to compete in high-volume display markets. Products based on its technology were too expensive for consumer markets.

When the U.S. government decided to consolidate a number of R&D programs relating to display technology in 1994, there was a flurry of research activity connected with the new emphasis on advanced displays, particularly on the part of the Defense Advanced Research Projects Agency (DARPA). Some important technological developments came out of these efforts. For example, the deformable mirrors that eventually became Texas Instruments' digital light processor technology now found in many projection televisions and data projectors were partially funded with the aid of DARPA grants and contracts (most of the funding came from Texas Instruments itself). DARPA provided some funding to firms like Photon Dynamics for TFT-LCD testing equipment. Firms like IBM, AKT, and Corning, in contrast, did not participate in these programs, except as observers, and did not receive major funding for further development of their core technologies.

A good example of decline in U.S. R&D capability in FPDs was the closing of a major government-funded display laboratory at the University of Michigan in the mid-1990s that had been started with DARPA funds but ended when the funding ran out.[16] The decline in capability was the result of lack of will on the part of the Republican-controlled Congress to fund FPD R&D efforts. Globalization played a key role in the evolution of the industry because the newer entrants in Korea and Taiwan were not able to match the technological resources that were available to Japanese firms and thus had to collaborate with firms in Japan, Western Europe, and the United States to solve some of the formidable problems of becoming globally competitive. This need to collaborate provided some U.S. supplier firms with opportunities to remain at the technological frontier even though no U.S.-owned firms were manufacturing TFT LCDs after the year 2001. However, since much of the innovation in the industry was connected with designing new commercial products and new manufacturing processes, U.S. firms who were not major suppliers to the industry and most U.S. laboratories and universities were increasingly unable to participate meaningfully in the industry.

CONCLUSIONS

In general, innovative activity has tended to follow investment in manufacturing in the FPD industry, but some important innovation continues to occur that is not necessarily located close to manufacturing. Scientists and engineers in East Asia have an important advantage over U.S. scientists and engineers because of the location of manufacturing there; nevertheless, some U.S. firms have remained key participants in the industry and their scientists and engineers have been able to contribute in very important ways to innovation in the industry. Without firms like IBM, Corning, Applied Materials, and Photon Dynamics, the FPD industry would not have been able to solve important scientific and technological problems. While the main benefit to date from innovative activity in this industry

---

[16]For further details, see Murtha et al. (2001, ch. 6).

has probably been captured mainly by firms and workers in Japan, Korea, and Taiwan, a not insubstantial number of beneficiaries can be found in the United States as well. The nearly 400 workers employed by Corning in its fusion glass facility in Harrodsburg, Kentucky, are an example.

A key lesson to be drawn is that U.S. supplier firms that are willing to establish service centers abroad and to work collaboratively with foreign firms wherever the latter are located can remain internationally competitive even in industries where manufacturing is primarily located abroad. Such willingness to collaborate does not necessarily imply the offshoring of formerly U.S.-based R&D, as the cases of Corning, AKT, and Photon Dynamics illustrate. On the contrary, the willingness to collaborate ensures that some important innovative activity will continue to occur in the United States. Any government policies that prevent firms from doing this are likely to be highly counterproductive. U.S. firms have many strengths that derive from the emphasis on government sponsorship of basic research, relatively strict enforcement of competition and intellectual property laws, the availability of venture capital for startups, and a generally favorable climate for entrepreneurship. If the United States wants to participate in dynamic, globalized industries like the FPD industry, it has to keep its economic nationalists on a short leash.

REFERENCES

Brody, T. P. (1996). The birth and early childhood of active matrix, a personal memoir. *Journal of the Society for Information Display* (April):117.

Cantrell, A. (2005). Corning's Rose-Colored Glass, *CNNMoney.com*, December 2. Available at http://money.cnn.com/2005/12/02/markets/spotlight/spotlight_glw/.

Castellano, J. A. (2005). *Liquid Gold: The Story of Liquid Crystal Displays and the Creation of an Industry.* Hackensack, N.J.: World Scientific.

Electronic News. (1993). Komatsu to invest $35M in Applied's display unit. September 20. Available at http://findarticles.com/p/articles/mi_m0EKF/is_n1981_v39/ai_14415110.

Guan, W. (2004). China Strategy of Corning's TFT-LCD Business: Interview with Mr. Nitin S. Kulkarni, President of Corning Display Technologies China. *Digi-Times*, December 23. Available at http://corning.com/displaytechnologies/ww/en/media_center/featured_articles/feature. aspx.

Hutchison, M. M. (1998). *The Political Economy of Japanese Monetary Policy.* Cambridge: MIT Press.

Johnstone, B. (1999). *We Were Burning: Japanese Entrepreneurs and the Forging of the Electronic Age.* Boulder, CO: Basic Books.

Kamiya, H., K. Tajima, K. Toriumi, K. Terada, H. Inoue, T. Yokone, N. Shimizu, T. Kobayashi, S. Odahara, G. Hougham, C. Cai, J. H. Glownia, R. J. von Gutfeld, R. John, and S.-C. Alan Lien. (June 2001). Development of one drop fill technology for AM-LCDs. Pp. 1354-1357 in *SID Symposium Digest of Technical Papers*, Vol. 32.

Lenway, S. A., T. Murtha, and J. A. Hart. (2000). Technonationalism and cooperation in a globalizing industry: The case of flat panel displays. In *Coping with Globalization*, A. Prakash and J. A. Hart, eds. New York: Routledge.

Los Angeles Times. (1989). Toshiba, IBM Set Plant for Large LCDs. p. A3.

Murtha, T., S. A. Lenway, and J. A. Hart. (2001). *Managing New Industry Creation.* Stanford, CA: Stanford University Press.

Newsday. (2006). Fired Corning, Inc., Employee Pleads Guilty to Selling Glass Secrets to Taiwanese Rival. January 19. Available at http://www.glassonweb.com/news/index/4187/.

Office of Technology Assessment of the U.S. Congress. (June 1990). *The Big Picture: HDTV and High-Resolution Systems.* Washington, D.C.: U.S. Government Printing Office.

Sharp Corporation. (2005). *Sharp Environmental and Social Report 2005.* Available at http://sharp-world.com/corporate/eco/report/2005pdf/sharp09e.pdf.

Softpedia News. (2005). The Transition to TFT LCD Could Cut Down Prices on Cell Phones. May 4. Available at http://news.softpedia.com/news/The-transition-to-TFT-LCD-could-cut-down-prices-on-cell-phones-1645.shtml.

Wood, C. (1992). *The Bubble Economy: Japan's Extraordinary Speculative Boom of the '80s and the Dramatic Bust of the '90s.* New York: Atlantic Monthly Press.

Young, R. (2005). Survival of the Fittest: Managing the Peaks and Valleys of Supply/Demand. Presented at the 7th Annual DisplaySearch U.S. FPD Conference, Austin, Texas.

5

Lighting

SUSAN WALSH SANDERSON
Rensselaer Polytechnic Institute
KENNETH L. SIMONS
Rensselaer Polytechnic Institute
JUDITH L. WALLS
University of Michigan
YIN-YI LAI
Rensselaer Polytechnic Institute

INTRODUCTION

Once a symbol of Edison's creative genius and the prowess of American innovation, the incandescent light bulb represents a mature technology, now mastered by new competitors and imported at pennies apiece from China. Lamp (the industry name for a light bulb) manufacturing was dominated for decades by a few firms, notably Philips, OSRAM, and General Electric (GE). Related industry segments have typically been more fragmented, with thousands of firms producing fixtures ranging from simple sconces to elaborate chandeliers. Increasingly both lamp and fixture manufacturing have been shifting to offshore locations, primarily in Asia.

Not only are North American and European lamp and fixture companies under the threat from low-cost imports, but solid-state lighting, a semiconductor- instead of bulb-based technology with greater potential energy efficiency and new capabilities, is poised to revolutionize the industry and change how we understand and use lighting—a change that will affect both traditional lamp and fixture producers. Solid-state lighting is challenging incumbents and throwing leadership in the future industry up for grabs. As innovative products composed of light emitting diodes (LEDs) are developed, new features like colors that change on command are expanding architectural possibilities. Other opportunities come from the convergence of lighting, information, and display technologies. In fiber optics light is data, and ordinary flat panel indoor lighting can serve as data transfer hubs, sending information to computers and appliances. Edison's lamp, and its successors, may soon be replaced with glowing ceiling panels or even lighting-enhanced wallpaper that changes patterns on command.

Which firms will successfully ride this new wave of innovation and what impact these changes will have on incumbents are not yet determined. Although the first wave of lighting innovation in the early 20th century spawned the development of global companies like GE, OSRAM, and Philips, these 21st-century innovations will create challenges for incumbents. New firms are emerging at all levels of the value chain to address the opportunity presented by solid-state lighting technologies.

In this chapter we contrast traditional lighting technologies with LED technologies. Traditional lighting technologies we define as incandescent, gas-discharge, and electric arc lighting (which includes fluorescent, high-intensity discharge, mercury and sodium vapor, metal halide, and neon lamps). We exclude lighting technologies such as chemiluminescence that yield insufficient light for illumination (such lights can be seen but not seen by). LED technologies (including organic and polymer LEDs) are the only nontraditional technology considered because LEDs are the only alternative lighting approach that has reached sufficient maturity to be considered commercially viable in the trade, technology, and technical literatures.

This chapter analyzes changes in lighting technology over the past two decades and its implications for U.S. industry competitiveness. We explore whether the rise of global competition is limited to low-cost manufacturing or whether strategic centers of decision making and research are moving away from the regions and firms that once dominated the industry. We examine the causes of these changes and what aspects of innovation in lighting, particularly in the arena of research and development (R&D), have changed since the 1990s. We speculate about the implications of these changes for firm strategy in the new era of intense global competition, we analyze how national policies have affected the development and diffusion of traditional and new lighting technologies, and we explore how public policy can best address the challenges and opportunities offered by solid-state lighting to aid countries in their struggles to conserve energy and reduce global warming.

We are entering an era of faster-paced competition as the lighting industry, which has been dominated by a few firms (at least in the lamp sector), faces competition from new technologies, firms, and regions. Asian firms, as well as firms headquartered in the United States and Europe, have performed strongly in patent invention for solid-state lighting and are making key contributions to these new technologies. Both new firms and incumbents are investing heavily in solid-state lighting technologies, and it remains to be seen which firms will predominate.

Public policy will likely play an important role in future developments by stimulating demand for energy-saving lighting, providing funding for R&D, and incubating startup companies as they seek to commercialize these new technologies. But retail firms like Wal-Mart are increasingly playing a role in the diffusion of energy-saving lighting technologies. We compare the policies of countries supporting development and diffusion of new lighting technologies and speculate about how these efforts may affect the location of R&D, manufacturing, and headquarters of surviving lighting producers.

EVOLUTION OF THE LIGHTING INDUSTRY

Globalization of Lighting Production

The global lighting market in 2004 was worth some $40 billion to $100 billion, about one-third of which represented lamps.[1] U.S. apparent consumption of lamps, fixtures, and other equipment totaled about $14.8 billion in 2004.[2] U.S. production of lamps grew steadily until the early 1970s, then fluctuated over the next 20 years, stabilized during the 1990s at about 1970 levels, and finally fell somewhat at the start of the 21st century, as shown in Figure 1.

The eventual leveling off and downturn in U.S. lamp production in the 1990s can be explained, in part, by a steady increase in imports over the past two decades. Total imports, as a percentage of U.S. apparent consumption, increased from less than 20 percent in 1989 to around 50 percent in 2004, as shown in Figure 2.[3]

About half of the imports come from China, Mexico, and Japan, with China representing the largest share as of 2004. In 1989, less than 3 percent of lamps were imported from China. By 2004, Chinese lamp imports represented 26 percent of all lamp imports, having grown more rapidly than imports from any other supplier nation, and 10 percent of apparent lamp consumption in the United States. Once concentrated in the hands of three large manufacturers, the incandescent bulb industry has new competitors, primarily low-cost manufacturers in Asia.

In the fixtures industry, broken down in Figure 2, these trends are more intense, with 86 percent of all fixtures imports in the United States arriving from China by 2004. Increased fixture imports are the result of both incursion of lower-cost Chinese manufacturers and shifting production abroad by U.S. firms that seek lower-cost manufacturing sites. An exception to this trend is Genlyte

---

[1]Hadley et al. (2004) cite the figure of $40 billion, one-third of which represents lamps, but publicly available estimates of the size of the global lighting industry vary greatly, and the firm Color Kinetics in a communication with us cites the figure of $100 billion based on data from Fredonia Marketing Research.

[2]Apparent consumption equals U.S. production plus imports less exports, where U.S. production is measured as value of shipments from Bureau of Economic Analysis data, and import and export data are from the U.S. International Trade Commission.

[3]Not shown in Figure 2 is an additional trade category "Other Lighting Equipment," for which imports increased from 38 percent in 1989 to 57 percent in 2004, with China's share of imports growing from 24 percent in 1996 to 32 percent in 2004.

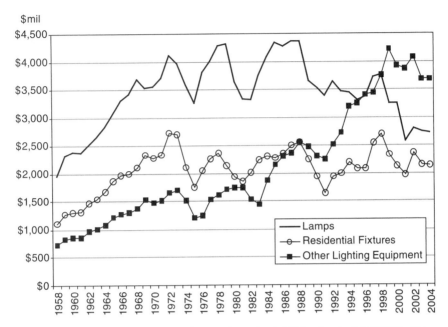

FIGURE 1 U.S. shipments of lighting products (real 2004 values). SOURCES: Shipment values: National Bureau of Economic Research (1958-1996), U.S. Department of Commerce, U.S. Census Bureau (1997-2001), Bureau of Economic Analysis (2002-2004). Producer Price Index from the Bureau of Labor Statistics.

Thomas, the largest lighting fixture and control company in North America, which manufactures 70 percent of its products in the North American region in order to keep close to its design centers and customers (Genlyte Thomas, 2005). Genlyte Thomas is introducing new energy-efficient light fixtures using compact fluorescent (CFL), high-intensity discharge (HID), and LED lamps and is conducting research on solid-state lighting to remain the premier fixtures company while the industry transitions to new lighting technologies.

The remaining area of growth for U.S. lighting production in the 1990s was in specialty lighting applications, such as Christmas decorations, underwater lighting, and infrared and ultraviolet (UV) lamps. This sector grew steadily throughout the second half of the 20th century and, as Figure 1 reveals, has surpassed the production value of lamps and of residential fixtures.[4]

---

[4]Figures 1 and 2 use the definition of traditional lighting, as defined in the beginning of this article, for "lamps" based on SIC (3641, 3648) and NAICS (33511, 335129), which includes all traditional lamp types including (regular and compact) fluorescent and HID lamps, but excludes LED lamps.

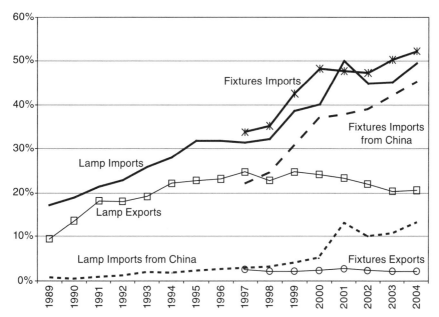

FIGURE 2 U.S. imports and exports of lamps and fixtures, total and imports from China, as percentages of U.S. consumption. SOURCE: U.S. International Trade Commission.

Big Three Lamp Producers

While there are hundreds of small lamp producers, which usually specialize in one type of lamp, the global lamp market is dominated by three big players: Koninklijke Philips Electronics (Philips), OSRAM-Sylvania (OSRAM), and GE.[5] All three firms produce a wide spectrum of lamps based on distinct technologies for most major commercial and residential markets. Philips has the largest global market share in lamps, and GE has the largest U.S. market share (Mintel, 2003).[6]

In the United States, GE has been a dominant player in lighting since the industry's inception (Leonard, 1992). As early as the mid-1890s, GE and Westinghouse controlled a 75 percent market share. GE eventually gained even greater market dominance, so that by 1927 GE and its licensees held 97 percent of the U.S. lamp market. Hygrade-Sylvania, whose lighting operations would much

---

[5]That most lamp producers specialize on a single type of lamp is apparent from industry directories such as www.lightsearch.com.

[6]Philips Lighting employs about 45,500 people and has 70 manufacturing facilities worldwide (Philips, 2006, p. 38).

later be bought by a German producer to form OSRAM-Sylvania, was GE's largest lamp licensee. Although GE's market dominance fell in the latter half of the 20th century, it remained the largest U.S. lamp producer.

In Europe, the lamp market also became concentrated early (Leonard, 1992). The leading firm was OSRAM, formed in a 1919 merger of the three leading German lamp producers, and now wholly owned by Siemens. Second in the European market was the Dutch company Philips. In part through cooperation with a European cartel set up in the 1930s under Swiss corporation Phoebus S.A., GE made substantial inroads in Europe and became the dominant worldwide producer.[7] The big three lighting firms all maintained leading positions in traditional lighting technology.

Traditional electric lighting patent applications during the period 1990-1993 were identified using data for the United States and Western Europe.[8] As noted earlier, we define traditional lighting to include incandescent, gas-discharge, and electric arc lighting (which includes fluorescent, HID, mercury and sodium vapor, metal halide, and neon lamps). All of the big three were leaders in these traditional electric lighting technologies, with 257.8 patent applications by Philips (credit is split equally in the case of multiple assignees); 232.1 applications by GE and by Thorn, whose lighting business GE acquired in 1991; and 219.4 applications by OSRAM, Sylvania, and OSRAM's parent firm Siemens. The big three each had more patent applications than any other firm.[9]

---

[7]GE's dominance varied substantially across nations. For example, in the United Kingdom in 1965-1967, the leading producer was British Lighting Industries, followed by Philips and OSRAM (Monopolies Commission, 1968, p. 8).

[8]Patents are included for international patent classifications H01J61-65, "Discharge lamps"; H01K, "Electric incandescent lamps"; and H5B31 and H5B35-43, which cover "Electric lighting . . . not otherwise provided for" excluding electroluminescent light sources (which provide sufficient light to see an object but not to see by). Patents are included for applications at patent offices of the United States (1,589 applications), Europe (976), Austria (190), Belgium (20), Denmark (49), Spain (976), Finland (79), France (121), Germany (1,798), Ireland (3), Italy (31), Netherlands (51), Norway (22), Portugal (5), Spain (194), Sweden (20), Switzerland (22), and the United Kingdom (218). Data are drawn from the European Patent Office's Worldwide Patent Statistics Database, version April 2006 (with the coverage of the Espacenet online database). Equivalent applications in multiple nations, detected by the fact that they share an identical set of priorities (as in Espacenet), were treated as a single application by weighting each application in inverse proportion to its number of equivalents (including itself).

[9]A German patent trust, Patra Patent Treuhand (possibly associated with OSRAM or Siemens), had 185.7 applications. The next three firms in number of patent applications were Toshiba with 70.2 applications, Motorola with 36 applications, and Matsushita with 33 applications. As in most areas of patenting, there were patent applications by many other individuals and companies (the total number of relevant patent applications was 3,236 during the period 1990-1993), and meaningful analyses are based on relative numbers, not on percentages of total applications. When figures are measured in terms of the number of patents actually granted from these applications (by the time of data collection), the conclusions are similar: Philips received 213.6 patents; OSRAM, Sylvania, and Siemens, 205; GE including Thorn, 181.5; Patra Patent Treuhand, 76.9; Toshiba, 52.6; Motorola, 35; and Matsushita, 31.

Evolving Technology

The lighting industry has developed several types of lamps. The incandescent lamp, little changed in form since the Edison era, is an evacuated glass tube (usually refilled with a gas) in which an electric current passes through a thin filament, heating it and causing it to emit light. Mercury vapor lamps, first patented in 1901 by Peter Cooper Hewitt, are high-pressure gas arc lamps and a forerunner to fluorescent lamps. Neon lamps were invented by Georges Claude 10 years later. Fluorescent lamps, first patented by Meyer, Spanner, and Germer in 1927, use a glowing phosphor coating instead of glowing wires to increase efficiency. Special types of incandescent lamps, such as bulbs filled with halogen gas to increase lifetime and efficiency, have also been developed.

Incandescent lamps account for a majority of household sales in the United States, but a smaller portion of total sales. In households, incandescent lamps represent 66.5 percent of sales revenues, whereas fluorescent and other lamps remain uncommon (Mintel, 2003). Residential sales, however, make up less than 10 percent of lighting demand measured in lumen-hours. Combining all economic sectors (residential, commercial, industrial, and outdoor), incandescent lamps represent 11.0 percent of lumen-hours of light output, as compared to about 57.5 percent for fluorescent, 31.0 percent for HID, and 0.01 percent for solid-state lighting (Navigant Consulting, 2003a, p. 7).

Each of these lamp types has experienced a steady march of small improvements in materials, design, light quality, energy efficiency, and manufacturing efficiency throughout the past century. While early improvements were made by independent inventors in the United Kingdom, more than three-quarters of these improvements originated in countries where the big three were headquartered—the United States, the Netherlands, and Germany.[10] In materials, for example, thorium oxide added to wires increased shock resistance, nonsag wire formulations made possible new configurations for brighter and more easily mounted incandescent filaments, and safer phosphors replaced the highly toxic beryllium coating in fluorescent lights. Examples of design changes include filling incandescent lamps with large-molecule gases to prolong filament lifetimes, new layouts of filament mounts to facilitate assembly and automated manufacture, and a proliferation of lamp varieties, shapes, and sizes. Light-quality changes were achieved by choosing appropriate filament and phosphor materials and sometimes by blocking part of the emitted light to attain, for instance, a look similar to sunlight.

Energy-saving lamps also progressed steadily but slowly. GE, for example, commercially introduced its first energy-saving incandescent lamp in 1913, but

---

[10]We catalogued 134 improvements in lamp technologies between 1905 and 2005. Sources: company websites of General Electric (2006), OSRAM (2006), Siemens (2006), Philips (2006), and Toshiba (2006); websites of Bellis (2006a,b,c), Williams (2005), and Arthur (2006), and Bowers (1982).

not until 1974 was the first energy-saving fluorescent lamp introduced. Manufacturing became increasingly efficient with machines and methods that allowed faster, higher-quality production with less manual labor. Automatic insertion and mounting of components, sealing, exhausting, basing, and flashing were key process technologies. Many of these and other improvements took place during the first half of the last century and are documented by Bright (1958, pp. 22-30). In the latter half of the century, improvements focused largely on improved efficiency and longer lamp lives. The discovery of substances such as narrowband phosphors led to the development of CFLs, gases such as xenon yielded brighter lamps such as those used in automobiles, and similar improvements had medical uses including UV lamps.

Whereas lowering manufacturing costs and streamlining production were the key lighting challenges of the late 20th century, saving energy is the new driving force for 21st-century development. Lighting accounted for about 22 percent of total energy used in residential and commercial sectors in the mid-1990s, as shown in Figures 3 and 4 (DOE, 1995, 1997). In 2001, 51 percent of the national energy consumption for lighting occurred in the commercial sector, 27 percent in residences, and 14 percent in industry; the remaining 8 percent was used in outdoor stationary lighting (Hong et al., 2005, p. 2). Almost half of the electricity used in commercial buildings is used in lighting, as Figure 5 indicates.

In the United States, residential homes largely use incandescent lamps (90 percent), whereas commercial and industrial sectors use mostly fluorescents (Hong et al., 2005). If residential homes in the United States replaced all incan-

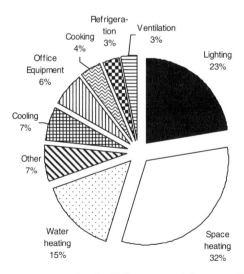

FIGURE 3 Energy consumption in U.S. commercial sector, 1995. SOURCE: DOE (1995).

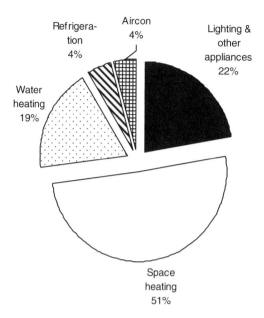

FIGURE 4 Energy consumption in U.S. residential sector, 1997. SOURCE: DOE (1997).

descent lamps with CFLs, they would save an estimated 35 percent of electricity used for all lighting applications (DOE, 1993).

   Although advances in energy-saving lighting technologies such as CFL have been an important part of the strategies of the big three lamp producers, the big three have had some difficulty getting residential customers to give up incandescent bulbs and replace them with the more energy-efficient but initially more expensive bulbs. The rate of adoption of CFLs in U.S. residential households has been low, particularly compared to that of Europe and Asia. Researchers attribute those differences to a variety of factors, including national coordination of promotional efforts, different cultural attitudes about resource consumption, and higher electricity prices (Calwell et al., 1999). U.S. residential consumers lack awareness of and knowledge about CFLs. Consumer buying habits, negative perceptions, and skepticism about fluorescent lighting and relatively low electricity prices have meant that the United States is behind the rest of the world in adoption of energy-saving lighting technologies (Sandahl et al., 2006). This may soon change; for example, Wal-Mart CEO H. Lee Scott, Jr., is committed to sell 100 million CFLs a year by 2008 and the firm is making a concerted effort to change consumer behavior (Barbaro, 2007).[11]

---

[11]Wal-Mart sold about 40 million CFLs compared to 350 million incandescent light bulbs in 2005.

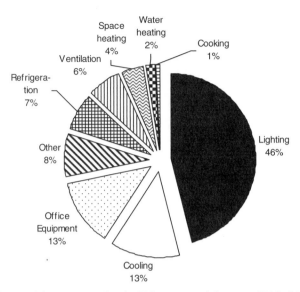

FIGURE 5 Electricity consumption in U.S. commercial sector, 1995. SOURCE: DOE (1995).

Since lamp efficacy is central to which lamp types dominate the market, it is important to understand efficacy and its role in purchasing decisions. Efficacy in lighting can be measured in terms of lumens produced per watt of electricity (lm/W). A standard 100-watt incandescent lamp, for example, lasts about 1,000 hours and produces 15 lm/W. By comparison, a standard 30-watt fluorescent lamp lasts 20,000 hours and produces 80 lm/W. A longer-lasting and more energy-efficient bulb is less costly over the long term but higher initial upfront costs and misconceptions about the efficacy of fluorescent lights (early fluorescents had poor color rendering and were noisy) have led to low adoption in residences. Optimal lamp choice involves not only energy efficiency but also replacement costs for burned-out lamps and labor costs to install lighting systems. In commercial and industrial settings, where life-cycle costs are important and companies can make upfront investments, fluorescents are usually chosen.

RADICAL INNOVATION IN LIGHTING: LEDS

Nature and Advantages of LEDs

An LED is a semiconductor diode. It is electroluminescent, emitting color that depends on the chemical composition of the semiconductor material or compound used and ranges along the spectrum from UV to infrared, as documented in Table 1.

TABLE 1 LED Color Spectrum Available from Alternative Materials

Semiconductor Material	Color
AlGaAs (aluminum gallium arsenide)	Red Infrared
AlGaP (aluminum gallium phosphide)	Green
AlGaInP (aluminum gallium indium phosphide)	Orange-red (bright) Orange Yellow Green
GaAsP (gallium arsenide phosphide)	Red Orange-red Orange Yellow
GaP (gallium phosphide)	Red Yellow Green
GaN (gallium nitride)	Green Pure green (emerald) Blue
InGaN (indium gallium nitride)	Bluish green Blue Near UV
SiC (silicon carbide) as substrate Si (silicon) as substrate, under development Al_2O_3 (sapphire) as substrate ZnSe (zinc selenide)	Blue
C (diamond)	UV
AlN (aluminum nitride) AlGaN (aluminum gallium nitride)	Far UV

SOURCE: Wikipedia (2006).

The first practical visible-spectrum LED was developed in 1962 by Nick Holonyak (*Inquirer*, 2004), and a variety of single-color LEDs followed. White LEDs have been a long-standing goal for researchers since they are most likely to replace traditional bulbs. White LEDs have been created by coating blue LEDs with a yellow phosphor, yielding a blue and yellow glow that appears white to the human eye. Another approach, taken by GE, uses UV LEDs driving phosphors, and a third approach is to use multiple colors of LEDs and combine them to create white light. Current white LEDs are cost-effective only for certain applications, such as backlighting and flashlights, and color LEDs remain more widely used.

Although incandescent and fluorescent lamps remain the predominant light sources, LEDs have several potential advantages. First, they use less energy: LEDs are three- to fourfold more efficient than incandescent and halogen sources. However, with the exception of laboratory devices, LEDs still fall short of fluorescent sources for many white light applications. Nevertheless, they are semiconductor devices and LED lighting is thought to follow an equivalent of Moore's law in computing, advancing rapidly and continually because of the pace of electronic circuit improvements.

Second, in contrast to incandescent lamps, LEDs use most of their energy in lighting (Herkelrath et al., 2005). LEDs also have a long life span, typically about 10 years of on-time—twice that of fluorescent lamps and twenty times that of incandescent lamps. In terms of luminous efficacy (lm/W), LEDs by 2004 were already about four times as efficient as incandescent lamps, and by 2020 they are targeted to be about 12 times as efficient as current incandescent lamps and more than twice as efficient as current fluorescent lamps (Hadley et al., 2004, p. 5; Tsao, 2002, p. 4). In addition, LEDs light up many orders of magnitude faster than incandescent lamps, and, rather than burning out abruptly, they do so slowly. LEDs require little maintenance and are cool to the touch, durable, and flexible. Furthermore, the technology is digitally compatible and, hence, can be integrated into digital networks, facilitating customizable electronic control.

LEDs come in many shapes and sizes and have multiple uses. Backlighting is one use, for cell phones, cars and other electronics, liquid crystal displays (LCDs), and specialized lighting applications. Specialty uses are possible since LEDs can be waterproofed, bent, shaped, multicolored, and dimmed.[12] LED applications are common in the entertainment industry, hotels, road signs, exit signs, pools, landscaping, and darkrooms.

The main drawback of LEDs is that they have not yet achieved the efficacy necessary for many white light applications. They are also still costly because they are expensive to produce. But production costs are expected to decline as volumes rise and the technology advances. For example, in 2002, the total cost of LED lamps (capital cost plus operating costs) was estimated at $16.00 per million lumen hours, compared to $7.50 for incandescent bulbs and $1.35 for fluorescent bulbs (Hadley et al., 2004, p. 8; Tsao, 2002, p. 8). However, by 2020, the total (capital plus operating) cost of LED lamps is targeted to be reduced to $0.63 per million lumen hours (Hadley et al., 2004, p. 8; Tsao, 2002, p. 8).

An additional limitation of LEDs, relative to incandescent lamps, is their imperfect color rendering, given the spectrum of light emitted. White light created by multiple color LEDs or by phosphors driven by LEDs involves a combination of wavelengths of light that differs from the color spectrum of traditional lamps and of sunshine, making objects with certain colors appear relatively dark.

---

[12]LEDs can also be designed to trap insects (through the use of insect-attracting colors) or to avoid attracting insects (since they do not generate UV light) (Bishop et al., 2004).

However, Ashdown et al. (2004, p. 8) indicate that color spectrum limitations are likely to be remedied as the technology progresses.[13]

Until recently, LEDs have been the only viable technology competing with the various types of traditional lamps as electricity-driven sources of illumination (Hong et al., 2005). A recently emerged technology that may become competitive with LEDs is microwave-driven light bulbs, for which claims of efficiency and low cost remain to be assessed; this technology emerged during the final stages of our work and is not assessed here (*Economist*, 2007a,b). Our analysis (including patent data presented later) includes two newer types of LEDs: organic light-emitting diodes (OLEDs) and polymer LEDs. OLEDs, which are LEDs involving organic (carbon-based) chemicals, are promising but still in an early development stage. Ching Tang and Steven Van Slyke of Eastman Kodak invented the first OLED in 1987 (Howard, 2004). The materials in OLED devices have broad emission spectra that provide an advantage over inorganic LEDs (minor changes in the chemical composition of the emissive structure can tune the emission peak of the device). It is believed that good-quality white light is achievable from OLEDs (OLLA, 2006a, 2006b).[14] An important focus of current OLED research is on improving operational life.

In particular, OLEDs are of interest to display firms since they are capable of producing true black colors, something LCDs cannot achieve since they require a backlight to function and are never truly "off." OLEDs can produce a greater range of colors, brightness, and viewing angles than LCDs because OLED pixels emit light directly. The display industry, with more than 70 companies including OLED pioneer Eastman Kodak, is set to commercialize OLED technology including OLED displays (Hong et al., 2005).[15] Kodak launched the first digital camera to use a full-color OLED display in 2003. The big three traditional lighting companies have all set up joint ventures to profit from OLED technology for the display market.

---

[13]Other limitations of LEDs are areas of active work. For example, LEDs driven with sufficient power for automotive headlights and taillights require heat sinks because heat degrades LEDs; relevant heat sinks are improving.

[14]White OLEDs by 2006 had achieved claimed efficacy of 25-64 lm/W (Burgess, 2006; Physorg. com, 2005; Ledsmagazine.com, 2006).

[15]In December 2007, Sony released the first commercially available OLED televisions, although they remain far from price-competitive with liquid crystal display (LCD) televisions (Eisenberg, 2007). Samsung has also been demonstrating prototype OLED displays (Gizmodo, 2008). LCDs, which are now dominant in televisions and computer monitors, are akin to (dynamically changing) stained glass windows behind which white light is projected, thus allowing some colors of light through the liquid crystals while blocking other colors. This means that light not allowed through the liquid crystals is wasted, turning electricity into heat, and that the liquid crystals require behind them a light source that adds to their thickness, weight, and cost. If OLED displays and their production methods can be improved sufficiently, therefore, they have potential advantages over LED displays: the desired color can be produced at each location on an OLED display with no wasted light and no liquid crystal layer.

LEDs as a Disruptive Technology: Diffusion Among Applications

Disruptive technology has been defined in several ways, and LEDs match at least two of the definitions. First, disruptive technology may be defined as a new technology that fills a long-standing need and for which the expertise and resources of incumbent firms do little to help them with this new approach; in this case solid-state lighting fills the need for illumination using a technology that differs totally from traditional lighting technologies. Second, disruptive technology may be defined as a technology in which new firms enter a market and threaten the market dominance of incumbent firms; in this case new firms have been entering the lighting market by creating products based on LEDs, and it has been unclear whether the leading existing lighting manufacturers can maintain strong market positions if purchases shift substantially to LEDs.

LEDs are a novel technology in lighting. LEDs are semiconductors and so manufacture of LEDs has little in common with traditional lamp production. The supply chain to produce LEDs as indicated in our discussions with industry experts is quite disaggregated, as is the case for other semiconductor technologies. Moreover, Kevin Dowling, Chief Technology Officer at Color Kinetics, stated to us that "the vertically integrated giants of the semiconductor world such as Intel and Applied Materials are becoming less numerous and rapidly becoming more the exception rather than the rule." Data on the participation of firms in each stage of the LED supply chain are available from solidstatelighting.net (2006). We catalogued the participation of each firm in each stage of the supply chain and found that most firms participate in only a single part of the supply chain, although a few large firms are involved in many parts of the supply chain. This supply chain is illustrated in Figure 6.

Semiconductor firms often specialize in specific stages of the supply chain, such as R&D, epitaxy, manufacturing, packaging, testing, and back-end processing. Each stage requires unique skills and equipment and significant capital expenditure, which is one reason why firms tend to specialize rather than integrate along the supply chain. Specialization is thought to drive down prices and improve performance, and this trend is similar among LED manufacturers. The development of this new technology will likely create opportunities at all levels of the value chain.[16]

The LED market in general lighting is still small compared to traditional lamps. LED applications command a high price, but relatively few units are sold and all are for specialty purposes. Traditional lamps (incandescent, fluorescent, and HID) are estimated by Navigant Consulting (2003a, p. 7) to have used 41,051 trillion lumen hours of electricity in the United States in 2005 compared to only

---

[16]As lighting moves to LEDs, traditional bulb manufacture may also be vertically disintegrating somewhat, judging from recent comments by Jim Campbell, President and CEO of GE Consumer & Industrial (General Electric, 2007).

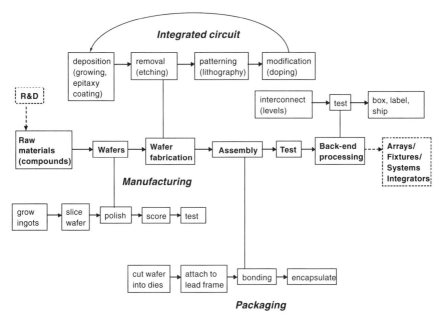

FIGURE 6 LED semiconductor supply chain.

5 for LEDs. Nevertheless, the LED market grew by 50 percent between 1995 and 2000 and has been forecast at $4.7 billion by 2007 (Ashdown et al., 2004, p. 9).

LED technology has some clear advantages over traditional lighting, which have allowed LED manufacturers to displace traditional lighting in niche markets (Griffiths, 2006). Indicator lights were one of the earliest uses, with color LED indicators predating the 1990s and white LED indicators used from about 2000 onward. In 2001, 30 percent of LED sales were for backlighting, 26 percent for automotive uses, 26 percent for signs and displays, 10 percent for electronic equipment, 4 percent for signals, and 4 percent for general illumination (Maccagno, 2002, cited by Ashdown et al., 2004, p. 9). By 2002, U.S. market penetration of LEDs was particularly high in exit signs (80 percent), truck and bus lights (41 percent), and traffic signals (30 percent) (Navigant Consulting, 2003b, p. xii; Hadley et al., 2004, p. 9).

Other niches that LEDs have entered include video screen backlights in the mid-1990s, decorative lighting in the late 1990s, and automobile lights including dashboard, interior, brake, and tail lights since about 2000. Other recent uses include architectural lighting, outdoor advertising, and long-lasting white-light flashlights. One example of the advantage of LEDs is in brake lights, where LEDs provide an extra 0.2-second response time and therefore help to prevent accidents. Although first introduced in luxury cars, LED brake lights are beginning to penetrate the more cost-conscious end of the market. Another example is Wal-

Mart's adoption of LED lighting for refrigerated display cases, an application once dominated by fluorescents. As well as lowering operating costs, the LED lights are amenable to added motion sensors so that lights come on only when shoppers are nearby. As a result, Wal-Mart is investing $30 million and expects a 66 percent reduction in freezer lighting energy costs (LIGHTimes, 2006).[17]

New LED Lighting Innovators

Advances in solid-state lighting offer an opportunity and a challenge for incumbent and startup firms. Although the big three lamp manufacturers have been making substantial investments in solid-state lighting, pioneering inventors in LED lighting come from universities, research labs, and companies, and R&D plays a vital role in development of these technologies.

Advances in red, yellow, and blue LEDs have been led by different research groups and companies. Several companies have "specializations" in one industry sector, due to a combination of strategy and luck in pioneering key product or process innovations. Nichia Corp. in Japan, for example, was one of the first companies to develop blue LEDs, a key advance when only red, green, and yellow were available. It also produced the first white LEDs in 1996 (Walker, 2004). The company is an illustration of how small firms have been able to penetrate the burgeoning industry. Nichia's key researcher, Shuji Nakamura, now a professor at the University of California, Santa Barbara, was largely responsible for the development of the blue LED. When Nakamura was hired in 1979, Nichia was a small firm in rural Japan with only 200 employees and Nakamura was assigned a project to synthesize a commercial-grade blue LED, needed to complete the color palette. At the time, large Japanese corporations were spending $85 million a year and Nichia had a very small research budget. Today Nichia controls 80 percent of the blue LED market with Cree and Toyoda Gosei (Cox, 2003). Nakamura's successful approach departed from the standard thinking in his field and in his company. He chose gallium nitride, a material most researchers thought would not yield significant results, as the basis for his research, and continued to work on blue LEDs for 10 years. Nichia's entry into the LED market was a lucky outcome of their hiring a particular employee and of that employee's actions.

The role of individuals in innovative companies in pioneering new lighting technologies is typical of the early stages of a technology cycle in which R&D efforts are lengthy and costly. Government grants have been instrumental to support startup companies and university research in the pursuit of emerging LED technologies. Government funding has filled key technology gaps, provided fund-

---

[17]Wal-Mart installations developed in collaboration with GE and Philips represent "the biggest investment to date in LED lighting for interior application [$30 million], and it is also the single largest installation of white LED lighting replacing fluorescent lighting in a display lighting application" (Griffiths, 2006).

ing to develop enabling knowledge and data, and advanced the solid-state lighting technology base. A team of researchers from Rensselaer Polytechnic Institute, for example, recently received $1.8 million in federal funding from the Department of Energy (DOE) to improve the energy efficiency of green LEDs, with a goal of doubling or tripling power output. The research was one of 16 projects selected for funding through the DOE's Solid-State Lighting Core Technologies Funding Opportunity Announcement, which supports enabling and fundamental solid-state lighting technology for general illumination.

"Making lighting more efficient is one of the biggest challenges we face," says Christian Wetzel, the Wellfleet Career Development Constellation Professor, Future Chips, and associate professor of physics at Rensselaer (RPI, 2006). To meet aggressive DOE performance targets that call for more energy-efficient, longer-lasting, and cost-competitive solid-state lighting by 2025, the team has partnered with startups such as Kyma Technologies and Crystal IS. Kyma, a North Carolina State University spin-off, specializes in gallium nitride substrates, while Crystal IS, founded by two Rensselaer professors, specializes in blue and UV lasers based on single-crystal aluminum nitride substrates. The DOE grant has funded these startups and researchers.

Government support has also been important for building demand and aiding firms to improve quality and reduce prices—keys to further diffusion. Such programs promote early diffusion of energy-saving technologies and are not unique to the United States. We will return to the role of national policies and government initiatives later in the chapter.

CORPORATE STRATEGIES TOWARD INNOVATION

The "Big Three"

The big three traditional lighting manufacturers, Philips, OSRAM, and GE, have responded to the opportunities in LED lighting by creating joint ventures with semiconductor firms that had preexisting expertise in these technologies. They later acquired these joint ventures outright. Philips established a joint venture in optoelectronics with Hewlett-Packard (HP) in 1999 (when HP split in two in 1999, the optoelectronics group was assigned to a new firm, Agilent Technologies) and acquired the venture Lumileds in 2005 for $950 million.[18] OSRAM established a joint venture with Infineon Technologies AG (formerly Siemens Semiconductors) in 1999, and acquired the venture in 2001, naming it OSRAM Opto Semiconductors GmbH.[19] GE established a joint venture, Gelcore,

---

[18]Philips' 2005 Annual Report states that Lumileds is the world's leading manufacturer of high-power LEDs.

[19]Siemens gradually spun off its semiconductor division as Infineon beginning in 1999, and sold its final 18.23 percent stake in the company in 2006.

with semiconductor maker Emcore in 1999, and acquired Gelcore in August 2006 for $100 million.[20] Additionally, Philips and OSRAM announced in January 2007 a cross-licensing agreement covering patents on LEDs and OLEDs (LIGHTimes, 2007), and Philips acquired the LED lighting controls firm Color Kinetics in August 2007.

Historically, the locus of innovation for traditional (i.e., non-LED) lamps originated in the primary R&D centers of the big three lighting firms in Germany, the Netherlands, and the United States. While these labs are still very important, in recognition of Asia's increasingly important role in the traditional lighting industry, the three firms have set up manufacturing, engineering, and R&D activity in other parts of the world, principally in Japan and Taiwan. Efforts are also being made to penetrate the Chinese market. GE, for example, began investing in China through joint ventures. The company combined a finished product purchasing center and an R&D center to form the GE Asia Lighting Center in Shanghai. By 2002, GE had four major plants in Shanghai and one in Xiamen, and had invested over $100 million in China for lighting, according to Matthew Espe, former president and CEO of GE Lighting (Zou, 2002).

Philips established an R&D campus with the Shanghai Science and Technology Commission with annual expenditures of $50 million, the majority of which is in lighting. Between 1988 and 2005, Philips Lighting established nine solely owned and joint ventures and five R&D centers, of which one conducts global level research while the other four mainly focus on the Asia-Pacific region (Chinesewings, 2005).

OSRAM China Lighting, Inc., 90 percent of which is owned by OSRAM, was formed in April 1995 with an investment of 49.7 million Euros. The company is located in Foshan, China, and has two factories in China, employing 6,000. In February 2006, OSRAM China Lighting announced it would acquire Foshan Electrical and Lighting Co. Ltd.[21]

Asian LED Producers

Beyond their expanding importance in the traditional lighting industry, Asian firms also play a significant role in the global LED market. Japan's LED industry leads with $918 million in sales, or a world market share of 47 percent, although a portion of these revenues are shared with some U.S. companies through joint ventures. Taiwan's industry holds second place at $712 million, or a market share of 25 percent (Taiwan Economic News, 2004a,b). LEDs are the largest type of

---

[20]Although Gelcore grew about 50 percent from 1995 through 2004, it nonetheless reported a net loss of $0.8 million in 2005. Thus, when Emcore sold its 49 percent stake in Gelcore, it traded possible future value for immediate cash gains.

[21]Source: "OSRAM China Lighting's Announcement to Acquire Foshan Electrical and Lighting Co. Ltd.," http://business.sohu.com/20060614/n243727208.shtml (in Chinese).

compound semiconductor production in Taiwan (Liu, 2003). Another source estimates the global LED market at $5.4 billion in 2004, with Japan's share at 51.3 percent, Taiwan's 22.7 percent, the United States 12 percent, and Europe around 9 percent (Ledsmagazine.com, 2005b). Data from www.solidstatelighting.net (2006) suggest that most LED R&D is conducted in the United States whereas Asia dominates manufacturing and packaging. For example, Taiwan, China, and Korea produced the majority of blue LEDs in the world, and more than 80 percent of the production of InGaAlP high-brightness (HB) LEDs in 2004. China boasts about 600 enterprises directly related to the LED industry in China, employing about 40,000 people (Ledsmagazine.com, 2005a).

Because of the youth of the industry, quality remains variable across firms and is sometimes difficult to assess. Quality for an LED includes a long operating lifetime before substantial fading occurs, controls to ensure against defective chips, and consistency of operating characteristics including a claimed color emission spectrum, light intensity, voltage drop, and viewing angle (Toniolo, 2006b). The LED industry has seen a surge of new players, which has flooded the market with low-quality LEDs (Toniolo, 2006a). The result of such commodity production is intense price competition and a "huge overcapacity" for lower-performance LEDs (Arensman, 2005). For high-performance LEDs, competition remains less intense. Among consumers, low-quality LEDs may give a bad reputation to all LEDs.

New Ventures

The global semiconductor market was worth $262.7 billion in 2006 (Gartner, 2007), considerably outstripping LED industry revenues of $3.7 billion in 2004 (www.gelcore.com). A large number of materials, substrates, epitaxy, packaging, and manufacturing companies have entered the LED market. In February 2006, lighting industry directory Lightsearch.com listed 71 companies that produced LED lamps.[22] Most companies operate at a single stage of the supply chain, illustrated in Figure 6. For example, companies that perform epitaxy do not usually do manufacturing or packaging. Likewise, most companies that focus on packaging do not produce raw materials or substrates. Moreover, companies that focus on basic R&D do not operate in the rest of semiconductor production.[23] Even on a national level, specialization sometimes occurs. For instance, Taiwan is strong in R&D and manufacturing of LEDs, whereas Korea specializes in packaging, and China, a late entrant, is setting up epitaxy, and wafer and chip production

---

[22]Recall that "lamp" as used in the traditional lighting industry means "bulb," and note that the former term is most appropriate for LED lighting as no glass bulb is involved. Lamp here means a light-producing device, not a fixture.

[23]Based on information on semiconductor companies in the LED industry gathered from solidstatelighting.net (2006).

(Wang and Shen, 2005). In addition, some countries specialize in the production of specific LED colors: Taiwan holds a majority market share for blue GaN LEDs at 34 percent, closely followed by Japan at 33 percent, whereas the United States and Korea lag with 19 and 12 percent, respectively (Wang and Shen, 2005).

Although LEDs are still a small subset of the semiconductor market, they have rapid growth in demand, making this an attractive market for new and existing semiconductor firms. LEDs offer opportunities for semiconductor firms to diversify into a new market that promises long-term growth potential. For example, Avago Technologies, the world's largest privately held semiconductor company, recently announced three new series of HB full-color LEDs for the outdoor electronic signs and signals market (*Business Wire*, 2006).

At the other end of the supply chain, LED "integrators" like Color Kinetics play an important role in LED lighting. Since its establishment in 1997, Color Kinetics has built an impressive patented portfolio of these technologies, which it uses in LED lighting systems. Color Kinetics has pioneered intelligent LED systems that are networked and has created a new niche as a "systems solutions" and lighting control technology provider. Color Kinetics has initiated several major projects that integrate LED lights with sound, movement, and rhythm through digital controls, and is working on white light systems. A subway tunnel in Chicago, for example, is bathed in several colors of LED light that periodically change (giving the impression of a sunset). The company leverages its strengths in innovation and engineering and works with selected Chinese manufacturers to assemble systems. In August 2007, Philips acquired Color Kinetics, which became Philips Solid-State Lighting Solutions, a business unit of Philips Lighting's Luminaires group.

LED LIGHTING R&D

The big three producers are dominant in traditional lighting technology, as shown earlier using data on patents for these technologies. In this section we analyze the R&D positions of these and other firms for LED lighting.

Methodology: Analysis of Patent Data

To assess trends in the global location of LED lighting R&D we use patent data. Patent data yield information on successful R&D outputs. Although the information is partial because many inventions and innovations are not patented, within an industry patents are highly correlated with R&D spending and are indicative of R&D success. Moreover, patents yield relatively defensible property rights and hence represent an important component of the value of firms' R&D outputs.

To analyze LED-related patents that pertain to lighting, a search criterion is needed to identify relevant patents. Choice of a criterion involves a trade-off between finding a subset of mainly relevant patents versus finding all relevant patents mixed with many more non-LED and non-lighting patents. We therefore chose a criterion to identify mainly relevant patents at the cost of excluding some LED lighting-related patents.[24] All patents were identified whose title contains the key word or phrase "LED(s)," "OLED(s)," "L.E.D.(s)," or "light emitting diode(s)," or the equivalents of the latter phrase in German, French, Spanish, Italian, or Portuguese.[25] Other languages, including Asia-specific languages, effectively are almost always included in our search because for almost all other languages the database we used has English translations of titles. The search criterion includes LED applications, including LED-type displays and (rarely) lasers, as well as LEDs whose glow is bright enough for general illumination. All patents granted are included regardless of whether they originated from firms, government programs, or university research labs.

Since both the traditional and the LED lighting industries are global in terms of the firms involved and startup efforts, we obtained data for patents issued in most nations worldwide, although we focus initially on patents granted in the United States and Europe.[26] For logistical reasons (we had to look up information by hand from actual copies of thousands of patents), we restricted the sample in the latter analysis to patents whose title included either "LED(s)" and "lighting," or the phrase "light emitting diode(s)." Our primary focus on U.S. and European patents addresses concerns that patents from other nations may face quite different approval requirements.

Patents are counted only once if the identical patent is filed multiple times

---

[24]International patent classification systems do not identify specific categories for LEDs or for LED lighting, and we draw on international data for which this limitation applies.

[25]The non-English terms used are "lichtemittierende Diode(n)," "Leuchtdiode(n)," "diode(s) luminescente(s)," "diodo(s) electroluminoso(s)," "diodo luminescente," "diodi luminescenti," "diodo(s) emissores de luz," and "diodo(s) emitindo-se claro(s)." The verb "led" sometimes appears in titles for reasons unrelated to light emitting diodes, so we read all patent titles and eliminated patents for which "led" was used as a verb.

[26]Data are obtained from the Espacenet worldwide patent database, maintained by the European Patent Office (EPO). The data include patents granted by the relevant patent authorities in almost all nations worldwide (a detailed listing is available from the Espacenet website), including not only the most developed world but also Eastern European and developing Asian, Middle Eastern, and African nations (or cooperating regions) with significant innovative activity. The European patent authorities for which LED patents appear in the primary sample are the European Patent Office plus the national offices of Austria, Finland, France, Germany, Italy, Sweden, and the United Kingdom. Patents from former Soviet-bloc nations are excluded. In our secondary sample of patents from patent authorities worldwide, patent authorities that actually granted patents in the data are (ordered by continent) the United States, Canada, Mexico; Germany, Great Britain, Belgium, France, Netherlands, Hungary, former Soviet Union, European Patent Office; Japan, Korea, Taiwan, Singapore, India, China, Hong Kong; Australia, New Zealand; and South Africa.

in different nations.[27] Patents granted in nations outside the United States and Europe are considered after our main analyses.

Analysis of LED Patenting

The analyses that follow compare the headquarters nationality of patenting firms and also the national R&D locations where invention was carried out. The headquarters location of a firm was identified as the international headquarters nation of the firm to which a patent was assigned. If a firm was owned by a "parent" firm, we use the headquarters nation of the parent firm. Rarely, patents were applied for by multiple firms or individuals, and assignee credit was divided equally among these applicants. The R&D location where invention was carried out was determined by the nation listed in the address of each inventor named on the patents. Since inventors' addresses are not generally available in electronic bibliographic data, we looked up the nation for each inventor using the original patent documents. Rarely, different inventors of a single patent had addresses in different nations, in which case credit for each R&D nation was divided in proportion to the number of inventors in each nation.

LED patent data are compared in two 4-year periods a decade apart, 1990-1993 and 2000-2003. These 4-year periods ensure an adequate-sized, representative sample of patent activity. Comparing between periods facilitates analysis of trends in R&D activity in LED lighting.

As LEDs have developed growing markets in new applications, so has LED R&D grown. Based on U.S. and European patents, the number of LED patents granted quintupled from 200 in 1990-1993 to 1,000 in 2000-2003, as shown in Figure 7.[28]

Globalization of LED R&D: U.S. and European Patents

The locations of inventors as reported on patent applications reflect where R&D occurred. We therefore assessed relative inventive activity in each nation, for 1990-1993 in Figure 8 and 2000-2003 in Figure 9, by determining the percent-

---

[27]Equivalent patents filed in multiple nations are identified as catalogued on the EPO's Espacenet patent web server, which defines equivalents based on identical priority claims.

[28]Substantial growth also holds using our worldwide data for a narrower range of patent titles, with LED patents granted growing from 438 in 1990-1993 to 1,114 in 2000-2003.

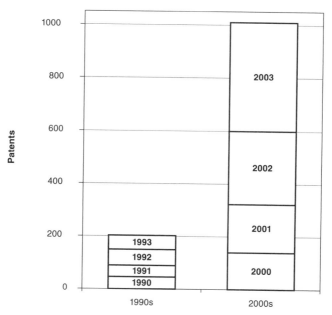

FIGURE 7 LED patents in sample granted during 1990-1993 and 2000-2003. SOURCE: Authors' analysis of patents granted in the United States and Europe (see text).

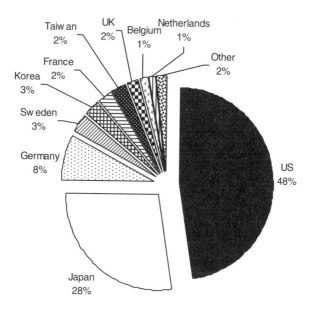

FIGURE 8 Invention locations of LED patents granted in 1990-1993. SOURCE: Authors' analysis of patents granted in the United States and Europe (see text).

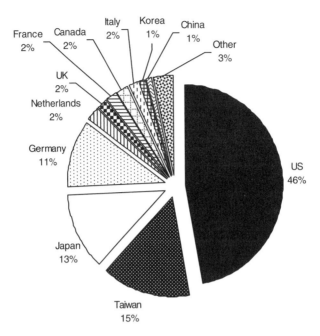

FIGURE 9 Invention locations of LED patents granted in 2000-2003. SOURCE: Authors' analysis of patents granted in the United States and Europe (see text).

ages of patents with inventors in each nation.[29] In the period 1990-1993, inventors located in the United States predominated with 47 percent of all of the LED patents. Inventors in Japan held second place with 27 percent of LED patents. Inventors in Germany ranked third with 8 percent of LED patents, followed by inventors in Sweden, Korea, Taiwan, France, and the United Kingdom, each with 2-3 percent of LED patents.[30]

---

[29]When computing the percentages of LED invention done in different nations, one concern is whether some nations' inventions might be of systematically poor quality, so that while those nations' inventors *appear* to accomplish a lot of R&D, in reality the value of their R&D output is much lower. One means to check whether this is the case is to examine only those relatively high-value patents for which firms went to the expense and trouble of obtaining equivalent patents in multiple countries or on multiple continents. Hence, we also examined patents using these criteria. These analyses yielded conclusions similar to the results reported in the text.

[30]Percentages in this section often represent small numbers of patents, so readers are cautioned that small variations in patenting activity can sometimes lead to substantial changes in the precise percentages reported. Our primarily conclusions, however, are robust to typical random variations, with statistical significance tests reported in footnotes. Differences in arrival rates of patents in 1990-1993 were statistically significant between the United States and Japan ($p = .0014$), between Japan and Germany ($p < .001$), and (marginally) between Germany and Sweden ($p = .0525$), using exact p-values from exact Poisson regressions. (Here and below, exact Poisson tests are actually close approximations

In the period 2000-2003, U.S. inventors' share of LED patents remained dominant at 47 percent. Rapid growth in inventions occurred in Taiwan, whose inventors achieved a 15 percent share of LED patents, ahead of Japanese inventors whose share had fallen to third place with 13 percent of LED patents. German inventors' share grew slightly to 11 percent of LED patents.[31] Other nations' share of LED patent invention remained at 2 percent or less.

Not only did LED innovation become somewhat more international, but non-U.S. companies became more international in the locations where they carried out research. Location of R&D (i.e., of inventors) is compared to locations of corporate headquarters in Table 2 for 1990-1993 and Table 3 for 2000-2003. During the period 1990-1993, the United States was the only country whose companies supported substantial LED R&D abroad.[32] LED patents were filed by inventors in Japan, the United Kingdom, and Germany for firms headquartered in the United States. Of the nine U.S. patents that had R&D locations abroad, Eastman Kodak was assignee for four, with inventors in Japan, and the remaining five had inventors in the United Kingdom (twice), France, Japan, and Malaysia.

By 2000-2003, more companies were supporting R&D across the globe. Overall, companies in 13 countries sponsored LED R&D abroad. U.S. companies, however, kept 95.1 percent of R&D within the United States. Asian LED invention sites for U.S. firms fell from 5.8 to 1.4 percent, and European sites for U.S. firms fell from 2.9 to 1.9 percent. In the period 2000-2003, U.S. companies' foreign-invented patents had inventors in Canada (1.6 percent) and nine other nations (each less than 1 percent). European companies began to support R&D in the United States, which now yielded 22.9 percent of European companies' LED patents. Asian companies also began to carry out R&D in the United States, yielding 3.6 percent of Asian companies' LED patents. Furthermore, Europe yielded 0.9 percent of Asian companies' LED patents.[33]

---

because they use rounded-to-integer values of weighted patent counts, rather than making assumptions about statistical (in)dependence between locations of inventors of some patents.)

[31]The differences between Taiwanese versus Japanese, and Japanese versus German, patent invention in 2000-2003 each involve 16-17 patents, and it is difficult to tell whether these reflect statistically meaningful differences in nations' propensities to generate LED inventions; these nations' rankings reported here should not be construed to indicate practically or statistically significant differences. (Exact significance tests are not feasible to compute in these cases, because the numbers of patents involved make exact Poisson calculations highly computationally intensive.) German inventors' arrival rate of patents was statistically significantly ($p < .001$) greater than the highest-ranked of nations with less invention, using an exact p-value from exact Poisson regression.

[32]This difference between the United States and other nations is marginally statistically significant, $p = .059$, using Fisher's exact test (two-tailed). (The test is approximate because it uses rounded-to-integer values of weighted patent counts, rather than making assumptions about statistical (in)dependence between locations of inventors of some patents.)

[33]Use of inventors outside the home country in 2000-2003 was statistically significantly less for United States patent applicants than for non-United States firms ($p < .001$ using Fisher's exact test as in the preceding footnote), with no significant difference between United States and Asian patent applicants.

TABLE 2 Location of Inventor vs. Firm Headquarters, 1990-1993

Location of HQ	Location of R&D			
	Asia	Europe	U.S.	Other
Asia	98.4%		1.6%	
Europe		100.0%		
U.S.	5.8%	2.9%	91.3%	
Other				100.0%

SOURCE: Authors' analysis of patents granted in the United States and Europe (see text).

TABLE 3 Location of Inventor vs. Firm Headquarters, 2000-2003

Location of HQ	Location of R&D			
	Asia	Europe	U.S.	Other
Asia	95.2%	0.9%	3.6%	0.3%
Europe	0.7%	76.0%	22.9%	0.4%
U.S.	1.4%	1.9%	95.1%	1.6%
Other				100.0%

SOURCE: Authors' analysis of patents granted in the United States and Europe (see text).

To a slight degree, the United States became an innovation hub for companies headquartered in other countries, mainly for Philips. Of the LED patents assigned to Dutch companies, 67.3 percent were filed by inventors located in the United States, as were only 1.8 percent of patents assigned to German companies. Germany also was an R&D source for 7.4 percent of Dutch firms' patents. These shifts largely reflect the internationalization of Philips as it acquired other firms and joint ventures.

There was a corresponding shift in the number of companies using inventors in other countries. In the period 1990-1993, six U.S.-headquartered companies sponsored LED research abroad, totaling nine patents, while only one European and one Japanese company sponsored LED research abroad with only one such patent each. In the period 2000-2003, 25 companies located in all parts of the world sponsored LED research abroad, totaling 95 patents. This a substantial shift, but far from complete globalization, as even in 2000-2003 only 5.2 percent of LED patents involved work outside companies' headquarters nations; the vast majority of patents still used inventors in firms' home countries.

As Table 4 shows, there was a shift from 1990-1993 to 2000-2003 in the dominant firms in LED patenting.[34] Dominant firms are ranked here in LED applications generally, including LED-type displays and LED backlights, not just

---

[34]The shift in firms' rank ordering from 1990-1993 to 2000-2003 is statistically significant ($p < .001$) using Wilcoxon's signed-rank test.

TABLE 4 Leading Firms in LED Patenting, 2000-2003 versus 1990-1993

Rank 2000-2003	Rank 1990-1993	Company	Headquarters Country	Patents 2000-2003
1	32	Philips/Lumileds[a]	Netherlands/U.S.	69.0
2	9	OSRAM/Siemens[b]	Germany	41.0
3	—	GE/Gelcore[c]	U.S.	26.5
4	—	United Epitaxy/Epistar[d]	Taiwan	17.2
5	3	Xerox	U.S.	14.5
5	13	Oki	Japan	14.5
7	1	Eastman Kodak	U.S.	14.0
8	2	HP[e]	U.S.	9.0
8	—	Para Light Electronics	Taiwan	9.0
10	—	Hon Hai Precision	Taiwan	8.0
11	—	IBM	U.S.	7.0
11	—	Lite On Electronics	Taiwan	7.0
13	8	Samsung	Korea	6.8
14	—	Nichia	Japan	6.0
15	10	Cree	U.S.	5.3

NOTE: Individuals and institutions were excluded from the table SOURCE: Authors' analysis of patents granted in the Untied States and Europe (see text).

[a]Includes all patents by Philips and its subsidiary Lumileds, Agilent as former joint venture partner of Lumileds, and Color Kinetics acquired in 2007.

[b]Includes all patents by OSRAM-Sylvania and its parent company Siemens, as well as 6.5 patents held by Patra Patent Treuhand.

[c]Includes all patents by GE and its subsidiary Gelcore.

[d]Includes all patents by United Epitaxy, Epistar with which it merged in 2005, and Epitech with which it merged in 2007.

[e]HP patents in the 1990s pertain to technology such as LED print heads that may not have been owned by HP's spin-off Agilent, which became part of Philips.

LEDs for general illumination. The listed firms are unlikely to include materials makers because of the search criteria used. Only 4 of the top 10 firms that filed LED patents in 1990-1993 remained in the top 10 a decade later (OSRAM plus its parent firm Siemens, Xerox, Eastman Kodak, and HP) and 4 entirely new LED patent assignees had appeared in the new top 10. All of the big three traditional lighting firms featured at the top of LED patenting in 2000-2003 (Philips ranked first, OSRAM second, and GE third). Of these big three, Philips and (through Siemens) OSRAM had LED patents in 1990-1993. The emergence of the big three as the dominant LED firms in 2000-2003 can clearly be attributed to their joint ventures that allowed them to enter fully into the semiconductor-based LED market.

Many LED patents came from firms not in the traditional lighting industry, including established semiconductor firms. Semiconductor firm United Epitaxy, founded in 1993, is one of several such Taiwanese firms. Firms' ranks are measured somewhat noisily here, because the sample of patents used does not cover every LED patent (as noted in our earlier description of patent methods).[35] Nonetheless, the evidence shows an important role in LED technology of Asian firms, representing half the firms listed in Table 4.[36]

Hence, the big three lighting firms have managed to established dominant positions in LED lighting technology. Nonetheless, if LED technology develops as anticipated there may be greater participation by firms in Japan, Taiwan, and perhaps Korea and other Asian nations in the global lighting industry.

Globalization of LED R&D: Worldwide Patents

The results differ somewhat when patents from Japan, Taiwan, and other nations are included. These patents were initially excluded because of concerns over whether patents are of comparable quality and value in different nations and because the U.S. and European markets have been two of the world's largest. However, focusing only on U.S. and European patents may introduce a bias because some applicants develop R&D competence but apply for patents only in their home countries. Also there may be international differences in propensities to patent in different markets. Filings by individuals (instead of companies) showcase the differences; in the period 2000-2003 almost four times as many patents were granted worldwide as in the United States and Europe, and almost five times as many individuals were granted patent rights, using our worldwide sample of data with a restricted definition of LED patents. The majority of individual filings originated in Taiwan (35.6 percent) and Korea (28.6 percent).

Among our sample of LED patent invention worldwide, in 1990-1993 Japanese inventors led with 78.3 percent of LED patents, the United States followed with 10.7 percent, and all other countries each invented less than 3 percent. A decade later in 2000-2003, Japanese invented 41.6 percent of LED patents, Taiwanese 22.0 percent, Americans 13.9 percent, and Koreans 12.9 percent of patents. Clearly by 2000-2003 more countries, notably Taiwan and Korea, were

---

[35]Errors in ranking could result if certain companies are less likely than others to include our search terms in their patent titles, and our purpose is not to compute exact rankings. However, the rankings computed confirm industry impressions and are approximately valid absent any odd variations in firms' choice of terminology in patent titles: the difference between Philips and OSRAM in their arrival rate of patents was statistically significant ($p = .010$), the difference between OSRAM and GE was not statistically significant at conventional levels ($p = .114$), the difference between OSRAM and United Epitaxy/Epistar was statistically significant ($p = .002$), and the difference between GE and United Epitaxy/Epistar was not statistically significant ($p = .174$), using exact p-values from exact Poisson regressions.

[36]If cumulative measures of LED patenting were considered, Asia would likewise emerge as playing an important role. Asian countries, particularly Japan, had strong early LED R&D, which is apparent in the 1990-1993 patent data.

locations for LED R&D.[37] Hence, these results indicate a possibly greater role of Asian inventors, when patents applied for in Asian nations are considered along with those in U.S. and European nations.

Other Indicators of R&D

Other indicators of globalization and Asian strength in LED innovation are international joint ventures and licensing agreements. Joint ventures in LEDs occurred between each of the big three traditional lighting firms and other international firms, as discussed earlier, all in 1999, with all three subsequently acquiring the joint ventures. International cross-licensing agreements, listed in Table 5, now exist between Philips (Dutch) and OSRAM (German), Philips and Nichia (Japanese), Philips and Toyoda Gosei (Japanese), OSRAM and Nichia, OSRAM and Seoul Semiconductor (Korean), OSRAM and Avago (American), Cree (American) and Nichia, and Cree and Seoul Semiconductor. The table also indicates other international licensing, disputes, and strategic research partnerships. The evidence indicates considerable global dispersion and a growing Asian contribution in LED innovation.

NATIONAL PROGRAMS AS INNOVATION DRIVERS

Promoting R&D

The development and market penetration of LEDs is closely linked with government policies and national programs. This is not uncommon in the semiconductor industry. For example, Japan saw extensive growth in semiconductor R&D, which displaced U.S. leadership in the DRAM market, following a mid-

---

[37]Japanese firms dominate the rankings when patents granted worldwide are considered. In the period 1990-1993, 9 of the top 10 ranked firms were Japanese. The five highest ranking firms by LED patents were Hitachi (35 patents), NEC (31), Toshiba (30), Mitsubishi (27), and Sanyo (23). Eastman Kodak (United States) was the only non-Japanese firm in the top 10 during the period 1990-1993. By 2000-2003, the top 5 ranking firms for LED patents were also Japanese: Nichia (38), Hitachi (30), Sharp (29), Showa Denko (26), and Citizen (25), and 8 of the top 10 were Japanese. Even among the Japanese firms, however, only Hitachi, Sharp, and Matsushita stayed in the top 10 ranking over the decade. The dominance of Japanese firms could, however, reflect differences in patent systems such as how often multiple claims are combined in one patent, coupled with the fact that firms are typically most likely to apply for patents in their home countries as well as possible international locations. The two non-Japanese firms in the top 10 during the period 2000-2003 were Taiwanese: Epistar (founded in 1996) held sixth place with 24.5 patents, and United Epitaxy (founded in 1993) held ninth place with 18.5 patents. The big three traditional lighting firms did not make the top 10 by this measure: Philips/Lumileds is ranked in 12th place, Osram plus Siemens in 15th place, and GE plus Gelcore in 36th place. Two Korean firms, LG and Samsung, were in the top 20. When only patents granted on multiple continents (a measure of high value) are considered, U.S. inventors are responsible for a slightly larger percentage of the sample than indicated here.

TABLE 5 Globalization of LED Patents: Licenses, Alliances, and Disputes

	Philips (Lumileds)	OSRAM	GE Lumination	Cree	Nichia	Seoul Semi-conductor
Philips (Lumileds)						
OSRAM	Cross-license					
GE Lumination						
Cree		Chip supply from Cree				
Nichia	Cross-license	Cross-license	Strategic alliance	Cross-license		
Seoul Semi-conductor		Cross-license		Cross-license	Lawsuits	
Epistar	Dispute				Dispute (resolved)	
Toyoda Gosei	Cross-license				Settlement	Strategic partnership
Citizen		Dispute (resolved)			White LED license	
ROHM		White LED license		License		
Lite-On		White LED license		License		
Kingbright		Lawsuit		License		
Others		White LED licenses; cross-license; dispute[a]		Licenses; lawsuit[b]	Investment[c]	

NOTES: Cross-licenses are agreements to share rights to large numbers of LED patents. A "license" indicates that a firm acquired the right to use a patent owned by a firm at the top of the table. White LED licenses pertain specifically to patents for white-light-producing LEDs. A "dispute" indicates a patent lawsuit brought by a firm at the top of the table against another firm. The "settlement" between Nichia and Toyoda Gosei involves patent infringement lawsuits that had been brought by each firm against the other. Strategic alliances and strategic partnerships are joint LED technology development efforts involving the two companies indicated.

[a]OSRAM has white LED patents licensed to Harvatek, Vishay, Samsung SEM, Everlight, Ya Hsin, Lednium (Optek); a cross-license with Avago; and a resolved dispute against Dominant.

[b]Cree has LED patents licensed to Cotco, and Stanley and a lawsuit against BridgeLux.

[c]Nichia has an investment in and chip-license agreement with Opto Tech.

SOURCE: Ledsmagazine.com (2007).

1970s research program (Macher et al., 2000). There appears to be a correlation between countries' national research programs for LEDs and innovative activity in those countries. Key LED programs exist in the United States, Japan, Taiwan, and South Korea, precisely those countries that dominate LED patenting. China recently announced programs targeted toward LED innovation and high-technology industries in general. Judging by the impact of similar research programs in other nations and import of U.S. and Taiwanese talent, China may become an additional key player in the LED industry.

While national programs collaborate extensively with universities and research labs, such institutions account for only about 4 percent of all LED patents, reflecting the limited funding available for commercializing their basic research. Nonetheless, university spin-offs have often created major companies such as Cree (with a market capitalization of $1.69 billion and revenue of $385 million in 2005). Universities and research institutions appear most innovative in Taiwan and Korea, accounting for about one-half and one-fifth, respectively, of all LED patents filed by universities and research institutions.[38] The remainder is split fairly evenly among the United States, the United Kingdom, Japan, and Belgium. Interestingly, China also features, filing 9 percent of all LED patents by universities and research institutions. With the exception of Belgium, each of these countries has a national program dedicated to development of LED lighting, with goals to improve energy efficiency and gain market share in general illumination, as outlined in Table 6.

Often the dedicated lighting program benefits from other supporting legislation or programs. For example, the U.S. initiative to develop LEDs may be partly driven by programs such as Vision 2020, an industry-led program to develop a technology roadmap for lighting, initiated by the U.S. Department of Energy. The program's goals are to develop standards for lighting quality; increase demand for high-quality lighting solutions; strengthen education and credentials of lighting professionals; provide R&D incentives to accelerate market penetration of advanced lighting sources and ballast technologies for superior quality, efficiency, and cost-effectiveness; and develop intelligent lighting controls and flexible luminaries/system delivery platforms (DOE, 2000). Apart from aims to establish integrated energy-efficient lighting systems, the program has also launched the Energy Star voluntary labeling program designed by the Environmental Protection Agency (EPA) and the National Appliance Energy Conservation Act that bans low-efficiency magnetic ballasts. Grants awarded by the DOE in 2006 totaled nearly $60 million, with a further $12 million provided by contractors

---

[38]Research output at universities has often been measured by journal publication rather than patents, but it would be difficult to use a publication-based measure here without possible language-related biases (non-English speakers frequently publish in non-English journals not catalogued in available publications databases). Our measure of patents rather than publications may be more pertinent to applied than to basic research. The numbers are based on all patents (including national patents) during the period 2000-2003.

TABLE 6 Major National Research Programs Pertaining to LEDs

Country, Funds[a]	Program	Phase & Objectives	Funding	Organizations
USA[b] estimated $42.1 mil/year (2003-2013); includes anticipated extension of funding	Energy Act of 2005: Next Generation Lighting Initiative (NGLI)	• 2007-2009: support research, development, demonstration, and commercial applications • 2010-2013 extension	$50 mil/year authorized from 2007 to 2009 Extended authorization to allocate $50 mil/ year from 2010-2013	Partnership DOE, industry, universities & laboratories
	SSL Project Portfolio (current projects in NGLI)	• Completed projects 2003-2005: six key research areas: quantum efficiency, longevity, stability and control, packaging, infrastructure, and cost reduction • Current projects through 2008: LED and OLED	Total: $70.9 mil	Partnership DOE, industry, universities & laboratories
Japan estimated $7.5 mil/year (1998-2008)	Light for the 21st century, New Energy and Industrial Technology Development Organization (NEDO)	• 1998-2002 (first phase): develop GaN-based LED technology for lighting applications • Develop 13% market penetration by 2010 (over 477 new patents already filed in one year) • Produce 120 lm/W and 80% efficiency by 2010	Yen 6 bil ($52 mil)	Japan R&D Center of Metals (JRCM) 13 companies and universities
	Ministry of Education, Culture, Science & Technology	• Financial year 2004: develop medical equipment and therapeutic techniques based on LEDs • Similar amounts of funding expected next 4 years (2005-2008): establish the Yamaguchi-Ube Medical Innovation Centre (YUMIC) • White HB-LEDs	Yen 500 mil ($4.6 mil) $4.6 mil/year 2005-2008	Several universities, more than 20 companies

continued

Country	Program/Project	Description	Funding	Notes
South Korea[c] estimated $59.4 mil/year (2001-2008); excludes equipment value, and half of LED Valley as "fiber-to-home"	Semiconductor Lighting National Program & KOPTI (Korea Photonics Technology Institute)	• 1993-1996 (R&D by LG, Samsung, universities, and Korea Research Institute): reduce use of glass, phosphors, heavy metals • 1999-2000 (business phase, JVs, production runs): meet environmental regulations July 2006 • Save $20 bil on energy • 2001 (activation phase, growth to more than 340 companies); produce 80 lm/W white LED by 2008	KOPTI receives $20 mil/year in funding KOPTI equipment value: $65 mil	KOPTI's costs covered 73.1% by government, 16.5% by Gwanju "City of Light," 10.4% by industry
	LED Valley Project	• 2005-2008: develop HB-LED • 2005-2008: second phase of a photonics industry project for HB-LEDs • Deploy fiber-to-the-home networks	$100 mil (first phase) $430 mil (second phase HB LED & fiber-to-home)	Gwanju "City of Light," mix of national and local government and private-sector investment
Taiwan estimated $4.0 mil/year (2002-2005)	Next Generation Lighting Project	• 2004-2005 (40 lm/W): improve performance of white LEDs • Second phase (60 lm/W): 100 lm/W output in labs	2004-2005: NT$383 mil ($11.5 mil)	Consortium of 11 companies
	National Science Council	• Circa 2002-2004: producing highly efficient LEDs • Led to 14 new patents and 20 new manufacturing process technologies	NT$12 mil ($0.4 mil)	NSC department of science & engineering

TABLE 6 Continued

Country, Funds[a]	Program	Phase & Objectives	Funding	Organizations
China[d] estimated $248.8 mil/year (2005-2010); only includes initial investment for five parks and the 5-year plan; not directly comparable to other nations' figures as this includes manufacturing site investments	Semiconductor Lighting Project	• Five parks established in Shanghai, Xiamen, Dalian, Nanchang, Shenzhen: establish industrial parks with up-, mid-, and down-stream products; first phase likely to be 2005-2010: collaboration with Taiwan and specialists from Taiwan and U.S.; anticipate $19 bil LED industry by 2010	Total investment: Yuan 10 bil ($1.2 bil), allocated as follows: Xiamen: $1.9 mil (with focus on opto-electronics), Dalian: $150 mil, Shenzhen: initial investment 3 bil Yuan ($375 mil); total 20 bil Yuan ($2.5 bil) over 3-5 years (2005-2010)	Xiamen: three companies and government & cooperation with Taiwan; Dalian: JV between companies and science & technology group; Shenzen: university, local city government support & 200 companies
	National Solid State Lighting project as part of 11th 5-Year Plan	• 2015 goals: savings from large-scale conversion to LED: 100 bil kW/h annually by 2015 • 150 lm/W LED and capture 40% of incandescent market • Reduce environmental pollution • Develop strong industrial base • International cooperation if necessary	2006-2010: $44 mil	15 research institutions & 2,500 companies
E.U. estimated $16.3-32.5 mil/year (2002-2006) est. assuming 5-10% dedicated toward LED	Sixth Framework program	• Strengthen science & technology base for international competitiveness	$1.3 bil earmarked for nanotechnology (with IST section)	

[a]The yearly fund flow was estimated as an annual mean of all funding programs over the entire time range of the programs. All figures in US$.
[b]Presidential budget for fiscal year 2006 includes request of $11 mil for SSL.
[c]Korea also has a national program for LCD and displays, from 2004-2008. Key players are LG and Samsung. No funding information.
[d]China's "863 Program," or National High Technology Research & Development Program, includes development of OLEDs as a focus.
SOURCES (in order of table): DOE (2005, 2006), Japan Research and Development Center of Metals' National Project (2000), Stevenson (2005), Compound Semiconductor (2004), Chiu (2004), Yahoo! News Australia & NZ (2004), Tang (2006), Ledsmagazine.com (2005c, 2005a), Steele (2006), European Commission Community Research (2002).

(DOE, 2006). Some 65 percent of the DOE grants were awarded to firms, with the remainder split about equally between research laboratories and universities.

Similarly, Japan has an LED association that promotes R&D and standardization in the LED industry. As well as aiming for energy-efficient lamps, the association has established a medical innovation center that conducts R&D on LED use in medical equipment and therapeutics. A 1979 Energy Conservation Law in Japan, updated in 1999, has been a key driver of energy conservation in factories, buildings, machinery, and equipment. Japan is the second-largest government supporter of R&D in general, after the United States, investing $90.3 billion in 1997. Of this budget, $6.8 billion was allocated toward national energy-related R&D—64 percent public sector and 36 percent private sector (Dooley, 1999).

South Korea's lighting program is supported by a government-backed organization, Korea Photonics Technology Institute, which aims to produce 80 lm/W white LEDs by 2008 and invests $20 million per year. Funding stems mainly from the government (73.1 percent), but also from industry (10.4 percent) and the "City of Light," Gwanju (16.5 percent). Gwanju is the center of the LED Valley project in Korea, aimed at penetrating LEDs into television backlighting by 2006, car lighting by 2008, and domestic lighting by 2010. Investment is significant at $100 million for the development of HB LEDs (plus $430 million partly for fiber-to-the-home) from 2005 through 2008. In addition, the Korean private sector, namely Samsung and LG, are investing in LCDs and OLEDs, using Korea's LED infrastructure as a platform. *Chaebols* such as Samsung and LG are doing so through their business units and research labs, as well as a partial spin-off in the case of LG, in which it still has a 60 percent equity stake. But there have also been new startups for epiwafer foundries, substrates/GaAs ICs, and fiber optic components—many set up by researchers from Samsung and LG or by university professors (Whitaker and Adams, 2002).

Taiwan has had support from the National Science Council for LED research. Together with a consortium of 11 companies, Taiwan invested $11.5 million in LED research and development during the period 2003-2005. The second phase, to produce high-efficiency LEDs, is expected to receive $0.4 million in funding. The goal is to produce 100 lm/W output efficiency of LED bulbs in laboratories. In addition, Taiwan has a 6-year national initiative on nanotechnology worth $700 million, some of which is dedicated toward LEDs (Liu, 2003).

China has budgeted $44 million to address solid-state lighting R&D needs as part of its 11th Five Year plan. The program will include 15 research institutions and university labs, and more than 2,500 companies involved in LED wafers, chips, packaging, and applications (Steele, 2006). The country expects to be the largest market for LEDs in the world, although it acknowledges a 6- to 20-year lag behind Japan, Europe, and the United States in LED device technology (Steele, 2006). The key driver behind the lighting project is energy savings. The goal is to penetrate 40 percent of the Chinese incandescent lighting market with 150 lm/W LEDs. The program was responsible for the establishment of five industrial parks in China during 2004 and 2005, backed by government, company, and university investment. The objective of the program is to save 30 percent of energy spent on lighting, the same as generated by the Three Gorges Project, in the next 15-20 years. An underlying national solid-state lighting project by the Ministry of Science and Technology aims to reduce environmental pollution and improve technology to develop a strong industrial base. Apart from its dedicated semiconductor lighting project, China is investing heavily in the semiconductor and advanced material industries in general. China is also focused on collaborating internationally to develop its semiconductor industry, recruiting talent particularly from Taiwan and the United States.

One aspect that stands out among national LED programs is that Europe appears to be lagging behind the United States and Asian countries. The European Union's Fifth Framework Program funds five research areas: nanotechnology, genomics and biotechnology, information technology, aeronautics and space, and food safety and health risk. The funding for the period 2002-2006 is $17.5 billion. Of this, $3.4 billion is assigned to the Information Society Technologies program, which includes research into semiconductor technologies and LEDs. The program funds research institutions, universities, and other organizations. The lack of specific initiatives for LED innovation may explain European countries' minor share of LED patents.

Some European countries have more specific programs dedicated toward LEDs. In September 2006, for example, the German Ministry of Education and BASF inaugurated a new research lab, the Joint Innovation Lab (JIL) (BASF, 2006). The JIL is a cooperative effort between 20 BASF experts and industrial and academic partners researching new materials in organic electronics, concentrated particularly on OLEDs for organic photovoltaics and appliances in the lighting market (OPAL). The German Ministry of Education and Research intends to invest around $800 million in the OPAL project. In addition, BASF spends over $1 billion on R&D each year. It is hoped that the projects will strengthen Germany's position in the emerging market of organic electronics and create the scientific and technological basis for initiating the production of OLED-based lighting (*A to Z of Materials*, 2006).

In the newer technology of OLEDs, much of the work is concentrated in research institutions and academia, both domestically and abroad. To be commercially viable, OLED research requires substantial infusion of capital. Foreign industry, heavily funded by their governments, could develop an insurmountable lead in the technology, making it very difficult for U.S. manufacturers to compete, if the U.S. government does not provide comparable support. With appropriate support from government and industry, commercialization could occur in as little as 5 to 8 years (Tsao, 2002).

A push is also being made to pursue good white LEDs, the "holy grail" of LED lighting. Analysis of PIDA data compiled by DigiTimes shows that each of the aforementioned countries is investing in white LEDs. The United States is investing $50 million over 10 years, Korea $23.4 million over 5 years, Japan $10.7 million over 5 years, Taiwan $4.6 million over 3 years, China $3.3 million over 3 years, and Europe $1.0 million over 4 years (Wang and Shen, 2005).

Demand Drivers

To spur innovation indirectly, regulations and incentives for energy-saving technologies can enhance demand for new lighting technologies. In a study comparing U.S. and Japanese lighting industry conservation measures, Akashi et al. (2003) found that conservation can be encouraged by regulation, incentives, and awareness campaigns. The U.S. Energy Policy Act of 1992 prohibited manufacturing and import of lamps that do not meet efficiency standards and mandated that lamp lumen output, efficiency, and life be printed on packaging, making it easier for consumers to compare and select more energy-efficient products. Nevertheless, consumers still experience considerable confusion in choosing lighting, particularly for residential settings. In new construction, builders have generally installed basic lighting packages that lack energy efficiency and other quality improvements in favor of lower capital costs (rather than lower operating costs). Bridging the gap between available lighting technology and consumer knowledge is a significant challenge and one that in Japan is met jointly by government and industry initiatives.

Future diffusion of LED lighting may reflect patterns now apparent for CFLs, which, although more efficient than incandescent and halogen lamps, have achieved low penetration in the U.S. market. Only about 2 percent of sockets nationwide, and 4 percent in California, now use CFLs. Flicker, color, upfront cost, and other drawbacks have contributed to their slow adoption, so that greater energy efficiency alone seems insufficient to penetrate much of the market, although the efforts of Wal-Mart to promote CFLs may result in a significant change in consumer behavior.

IMPLICATIONS AND POLICY RECOMMENDATIONS

This chapter has documented a shift taking place in the lighting industry. Traditional lamps are being replaced with CFLs. While the early traditional lighting industry was dominated by three big companies—GE, Philips, and OS-RAM—as production of lamps became commoditized competitive pressures in lighting increased. Lower prices and margins shifted production of traditional lamps to Asia, especially China, the largest source of lamp imports in the United States. Improvements in lamp efficiency led to the development of fluorescents and other types of lamps, which successfully penetrated commercial and industrial markets and are poised to enter U.S. residential markets after years of delay among consumers who lacked awareness and were unwilling to spend money up front for savings later on.

A new lighting technology, LEDs, is leading to a shift in how we view lighting. LEDs have already penetrated end-use markets for automobile brake lights, signs and displays, backlighting, and traffic signals. Investments in the development of white LEDs are setting the stage for the use of LEDs as general illumination and threaten the traditional lighting industry and its three big players. LEDs are a disruptive technology that has allowed many new players to enter the lighting market. While Japan and the United States dominate the LED market in terms of R&D and revenue, their market share is being eroded by fast-growing entrants, especially from Taiwan. Taiwan leads global production of blue (GaN) LEDs (Wang and Shen, 2005), had the second- or third-largest amount of LED patents by our counts in 2000-2003, and has two firms high on our LED patent ranking tables.

Philips, OSRAM, and GE were not involved in the early stages of LED technology development. It was only in 1999 that the big three decided to enter the LED market through a series of joint ventures that the companies later acquired. These firms may further build their strengths in this technology through acquisition. In mid-June 2007, Philips acquired Color Kinetics for approximately $791 million.[39] These big three firms appear to have established dominant positions in LED technology, judging from our patent analyses. However, it remains to be seen whether the big three will replicate the tight oligopoly they held in the traditional lighting industry in most of the 20th century. Partly this is because the semiconductor supply chain is fragmented as firms in this sector are typically not vertically integrated; by specializing, companies are able to keep costs down. Our analysis indicates that LED producers likewise operate at various stages of the supply chain and do not integrate vertically. This means that the LED market has witnessed many new entrants and has also created opportunities for new ventures in areas such as system controls and integration.

Although LEDs have some clear advantages over traditional lamps, such as added flexibility, integration with digital systems, and higher energy savings, they are also still costly to produce. The question remains when (indeed whether) white LEDs will successfully displace traditional general illumination technologies, especially among residential buyers. Evidence from CFL, HID, and other efficient traditional technologies shows low penetration rates among consumers. To aid success of LED lighting, therefore, governments might not only fund basic R&D but also promote awareness among consumers so that LED lighting products diffuse in the residential market. Governments worldwide are making significant investments into LED R&D and promotion of the technology. Government programs, such as the one in the United States, have allowed small startups

---

[39]At the time of the announcement Color Kinetics had 71 patents granted and over 15,000 installations. The merged entity will operate under the name Philips Solid-State Lighting Solutions, with intelligent and premium LED product lines ultimately co-branded Philips/Color Kinetics (Color Kinetics, 2007).

and university research labs to make progress on LED R&D and gain a foothold in this new market.

U.S. and Asian government programs, in particular, have made the largest investments. China, which is still at the early stages of ramping up capacity and technology to produce LEDs and therefore lagging behind other countries, is addressing R&D in solid-state lighting as part of its 11th Five Year Plan and is setting up five business parks dedicated to these new lighting technologies. China has a strong interest to meet its own energy-efficiency needs. Already, there is a trade imbalance between China and the United States for semiconductors generally. In 2002, the United States imported $6.4 billion worth of semiconductor products from China, while exporting only $2.2 billion worth (Holtz-Eakin, 2005). Given its investments in R&D, China might become an important player in the global LED market.

Analysis of these trends indicates that Asian countries such as Japan and Taiwan, and possibly China and Korea, are poised to take an increased role in R&D, production, and diffusion of LED technology. Evidence provided by the patent analysis suggests a potential shift toward these Asian nations. Extensive public and private investment will help if the United States is to keep up with the opportunities presented by these new technologies. Moreover, efforts to encourage consumers to use solid-state lighting as it becomes efficacious in new applications may help domestic markets to grow and support the commercialization of these important energy-saving technologies.

ACKNOWLEDGMENTS

Partial funding is gratefully acknowledged from the Alfred P. Sloan Foundation and from the Center for Future Energy Systems at Rensselaer Polytechnic Institute. Insightful comments on an earlier draft were kindly provided by Kevin Dowling, Jeffrey Macher, Bill McNeill, Klaus Minich, David Mowery, and David Simons. We also thank our discussants and anonymous referees at each stage of development of this work.

REFERENCES

Akashi, Y., L. Russell, M. Novello, and Y. Nakamura. (2003). Comparing Lighting Energy Conservation Measures in the United States and Japan. Working Paper, Lighting Research Center, Rensselaer Polytechnic Institute.

Arensman, R. (2005). LED market lights up. *Electronic Business* 31(4):24.

Arthur, A. (2006). A unique history of the light bulb. Available at http://www.contentmart.com/ContentMart/content.asp?LinkID=19298&CatID=328&content=1. Accessed December 2006.

Ashdown, B. J., D. J. Bjornstad, G. Boudreau, M. V. Lapsa, B. Shumpert, and F. Southworth. (2004). Assessing Consumer Values and Supply-Chain Relationships for Solid-State Lighting Technologies. Report ORNL/TM-2004/80, Oak Ridge National Laboratory, Oak Ridge, Tenn.

A to Z of Materials. (2006). BASF Open Organic Electronics Research Laboratory, 11 September. Available at http://www.azom.com/details.asp?newsID=6627. Accessed September 22, 2006.

Barbaro, M. (2007). Power-sipping bulbs get backing from Wal-Mart. *New York Times*, January 2.

BASF. (2006). Joint Innovation Lab–Organic Electronics: Research with customers and academic partners. Press release, September 11.

Bellis, M. (2006a). The history of fluorescent lights. Available at http://inventors.about.com/library/inventors/bl_fluorescent.htm. Accessed December 2006.

Bellis, M. (2006b). Timeline of electrical lighting. Available at http://inventors.about.com/od/lstartinventions/a/lighting_2.htm. Accessed December 2006.

Bellis, M. (2006c). The history of the incandescent lightbulb. Available at http://inventors.about.com/library/inventors/bllight2.htm. Accessed December 2006.

Bishop, A. L., R. Worrall, L. J. Spohr, H. J. McKenzie, and I. M. Barchia. (2004). Response of *Culicoides* spp. (Diptera: Ceratopogonidae) to light-emitting diodes. *Australian Journal of Entomology* 43(2):184-188.

Bowers, B. (1982). *A History of Electric Light & Power.* Peter Pergrinus.

Bright, J. R. (1958). *Automation and Management.* Boston, MA: Harvard University.

Burgess, D. S. (2006). Efficient white OLED employs down-conversion. LED Focus, May 2006. Available at http://www.photonics.com/content/spectra/2006/May/LED/82586.aspx. Accessed December 2006.

Business Wire. (2006). Avago Technologies introduces new high performance extra bright oval LEDs for outdoor electronic sign and signal applications. September 19.

Calwell, C., C. Granda, L. Gordon, and M. Ton. (1999). Lighting the Way to Energy Savings: How Can We Transform Residential Lighting Markets? Volume 1: Strategies and Recommendations. Prepared by Ecos Consulting for the Natural Resources Defense Council, San Francisco, Calif.

Chinesewings. (2005). Philips Celebrate the Shanghai Innovation and Technology Park Lighting Project. August 1.

Chiu, Y. T. (2004). Optical devices to replace toxic lighting mechanisms. *Taipei Times.* April 16.

Color Kinetics. (2007). Color Kinetics Announces Acquisition by Philips. Press release, June 19.

Compound Semiconductor. (March 2004). Taiwanese suppliers target white LED improvements. *Compound Semiconductor.*

Compoundsemiconductor.net. (2004). Japanese LED project targets medical uses. Available at http://compoundsemiconductor.net/articles/news/8/7/26/1.

Cox, J. B. (2003). Semiconductor-maker Cree gets boost from Taiwan court ruling on patents. *Knight Ridder Tribune Business News.* September 26, p. 2.

DOE (U.S. Department of Energy). (1993). Housing Characteristics 1993. Energy Information Administration, Office of Energy Markets and End Use, Report DOE/EIA-457.

DOE. (1995). Commercial Buildings Energy Consumption Survey. Energy Information Administration, Office of Energy Markets and End Use, Report DOE/EIA-871.

DOE. (1997). Commercial Buildings Characteristics 1995. Energy Information Administration, Office of Energy Markets and End Use, Report DOE/EIS-E-0109.

DOE. (2000). Vision 2020, The Lighting Technology Roadmap. A 20-Year Industry Plan for Lighting Technology. Office of Building Technology, State and Community Programs, Energy Efficiency and Renewable Energy.

DOE. (2005). Energy Act Authorizes Next Generation Lighting Initiative. Available at http://www.netl.doe.gov/ssl/083105.html. Accessed December, 2006.

DOE. (2006). 2006 Project Portfolio: Solid State Lighting. D&R International Ltd. for Building Technologies Program, Lighting Research and Development, Office of Energy Efficiency and Renewable Energy.

Dooley, J. J. (1999). Energy R&D in Japan. Prepared for U.S. Department of Energy, contract DE-AC06-76RLO 1830. Pacific Northwest Laboratory, Battelle Memorial Institute. Document no. PNNL-12214.

Economist. (2007a). An environmentally friendly bulb that may never need changing. *The Economist* June 19.

Economist. (2007b). Everlasting light. *The Economist*, Technology Quarterly insert, September 6.

Eisenberg, A. (2007). The television screen, sliced ever thinner. *New York Times.* December 23.

European Commission Community Research. (December 2002). The Sixth Framework Programme in Brief. Brochure available at http://ec.europa.eu/research/fp6/pdf/fp6-in-brief_en.pdf. Accessed December 2006.

Gartner. (2007). Gartner's Final Semiconductor Market Share Results Show Industry Grew 10 Percent in 2006. Available at http://www.gartner.com/it/page.jsp?id=503221. Accessed January 2008.

General Electric. (2006). Lighting: The last 125 years. Available at http://www.gelighting.com/na/business_lighting/education_resources/learn_about_light/history_of_light/last_years.htm. Accessed December 2006.

General Electric. (2007). Consumer & Industrial Announces Intention to Restructure Its Lighting Business to Capture Another Century of Growth and Leadership. Press release, October 4. Available at http://www.genewscenter.com/Content/Detail.asp?ReleaseID=2704&NewsAreaID=2. Accessed November 1, 2007.

Genlyte Thomas. (2005). Annual Report, 2005.

Gizmodo. (2008). Samsung's 31-inch OLED is biggest, thinnest yet. Available at http://gizmodo.com/342912/samsungs-31+inch-oled-is-biggest-thinnest-yet. Accessed February 20, 2008.

Griffiths, T. (2006). Solid state lighting success is niche to niche. *LIGHTTimes Online*, December 11, p. 1.

Hadley, S. W., J. M. MacDonald, M. Alley, J. Tomlinson, M. Simpson, and W. Miller. (2004). Emerging Energy-Efficient Technology in Buildings: Technology Characterizations for Energy Modelling. Prepared for the National Commission on Energy Policy. Oak Ridge National Laboratory, U.S. Department of Energy.

Herkelrath, M., A. Laksberg, and L. Woods. (2005). A Brighter Future: Advances in LED Energy Efficient Lighting Technology. Manuscript, University of Washington.

Holtz-Eakin, D. (2005). Economic Relationships Between the United States and China. Congressional Budget Office Testimony, Statement before the Committee on Ways and Means, U.S. House of Representatives, April 14.

Hong, E., L. A. Conroy, and M. J. Scholand. (2005). U.S. Lighting Market Characterization, Volume II: Energy Efficient Lighting Technology Options. Technical report prepared for Building Technologies Program, Office of Energy Efficiency and Renewable Energy, U.S. Department of Energy by Navigant Consulting.

Howard, W. E. (2004). Better displays with organic films. *Scientific American* 290(2):76-81.

Inquirer. (2004). Inventor of LED honoured in his own day, strike a light. April 27.

Japan Research and Development Center of Metals' National Project. (2000). Light for the 21st Century: The Development of Compound Semiconductors for High Efficiency Optoelectronic Conversion, Year 2000 Report of Results. (English translation, 2002).

Ledsmagazine.com. (2005a). China promotes benefits of solid-state lighting. Available at http://leds-magazine.com/features/2/2/5/1. Accessed June 2005.

Ledsmagazine.com. (2005b). LED industry in Taiwan set to grow by 16 percent. Available at http://ledsmagazine.com/news/2/3/36/1. Accessed April 2007.

Ledsmagazine.com. (2005c). Shenzhen starts to build LED manufacturing base. Available at http://ledsmagazine.com/news/2/6/22/1. Accessed April 2007.

Ledsmagazine.com. (2006). OLED focus for Osram Opto, efficacy reaches 64 lm/W. Available at http://www.ledsmagazine.com/news/3/8/8. Accessed February 2008.

Ledsmagazine.com. (2007). Seoul and Osram sign LED patent cross-license deal with Osram Opto Semiconductors. Available at http://ledsmagazine.com/news/4/8/33. Accessed October 2007.

Leonard, R. S. (1992). Lighting the path to profit: GE's control of the electric lamp industry, 1892-1941. *Business History Review* 66(2):305-335.

LIGHTimes. (2006). Wal-Mart sees cold cash in LED holiday lighting and LED lighting for refrigerator cases. November 21, p.1.

LIGHTimes. (2007). Quiet big news announces cross licensing of all Osram and Philips inorganic and organic LEDs. Available at http://www.solidstatelighting.net/lightimes/?date=2007-02-02. Accessed February 2, 2007.

Lightsearch Directory. www.lightsearch.com. Accessed June 2006.

Liu, Y. S. (2003). III-V Compound Semiconductor Industry and Technology Development in Taiwan. International Conference on Compound Semiconductor Manufacturing. GaAsMANTECH, Inc.

Maccagno, P. (2002). Overview of the high brightness LED market. In *Light Emitting Diodes 2002: The Strategic Summit for LEDs in Illumination*, conference proceedings, San Diego, Calif., October.

Macher, J. T., D. C. Mowery, and D. A. Hodges. (2000). Semiconductors. Pp. 245-285 in *U.S. Industry in 2000: Studies in Competitive Performance*, D. C. Mowery, ed. Washington, D.C.: National Academy Press.

Mintel. 2003. Lightbulbs—U.S.—May 2003. Mintel International Group.

Monopolies Commission. (1968). *Second Report on the Supply of Electric Lamps.* London: Her Majesty's Stationery Office.

Navigant Consulting. (2003a). Energy Savings Potential of Solid State Lighting in General Illumination Applications. Technical report prepared for Building Technologies Program, Office of Energy Efficiency and Renewable Energy, U.S. Department of Energy.

Navigant Consulting. (2003b). Energy Savings Estimates of Light Emitting Diodes in Niche Lighting Applications. Technical report prepared for Building Technologies Program, Office of Energy Efficiency and Renewable Energy, U.S. Department of Energy.

OLLA Project Report. (2006a). Demonstrate a white p-i-n type OLED. High brightness OLEDs for ICT & next generation lighting applications funded under the IST priority (contract nr 4607) of the European 6th Framework Programme. February 2006.

OLLA Project Report. (2006b). White pin-OLED with improved polymer injection layer and efficiency above 10 lm/W. High brightness OLEDs for ICT & next generation lighting applications funded under the IST priority (contract nr 4607) of the European 6th Framework Programme. August 2006.

OSRAM. (2006). Highlights from 100 years of the OSRAM brand. Available at http://www.osram.com/cgi-bin/press/archiv.pl?id=450. Accessed December 2006.

OSRAM China Lighting Ltd. Available at http://www.osram.com.cn/aboutus/index.jsp (in Chinese). Accessed December 2006.

Philips. (2006). Annual Report 2005.

Philips. (2006). A history of Philips and Lighting. A century of innovation at Philips Lighting. Available at http://www.lighting.philips.com/gb_en/about/sub_feature_4. php?main=gb_en&parent=1&id=gb_en_about&lang=en. Accessed December 2006.

Physorg.com. (2005). Universal Display First to Achieve 30 Lumens Per Watt White OLED. August 5.

RPI (Rensselaer Polytechnic Institute). (2006). Rensselaer Researchers Aim to Close "Green Gap" in LED Technology. Press release, August 23.

Sandahl, L. J., T. L. Gilbride, M. R. Ledbetter, H. E. Steward, and C. Calwell. (2006). Compact Fluorescent Lighting in America: Lessons Learned on the Way to Market. Research Report Prepared for the U.S. Department of Energy under Contract DE-AC05-76RLO 1830. Pacific Northwest National Laboratory, Richland, Wash.

Siemens. (2006). Lighting. Available at http://w4.siemens.de/archiv/en/innovationen/licht.html. Accessed December 2006.

Solidstatelighting.net. (2006). Directories. Available at http://www.sslighting.net/cts/. Accessed June 2006.

Steele, B. (2006). China pours millions into solid-state lighting program. Ledsmagazine.com, July 25. Available at http://ledsmagazine.com/news/3/7/21/1. Accessed December 2006.

Stevenson, R. (2005). Oil-free Korea prioritizes solid-state lighting project. Available at http://compoundsemiconductor.net/articles/magazine/11/4/4/1. Accessed December 2006.

Taiwan Economic News. (2004a). Taiwan poised to retain no. 2 place in world LED production. March 24.

Taiwan Economic News. (2004b). Taiwanese led suppliers threatened by Japanese rivals' bold expansion plans. April 29.

Tang, Y. K. (2006). Lighten up: China's resolve to embrace energy-efficient lighting unleashes a powerful market potential. *Beijing Review*, March 2, updated December 13, 2006. Available at http://www.bjreview.com.cn/science/txt/2006-12/13/content_50654_2.htm.

Toniolo, T. (2006a). Shining a light on quality. *Product Design & Development* 61(11):14-15.

Toniolo, T. (2006b). What determines an LED's quality? *Electronic Products* (June).

Toshiba. (2006). The History of Toshiba lighting. Available at http://www.tlt.co.jp/tlt/english/company/company.htm and http://www.toshiba.co.jp/worldwide/about/ history.html#1875. Accessed December 2006.

Tsao, J. Y., ed. (2002). *Light Emitting Diodes (LEDs) for General Illumination: An OIDA Technology Roadmap Update 2002*. Optoelectronics Industry Development Association.

Walker, R. C. (2004). Cutting through the buzz production of HB-LEDs in Taiwan, South Korea and China. CompoundSemiNews. Available at: http://www.compoundsemi.com/documnts/article/news/3850.html. Accessed May 2006.

Wang, M., and J. Shen. (2005). LEDs lighting the future: Q&A with Biing-jye Lee, president of Epistar. Ledsmagazine.com, November.

Whitaker, T., and B. Adams. (2002). Korea is the new compound semiconductor boom region. Compoundsemiconductor.net, April.

Wikipedia. (2006). Light-emitting diode. Available at http://www.wikipedia.org/wiki/LED. Accessed August 30, 2006.

Williams, B. (2005). A history of light and lighting. Edition: 2.3. Available at http://www.mts.net/~william5/history/hol.htm. Accessed December 2006.

Yahoo! News Australia and New Zealand. (2004). China launches 50 semiconductor lighting projects. Available at http://au.news.yahoo.com/040722/3/q08l.html, July 22. Accessed December 2006.

Zou, H. (2002). Envoy of brightness. *Shanghai Star*. January 24.

6

Pharmaceuticals

IAIN M. COCKBURN

Boston University and National Bureau of Economic Research

INTRODUCTION

Pharmaceuticals is a highly globalized industry, dominated by multinational companies that engage in significant business activity in many countries and whose products are distributed and marketed worldwide. Historically, the industry has been dominated by vertically integrated firms performing almost all of the activities in the value chain, from basic research through to sales and marketing. But it is far from clear that these activities should necessarily be geographically co-located; these firms have generally operated globally, with many aspects of their operations spanning several countries. In recent decades the industry has undergone dramatic structural changes, with the rise of the biotechnology sector, substantial growth in demand driven by demographics and substitution away from other therapeutic modalities such as surgery, and increased competition from globally active generic manufacturers. These changes have led to some degree of dis-integration and geographic dispersion, but innovative activity is nonetheless highly geographically concentrated, reflecting the economic significance of factors such as localized knowledge spillovers and the strength of patent protection, as well as the influence of government policies such as price regulation, state procurement of drugs, and health and safety regulation. Rising research and development (R&D) expenditures in the face of health care cost containment pressures and apparently slowing research productivity give pharmaceutical companies a powerful incentive to seek out cost savings and new models for innovation, and to the extent that "offshoring" can raise research productivity it will generate substantial benefits to U.S. consumers and taxpayers. However, although we see some evidence of cost-driven geographic redistribution of R&D into new low-cost locations, this process has thus far been limited.

Global Distribution of Activity

Many, though not all, new pharmaceutical products are marketed world-wide. "Blockbuster" products—the relatively small fraction of drugs that realize global sales in the range of billions of dollars per year—are normally sold in most middle- and high-income countries. Patients in Organisation for Economic Co-operation and Development (OECD) countries ultimately have access to 80-90 percent of these products, albeit with longer delays in countries with more stringent price controls or weaker intellectual property (IP) protection. "Minor" products—those products with global sales below $1 billion per year—are typically launched in fewer countries; patients in the average country ultimately gain access to well under half of the total number of drugs introduced worldwide within 10 years of their first sale. New drugs are significantly less likely to be launched in poorer countries and, even if they do ultimately become available, it can take many years. Some countries are noticeably different in these respects; for example, Japan and Italy have much higher frequencies of single-country products.[1] Country-specific regulatory requirements and differences in health care delivery systems may require drug companies to make significant investments in local capacity in regulatory affairs and sales and marketing, even where promotion of pharmaceutical products is highly restricted.

Drug manufacture is also a multinational phenomenon, with an active global trade in intermediates (specialty chemicals), active pharmaceutical ingredients, and finished products. Stringent regulatory requirements for manufacturing imposed by government agencies in major markets such as the United States have extended quality standards worldwide, and several countries have become major loci of manufacturing activity that supplies global markets, notably Ireland and Puerto Rico, as well Israel and India for generic products.

R&D, by contrast, is much more geographically concentrated; the bulk of all R&D expenditure occurs in the United States, a handful of European countries, and Japan. Table 1 provides one measure of the global allocation of aggregate industrial pharmaceutical R&D expenditures. Unfortunately, reliable nationally comparable statistics on R&D spending in pharmaceuticals are difficult to obtain. National trade associations for the industry often report global rather than national spending by their members, and use varying definitions of R&D. For smaller countries, particularly emerging economies, data are simply unavailable, intermittently available, or of very questionable quality. With these caveats, aggregate statistics based on government censuses may nonetheless be informative, and they suggest a relatively stable share of the allocation of total industry R&D expenditure among developed countries, though even these numbers are difficult to compare due to differences in industry definitions, reporting standards, and data collection methods as well as exchange rate issues.

---

[1]See Lanjouw (2005), Kyle (2006, 2007), and Danzon et al. (2003).

TABLE 1 Business Expenditure on Pharmaceutical R&D by Country

	1990	1995	2000	2004
Total BERD at PPP (current million $), of which:	16,853	24,587	33,781	46,216
USA	37.3%	41.5%	38.3%	36.5%
EU15	39.8%	36.3%	40.4%	39.0%
UK	12.1%	11.8%	13.3%	11.1%
France	6.4%	8.5%	7.8%	7.6%
Germany	8.1%	5.0%	6.7%	7.5%
Italy	5.5%	2.5%	1.9%	1.5%
Sweden	2.1%	2.7%	3.7%	3.6%
Japan	16.2%	14.9%	14.3%	14.8%
Other developed countries[a]	6.7%	6.3%	5.8%	8.0%
"New Europe"[b]	—.	0.8%	0.9%	1.2%
Other emerging economies[c]	—.	0.1%	0.4%	0.6%

NOTES: In 2004, data for Australia, France, Greece, Japan, Mexico, Sweden, and Turkey are inferred from 2003 values and the average annual growth rate (AAGR) of BERD over the past 5 years in that country. In 2003 data for Austria, Denmark, Greece, and Iceland are inferred from adjacent year values and the 5-year AAGR of BERD in that country. The same applies to Austria in 1990 and 1995, and Belgium in 1990. Data for Switzerland may not be consistent over time.

[a]Australia, Canada, Iceland, Korea, Norway, Singapore, and Switzerland.
[b]Czech Republic, Hungary, Poland, and Slovenia.
[c]Taiwan, Mexico, and Turkey.
SOURCES: OECD Main Science and Technology Indicators Vol. 2006 release 02, and UK Pharmaceutical Industry Competitiveness Task Force: Competitiveness and Performance Indicators 2005.

Data published by OECD on business expenditure on R&D in pharmaceuticals are one basis for examining the global distribution of research effort in the industry. These data are converted to U.S. dollars using purchasing power parity (PPP) exchange rates and are based on the geographical location of spending rather than the "nationality" of the parent corporation. As shown in Table 1, there was almost a threefold increase in total nominal R&D between 1990 and 2004. However, shares by country or region were relatively constant, with both the United States and the European Union (EU) countries accounting for about 40 percent each of "world" R&D expenditure, Japan about 15 percent, and other developed countries (principally Switzerland) about 7 percent. The emerging economies and former Soviet Bloc countries present in this database have a small but increasing share of this total. It is important to note, however, that there is some R&D activity in countries that are not included here. India and China may account for as much an additional $1.5 billion or more of R&D spending in 2004,[2] and other indicators suggest small, though rapidly growing, levels of

---

[2]OECD reports over $1 billion in pharmaceutical R&D in China in 2000 but other sources suggest much lower figures. The *China Statistics Yearbook on High Technology Industry 2006* published by China National Bureau of Statistics, National Committee of Development and Reform, Ministry

activity in the Russian Federation and other Eastern European countries such as Romania and the Slovak Republic, in some parts of Latin America, as well as countries in Southeast Asia.

Note that Table 1 excludes contributions to the development of new drugs from publicly financed R&D and research supported by private nonprofit organizations. Noncommercial R&D is critically important to the industry and is a major source of "upstream" technology in the form of knowledge externalities from basic research, or spin-off products and entrepreneurial companies. Internationally comparable data on these forms of R&D expenditure are not available for many countries; however, data collected by the OECD suggest that medical sciences account for 20 to 30 percent of academic R&D expenditure in most developed countries. Given that the United States accounts for at least one-third of global public-sector research, it seems clear that the U.S. share of the total global research expenditure in pharmaceuticals from all sources would be significantly higher.[3]

Focusing on U.S. pharmaceutical companies, self-reported data compiled by PhRMA, the U.S. trade association, indicates that a significant share of R&D spending by U.S.-based companies was incurred outside the United States. (PhRMA members are largely though not exclusively U.S.-headquartered companies.) In 2005, these companies spent just under $9 billion outside the United States (21.5 percent of their total R&D spending), almost all of which was in Western Europe and Scandinavia. Table 2 summarizes these data.[4]

Interestingly, the ex-U.S. share of total R&D spending has been quite stable: while total non-U.S. R&D spending reported by these companies increased by 633 percent in real terms between 1980 and 2005, over this period the "abroad" share fluctuated between 17 and 22 percent, with no obvious trend over time.

Innovation in the Pharmaceutical Value Chain

The pharmaceutical value chain encompasses many activities, ranging from basic scientific research to marketing and distribution. Innovation in the industry is tightly linked to basic biomedical science, and many companies participate actively in basic scientific research that generates new fundamental knowledge, data, and methods. R&D activity is conventionally divided into two phases:

---

of Science and Technology, for example, reports just under 1.4 billion yuan of intramural R&D in pharmaceuticals in 2000—about $700 million at PPP. (However, more than 20 percent of this was for "processing of traditional Chinese herbal medicine.") The same source gives total pharmaceutical R&D in 2005 at just under 4 billion yuan, a remarkable increase.

[3]In 2004, for example, U.S. expenditure on academic R&D was more than $42 billion out of a total for OECD member countries of $125 billion. In addition, the United States has one of the highest shares of medical sciences in total academic R&D. SOURCE: OECD Main Science and Technology Indicators database.

[4]Comparable data are not available for non-U.S.-based companies.

TABLE 2 Ex-U.S. R&D Spending by PhRMA Members, 2005

Area	$ Million	Share (%)
Africa	28.0	0.3
Canada	479.3	5.4
Latin America and Caribbean	174.9	2.0
Asia-Pacific (except Japan)	117.5	1.3
India and Pakistan	10.9	0.1
Japan	1,025.4	11.5
Australia and New Zealand	144.6	1.6
Europe	6,524.7	73.4
Central and Eastern European nations, including Russia	244.6	2.8
Middle East	37.7	0.4
Uncategorized	101.3	1.1
Total	8,888.9	100.0

SOURCE: PhRMA Profile 2007, Table 6.

discovery and development. Drug discovery includes basic science and research on disease physiology, identification and validation of "druggable targets" in the body where therapeutic molecules may affect disease processes, identification and optimization of drug candidates, and preclinical testing. The development phase of research focuses on testing in humans, from the first small-scale trials directed at establishing basic physiological data in healthy volunteers through to large-scale trials on patients with the disease, which are designed to provide data on safety and efficacy to support applications for regulatory approval of the drug. Following marketing approval, research often continues to develop improved formulations of the product and to establish safety and efficacy in treatment of additional diseases or patient populations. Reflecting extraordinary advances in biology since the 1970s, the industry has become progressively more science-intensive, relying closely on fundamental advances in physiology, biochemistry, and molecular biology rather than "brute force" application of large-scale resources. If anything, this process has accelerated over the past decade as the industry has focused on complex and systemic diseases such as cancer, autoimmune diseases, and psychiatric conditions. Particularly in drug discovery, industrial and publicly funded research efforts are deeply intertwined.[5]

Though the bulk of innovative activity is concentrated in drug discovery and development, some R&D is also directed at manufacturing technologies and process improvement. However, stringent regulation of manufacturing processes inhibits experimentation and innovation, and for many drugs manufacturing costs are a small fraction of sales. This limits returns to investing in process innovation

---

[5]See Gambardella (1995) and Cockburn and Henderson (1998).

as opposed to investing in developing new products. In some cases advances in manufacturing technology can be important to allow use of advanced delivery systems or new formulations, but for many products innovation in manufacturing and production processes is largely confined to generic producers who compete fundamentally on costs.[6] By contrast, for other drugs, particularly the "large-molecule" therapeutic proteins based on biotechnology, manufacturing costs are substantial and production processes are more tightly linked to research activity (Grabowski et al., Forthcoming).

While the industry continues to be dominated by large integrated firms that conduct much of this innovative activity in-house, recent decades have seen significant vertical restructuring of the industry and these firms increasingly rely on externally sourced R&D in both the discovery and the development phases of research. In drug discovery, an active entrepreneurial sector that bridges academic and publicly funded research and industrial science has become a very important supplier of drug candidates and tools for performing R&D. In the development phase, specialist firms (contract research organizations) now play a significant role in conducting clinical trials on behalf of the sponsor of a drug. The causes of this restructuring of R&D activity are complex, ranging from changes in patent law and practice that have extended exclusionary IP rights into "upstream" science, financial market innovations that have eased access to capital for early-stage companies, and the development of institutions that have encouraged universities and public laboratories to actively promote commercialization.[7]

One consequence of these changes is that pharmaceutical innovation now relies heavily on a complex web of contractual agreements linking a variety of actors at various stages of the drug development process. Danzon et al. (2003) found that over one-third of new drugs approved between 1963 and 1999 originated in alliances between industry participants. Data on strategic technology alliances show an explosion of collaborative activity in the biomedical sector since the early 1990s. While the total number of new industrial technology alliances captured per year in the CATI-MERIT database grew by 76 percent between 1990 and 2003, the number of new alliances per year in "biotechnology"—which likely captures much of the external sourcing of drug discovery by pharmaceutical companies—increased by 818 percent.[8] Furthermore, many of these alliances spanned national boundaries: between 1990 and 2003, 45 percent of the alliances in this database in which U.S.-headquartered companies participated were "U.S.-only"

---

[6]This is not to say that process innovation cannot generate large aggregate savings for the industry. Macher and Nickerson's (2006) benchmarking study of pharmaceutical manufacturing suggests that adoption of process-analytical technologies, greater use of IT, and complementary work practices could generate cost savings of up to $50 billion per year.

[7]For a lengthier discussion, see Cockburn (2004).

[8]See National Science Foundation Science Indicators 2006, Table AT04-37. Note that many of these alliances may be relatively short-lived, and the "coverage" of alliance formation in this database is difficult to assess.

(i.e., all participants had U.S. headquarters), 23 percent of the alliances in which European-headquartered companies participated were Europe-only, and only 8 percent of the participated alliances in which Japan headquartered companies were "domestic."[9]

FORCES DRIVING LOCATION OF INNOVATIVE ACTIVITY

Pharmaceutical companies have always been able to operate R&D facilities largely independently from other activities: though a typical large pharmaceutical firm operates as an integrated economic entity, it normally conducts R&D in multiple locations around the world. The nature of the product development process, along with historically strong IP rights, and relatively straightforward licensing practices, has allowed pharmaceutical companies to "decouple" manufacturing and marketing from R&D. This has been the case for many decades, but increased vertical dis-integration in R&D activities since the mid-1980s has further relaxed organizational constraints on the location of research activity, permitting extensive geographic reorganization of R&D across countries and regions as well as vertical reorganization within firms. In the United States, for example, "upstream" firms specializing in new technologies for drug discovery are now often located in different locations (such as Boston and the San Francisco Bay area) than those historically used by the "big pharma" firms concentrated in Philadelphia, New Jersey, Connecticut, and the Midwest.[10]

Many factors drive these R&D location decisions, and the observed geographical distribution of research reflects complex trade-offs among them. One the one hand, economies of scale and scope in performing R&D, the presence of internal knowledge spillovers, and costs of coordinating activity across dispersed units suggest that, all else equal, firms should limit geographic dispersion of R&D. Furthermore, some locations may be more intrinsically economically attractive because of lower costs, access to government subsidies, or favorable tax treatment of R&D. Proximity to centers of academic excellence and other forms of noncommercial research also appears to convey benefits such as raised research productivity (see Furman et al., Forthcoming). On the other hand, these economic factors, which tend to concentrate R&D, are offset by political considerations. In some countries, pharmaceutical companies face strong political pressure to maintain domestic R&D. Some countries, such as the United Kingdom, have explicitly linked the stringency of price regulation to local R&D spending

---

[9]These marked differences in alliance activity are presumably driven by the relative concentration of potential partners (i.e., biotech companies in the United States).

[10]It is unclear whether the extent of vertical specialization and accompanying geographic reallocation of effort are more or less pronounced in the United States than elsewhere. The United States has attracted the lion's share of investment in new enterprises in biotechnology and pharmaceuticals, but there is no evidence that U.S.-headquartered firms rely more or less heavily on external R&D than, say, those headquartered in Europe.

levels; in other cases, such as in Canada, local R&D spending reflects a political bargain to avoid compulsory licensing.

Historically, the United States has been perceived by the industry as a very attractive location for pharmaceutical R&D because of its very limited use of price regulation and government purchasing, and its strong patent rights.[11] In contrast, in the late 1990s, EU governments became very concerned that overly aggressive price controls and hard bargaining by state purchasers were driving away investment in pharmaceutical R&D and adversely affecting the competitiveness of EU-based companies, though there is little evidence (see Table 1) of any major shift in R&D spending away from Europe. Episodes such as Canada's experience with compulsory licensing of pharmaceuticals in the 1970s and 1980s, or more recent examples such as the periodic heated disputes between OECD-based companies and governments of developing countries over pricing of antiretroviral drugs, suggest that R&D location decisions can be quite sensitive to government policies directed at lowering the cost of acquiring pharmaceuticals. Notwithstanding its long tradition of excellence in medical and pharmaceutical research, and substantial historical investments by multinational drug companies, Canada experienced a steep decline in domestic R&D activity in pharmaceuticals when it introduced its compulsory licensing regime. Only when full patent rights were restored, and a relatively loose drug price regulation scheme was instituted, did commercial R&D spending return to previous levels. Countries such as Australia, which have relatively stringent drug price controls, continue to face major challenges in attracting significant R&D investment by multinational drug companies, in spite of strong academic research capabilities, an attractive business environment, and substantial public support of commercial biomedical research.

Beyond these "price" drivers, several other factors have been identified as influencing R&D location decisions. These often work through indirect, or unpriced, effects such as knowledge spillovers that are conveyed by "open" publications, geographic proximity, or communication through informal professional networks rather through economic transactions. For example, drug discovery labs sites tend to specialize in therapeutic areas or scientific disciplines[12] and, since

---

[11]The Hatch-Waxman Act lowers the costs of generic entry and provides incentives to challenge pharmaceutical patents, but it also provides certain protections to patent holders. The United States also has provisions that extend the duration of pharmaceutical patents to offset time lost waiting for regulatory approval.

[12]For example, in 2000 Hoffman La Roche operated six major drug discovery facilities: Kamakura, Japan; Penzberg, Germany; Basel, Switzerland; Welwyn Garden City, United Kingdom; Nutley, New Jersey; and Palo Alto, California. A research center in Basel focused on basic research in genomics, proteomics, and bioinformatics, while the Kamakura, Penzberg, and Nutley labs specialized in oncology, Basel and Nutley in metabolic disorders and vascular diseases, Basel and Palo Alto in central nervous system disorders, Welwyn Garden City in virology, and Palo Alto in inflammation and genitourinary diseases.

proximity to publicly funded science appears to be an important determinant of research productivity, these often reflect local academic centers of excellence in particular fields. Furman et al. (Forthcoming) show that patenting by pharmaceutical companies is positively correlated with the volume of academic publications by "local" public-sector scientists.[13] The very substantial levels of publicly funded biomedical research in the United States, the United Kingdom, and some other countries has therefore played an important role in sustaining similarly high levels of commercial investment in drug discovery in these countries.

More generally, like other knowledge-intensive activities, discovery research appears to display substantial agglomeration externalities. Drug discovery activity tends to "cluster" in a small number of locations around the world: many major discovery labs are located in the New York/New Jersey/Connecticut standard metropolitan statistical area, Boston, the San Francisco Bay area, the suburbs of Philadelphia, Research Triangle in North Carolina, the Rhine Valley, the suburbs of London, Stockholm, and Tokyo/Kansei. These are conspicuously not low-cost locations, so this clustering suggests substantial offsetting economic benefits derived from being co-located with other firms. Beyond the role of localized knowledge spillovers, benefits from co-location with other pharmaceutical firms include access to skilled labor and "infrastructure" in the form of specialized services and suppliers, and efficient interaction with collaboration partners.[14]

A final factor that may affect R&D location decisions is the strength of IP protection. Though there is no obvious connection between the degree of patent protection in the local product market and the productivity of R&D conducted in any given country, the nature of a country's IP regime appears to affect multinationals' willingness to conduct R&D activities there. This may be because weak patent protection for products often correlates with weak legal protection of other forms of IP such as trade secrets and associated contractual agreements with employees and suppliers, and limited avenues to enforce these rights. Both patent and nonpatent protection of IP play an important role in maintaining exclusive access to, and control over, proprietary knowledge, and in countries with weak IP companies may have well-founded concerns about "leakage" of valuable information to local competitors. Zhao (Forthcoming) argues that weak

---

[13]To some degree, drug discovery labs operated by different firms within the same region appear to specialize in particular therapeutic classes or scientific disciplines. Cockburn et al. (2002) report, for example, that for their sample of firms commercial R&D in New Jersey was primarily, though not exclusively, focused on cardiovascular therapies, whereas that conducted in the suburbs of London was primarily in antipsychotics.

[14]Returning to the example of Hoffman La Roche in 2000, of nine important collaborators identified by the company, Tularik, Affymetrix, Clontech, and Incyte (plus its majority-owned subsidiary Genentech) were geographically proximate to its Palo Alto lab, CuraGen and Progenics were close to its Nutley lab, and Vernalis, Imperial College, and Oxford University were close to Welwyn Garden City. In a study of cross-regional collaborations by a broad set of biomedical technologies, Zhao and Islam (2007) document increasing geographic dispersion of R&D activities of large firms, but also increased internalization of knowledge spillovers within these firms.

IP regimes need not deter R&D investment by multinationals: absent strong IP rights, companies can nonetheless develop alternative mechanisms for realizing returns on innovation and IP. These mechanisms include rapid "internalization" of knowledge through efficient internal organizational processes and control of complementary assets and may make it possible to profitably exploit low prices of R&D inputs and underutilized domestic innovation capabilities. However, this argument is most appealing for technologies that have a substantial tacit component, are strongly complementary to other protected assets held by the firm, and have rapid development cycles. This is not the case for pharmaceutical R&D, where results from R&D are often easy to "externalize" and imitate, and product life cycles are measured in decades.

Not surprisingly, therefore, R&D activity in pharmaceuticals has historically been concentrated in countries with strong and enforceable IP and has only just begun to grow in countries that have recently adopted OECD-style patent systems under the provisions of the Trade-Related Aspects of Intellectual Propoerty Rights (TRIPS) agreement. Compliance with TRIPS requires all World Trade Organization (WTO) members to (ultimately) adopt key features of the patent systems of wealthy industrialized countries, such as a 20-year term, nondiscrimination across fields of technology and nationality of applicants, and effective enforcement procedures. Strong patent protection for pharmaceuticals is controversial in many of these countries (see discussion of India which follows), and the degree to which domestic political pressures will limit the enforceability of patents, or push the limits of the TRIPS agreement by, for example, instituting compulsory licensing of drugs, remains to be seen. Patent rights obtained by multinationals in countries such as India give these companies the ability to exclude generics and to set prices above marginal cost. But patents also provide protection for domestic firms conducting R&D, and political choices to weaken or limit patent protection on the products of multinationals may have serious consequences for nascent research sectors in these economies.

Impact of Industry Restructuring on Innovation

Structural change in the pharmaceutical industry has given pharmaceutical companies more opportunities and much greater flexibility to improve R&D performance by reallocating R&D effort between internal and external projects, and across different locations both within countries and around the world. Whether greater globalization of R&D has been caused by this vertical disaggregation of the industry—or vice versa—is an open question. Clearly the two phenomena are closely linked and, beyond the industry-specific factors that have driven vertical disaggregation discussed earlier, more general phenomena affecting many industries (such as improvements in communication technologies, greater international mobility of labor and capital, innovation in capital markets, and international harmonization of IP rules following the TRIPS agreement) have also played a role in creating new opportunities for collaborative R&D and specialist providers of R&D inputs.

There are many reasons to believe that industry restructuring and globalization may generate substantial gains in R&D performance. R&D outcomes depend critically on resource allocation (which projects to pursue, how much to spend, which to shut down), which at one time was done almost entirely through internal decisions of large vertically integrated firms. In today's industry, market transactions and the price mechanism play a much greater role in resource allocation, with specialization and competition in the supply of research inputs and services. Capital markets play an important role in pricing risk and provide high-powered incentives to entrepreneurial firms, and strong IP rights support a global "market for technology." These powerful economic forces may well result in significantly faster/better/cheaper drug development.

On the other hand, industry restructuring and globalization may be responsible for some inefficiencies that limit any gains in R&D performance. There is no guarantee that the market for technology (i.e., licensing and collaboration deals) creates reliable price signals, and market-driven resource allocation may therefore generate worse results than those obtained by the internal capital markets of large firms.[15] The struggle between entrants and incumbents in the industry may also be wasting significant resources in bargaining costs; payments to intermediaries such as lawyers and bankers; extra organizational overhead dedicated to seeking out, structuring, and operating collaborative ventures; or in defensive investments to improve bargaining and so forth.[16]

While the ultimate impact of restructuring and globalization on R&D performance will surely take decades to become apparent, it is clear that this more open competitive environment presents severe challenges. Heightened competitive pressure, greater cost transparency, and global competition have contributed to an extraordinarily high failure rate among would-be entrants to the industry. Of the many thousands of well-financed entrants with strong patent portfolios and exciting science that have attempted to gain a foothold in the industry as a supplier of technology or competitor to the established multinational incumbent firms, only a few hundred have survived. The prospects for new players in the industry based in emerging countries are therefore mixed. Success in the global pharmaceutical industry requires (among other things) substantial and sustained investments in R&D capacity, IP portfolios, and access to leading-edge science. It will likely take many years before new competitors appear on the world stage to present a serious head-to-head challenge to existing OECD-based firms.

---

[15]Markets for "knowledge goods" have long been understood to be subject to numerous forms of market failure such as imperfect and asymmetric information, nonexcludability, limited numbers of buyers and sellers, exernalities, and so forth.

[16]See Cockburn (2004, 2007) for further discussion.

Evidence on Offshoring of Pharmaceutical R&D

Rapid growth in technological capabilities in low-cost emerging economies is presenting new opportunities and challenges for pharmaceutical companies. Some geographic redistribution of R&D activity does appear to be taking place. On the one hand, companies located in countries such as India and China are performing more in-house R&D oriented toward developing new drugs, rather than reverse-engineering existing products or improving production efficiency. On the other hand, reflecting the general trend in the industry toward greater specialization and external sourcing of R&D services, OECD-based companies are beginning to look to low-cost countries as suppliers of contract research services, and growing numbers of clinical trials are being conducted in emerging economies. India and China are the two countries most frequently mentioned in this regard; however, by some indicators significant growth in activity also appears to be taking place in some Eastern European countries, Argentina, Brazil, Taiwan, South Africa, and Israel.

Data on this activity are limited. As discussed earlier, internationally and intertemporally consistent aggregate statistics on R&D expenditure in this industry are often not available for low-cost locations, and their reliability is difficult to assess. The experience of specific countries may nonetheless be informative, and the case of India is presented in the next section.

Case Study: India's Pharmaceutical Industry

Indian pharmaceutical companies have attracted much attention. Chaudhuri (2005) discusses the development of the Indian pharmaceutical industry in detail, charting the rise of companies such as Ranbaxy and Dr. Reddy to their current position as significant global players in generic drug manufacturing. Many of these companies have developed advanced capabilities in low-cost manufacturing, reflecting local expertise in chemical engineering and a historically process-oriented patent regime. (The success of these companies also reflects their mastery of the regulatory approval process for generic drugs, and associated litigation, in foreign markets.) Sustained profitability, together with the introduction of domestic product patents for pharmaceuticals in 2005 as part of the TRIPS agreement,[17] has prompted some of these companies to increase R&D spending and take on drug development projects. Chatterjee (2007) analyzes financial statements of Indian pharmaceutical manufacturers and finds very high growth rates of R&D spending for the 40-50 firms that have obtained U.S. patents and/or Food and Drug Administration (FDA) approval of manufacturing processes, and attract the

---

[17]The TRIPS agreement was reached as part of the Uruguay Round of multilateral trade negotiations in the mid-1990s. India and other countries with weak or nonexistent IP regimes agreed to put in place OECD-style patent protection by 2005 as a condition of retaining full participation in the WTO.

attention of domestic stock analysts. These firms increased their R&D-to-sales ratio by a factor of 10 to 20 between 1990 and 2000, and doubled it again between 2000 and 2005 to reach respectable average levels of 5-6 percent of sales. Among these, a handful of "elite" firms have raised R&D spending to 10 percent per year or more in 2005, with compound average growth rates of R&D spending over the past decade of 30-50 percent per year.[18]

Some of these firms have made significant investments in product-oriented research since the mid-1990s, with some proportion of increased R&D expenditure directed toward generating novel compounds to be tested as drug candidates (often referred to as New Chemical Entities [NCEs]). As of early 2006, more than 60 molecules developed by Indian companies were reported to be in late stages of preclinical research, and 16 had reached the clinical trials stage, though none have yet completed Phase III or been approved by the FDA.[19] Indian firms have also entered into a small number of highly publicized product development partnerships with multinational R&D-oriented firms: examples include Dr. Reddy's outlicensing of diabetes candidates to Novo Nordisk and Novartis in 1997, Glenmark's 2004 outlicensing agreement with Forest Labs to develop a candidate drug for asthma, and Nicholas Piramal's 2007 agreement with Eli Lilly to conduct clinical trials and regional marketing for various candidates for treatment of metabolic diseases. As with such deals in OECD countries, many of these agreements have been terminated or otherwise failed to generate significant licensing revenue for the Indian partner (Frontline, 2002). Note also that all of these NCEs appear to be "small-molecule" chemistry-based drug candidates rather than the "large-molecule" products characteristic of leading-edge biotechnology research.

Indian pharmaceutical companies are also reported to be playing a growing role as contract research providers to research-based multinationals. Indian companies have strong capabilities in medicinal and analytical chemistry, process engineering, and organic synthesis developed originally to support generic manufacturing. Though the nature of the R&D activities covered by these contract research agreements are unclear, expertise in chemistry-driven R&D activities positions these companies well to support the sourcing company's core product innovations by developing efficient manufacturing processes, "tweaking" candidate molecules for better bioavailability, or developing formulations optimized for specific marketing purposes. Indian companies are also developing expertise in toxicology and animal studies to support preclinical research, and total expenditure on clinical research in India has been estimated as $100 million per year for 2005. Small but significant investments are being made by OECD-based compa-

---

[18]Some caution may be warranted in assessing these data. Accounting standards for defining R&D may vary over time, some of the growth in expenditure may reflect cost inflation, and a certain amount may reflect "window dressing" to attract attention from outside investors or potential joint venture partners.

[19]"Pharmaceuticals" Report by Ernst & Young for the India Brand Equity Foundation, 2006.

nies in captive research facilities located in India and in long-term collaborative discovery partnerships with Indian companies.

But though these ventures may signal a new phase in the development of India's pharmaceutical industry, with an increasingly significant role in global innovation, it is important to note that the scale of this activity is currently very small. To put India's current scale of activity in drug discovery and development (at most 100 early-stage candidates since 2000[20]) in perspective, it is worth noting that worldwide several thousand molecules enter preclinical research and Phase I trials every year. The total expenditure on pharmaceutical R&D in India from all sources is growing rapidly but is unlikely to exceed $500 million per year in the near future, which is less than 1 percent of expenditure in OECD countries. Notwithstanding substantially cheaper inputs to R&D in India (labor costs for skilled scientists are claimed to be as little as one-seventh of U.S. levels), the scale of India's R&D effort is still very small.

Complying with the TRIPS Agreement, India has now implemented an OECD-style patent system. Effects on domestic prices are as yet unclear, but availability of product patents appears to be increasing the number of drugs available to Indian consumers and decreasing the amount of time elapsed between their first worldwide launch and availability in the Indian market.[21] Patent protection may also be playing a role both in stimulating R&D by domestic firms and in supporting multinational companies' participation in contract research agreements and licensing deals. But drug patents are politically controversial in India, and it remains to be seen whether the operation of the Indian Patent Office and enforcement in domestic courts will provide adequate IP protection for product innovators.[22]

Other Evidence on the Scope and Volume of Global R&D

Though comprehensive and reliable data on global R&D expenditure are not available, the location of discovery and development activity can be tracked using proxy indicators such as patent applications, academic publications, or databases that document clinical trial sites.

---

[20]HSBC Securities, cited in IBEF (2006).

[21]See Lanjouw (2007).

[22]Novartis is currently fighting a closely watched legal battle to secure Indian patent rights on Gleevec, its breakthrough cancer drug. Novartis' application for patent rights in India was rejected by the Indian Patent Office in 2006. "The patent office ruling in the Gleevec case has sent two far-reaching signals on the new TRIPS-compliant law: First, that India will not grant product patent [sic] for any drug unless it was invented after January 1, 1995. Second, that the standards set by the Indian law for grant of product patent for drugs could be more exacting than even those of advanced countries" (*Times of India*, February 24, 2006).

Discovery

Patent applications provide some information on the location of drug discovery activity. Because the United States is largest single pharmaceutical market, U.S. patent protection is likely to be sought for most promising drug candidates. Therefore, both the location of the patent assignee and the address of the inventors listed on U.S. patents are useful indicators of the location of drug discovery activity.[23] Table 3 shows the geographic distribution over time of granted U.S. applications for pharmaceutical patents assigned to corporations. These data are not comprehensive, since they cover only a single class of patents (IPC A61K), and the sample excludes patents where the country of the assignee cannot be determined. Data are tabulated by the date of application, which induces truncation of the sample in later years to due the application/grant lag.

Patenting in this sample is dominated by U.S. and European companies. Although the number of patents in this sample that are assigned to corporations based in India and China has grown quite rapidly, as the table shows this growth is from such a small base that the share of companies based in these countries remains very small.

Table 4 shows results from a slightly different exercise, breaking down the regional share of drug discovery activity based on the location of the inventors listed on drug patents. Again, although the volume of activity in India and China has grown very rapidly since 1990, it still represents a tiny share of total activity. In 1990, for example, France alone accounted for 661 instances of "inventorship," whereas India had 20 and China had 22.

These indicators provide little evidence of substantial global relocation of activity in drug discovery. As more current data become available, we are likely to see substantially increased levels of activity in countries like India and China. But although activity in these countries is growing quite quickly, it will be many years before the share of the locations where drug discovery has traditionally been concentrated is materially affected.

Development

A somewhat different picture emerges, however, for clinical development. For this aspect of innovative activity in pharmaceuticals, the location of clinical trial sites provides some insight into the global distribution of activity. Berndt, Cockburn, and Thiers (2007) report results from tabulating the location of more than 65,000 trial sites participating in the clinical trials that have been registered on clinicaltrials.gov since 2001. Table 5 summarizes this measure of clinical development activity. Because registration is not compulsory for all trials, the

---

[23]"Home bias" may inflate the U.S. share, and language barriers or lack of experience with the U.S. patent system may result in underrepresentation of some countries.

TABLE 3 Location of Corporations Obtaining U.S. Pharmaceutical Patents

Application Year	1990	1995	2000	2002
Total number of patents in sample	3,414	9,277	7,073	3,257
Regional share of patents based on location of assignee (%)				
United States	55.1	63.3	58.3	57.2
EU15	24.6	21.7	22.3	22.8
Japan	15.3	7.5	9.0	9.5
Other OECD	2.8	5.0	6.2	6.5
India	0.0	0.1	0.7	1.3
China	0.1	0.0	0.1	0.2

NOTES: Table entries based on the count of U.S. patents in IPC class A61K assigned to corporations whose country can be identified. SOURCE: Author's calculations based on U.S. Patent and Trademark Office (USPTO) data.

TABLE 4 Location of Inventors on U.S. Pharmaceutical Patents

Application Year	1990	1995	2000
Total pharmaceutical inventorships[a]	10,582	30,135[b]	23,923
Regional share (%)			
United States	42.8	56.7	54.4
EU15	28.6	25.0	24.8
Japan	21.2	10.4	10.7
Other OECD	3.1	5.3	6.0
India	0.2	0.2	1.1
China	0.2	0.1	0.2
Other	4.0	3.0	2.8

[a]An inventorship is an instance of an inventor being listed on a patent application; therefore, a single patent with three inventors will generate three observations.

[b]1995 saw a surge of applications at the USPTO in order to secure various procedural advantages before the passage of patent reform legislation.

SOURCE: Author's calculations based on USPTO data. Table entries are based on the number of instances of inventorship for patents falling in IPC class A61K, by country of the inventor and date of application for the patent.

TABLE 5 Clinical Trial Sites by Region

Year	2000	2001	2002	2003	2004	2005	2006
Total sites	2,385	4,139	6,677	8,034	14,224	23,536	33,045
Share of (%)							
United States	51.4	48.8	50.9	49.1	54.6	49.5	45.2
EU15	32.2	29.4	28.9	25.7	23.1	26.3	26.8
Japan	0.0	0.0	1.6	2.1	2.0	2.5	2.7
Other OECD	7.5	8.2	10.3	8.0	7.4	8.0	7.2
India	0.0	0.1	0.3	0.8	0.7	0.9	1.0
China	0.0	0.4	0.2	0.5	0.3	0.5	0.9

NOTES: Table entries are based on the number of sites participating in clinical trials registered on clinicaltrials.gov. The average number of sites per trial is 7.6.

extent to which these data are representative of all trials is unclear. (In 1997 the FDA began requiring ex ante registration of trials for life-threatening diseases, and community norms have encouraged participation in trial registries, but it was only when a consortium of editors of the major medical journals [ICMJE] agreed in 2004 to require ex ante registration of trials as a condition of publication that the volume of registrations appears to have begun to approach full coverage.) Growth in the total number of trials, particularly in the early years of this sample, largely reflects increases in the coverage rate rather than increases in the volume of activity. But provided nonregistration does not vary systematically across countries, shares of activity are nonetheless a reasonable metric for the extent of a country or region's involvement in clinical research.

As Table 5 shows, the United States, the EU, and Japan continue to account for the bulk of clinical trial activity, but India and China had a significant and rapidly growing share of activity.

Overall, emerging economies and low-cost locations are a relatively small share of global activity, but this is changing rapidly. Table 6 shows the share of global trial sites by geopolitical region and by traditional versus emerging countries. In 2002, more than 90 percent of trial sites were located in "traditional" countries (North America, Western Europe, and Scandinavia), but this proportion has fallen rapidly in recent years, with the share of nontraditional countries in the total number of trial sites rising from 7 to 17 percent. Growth in this form of R&D activity has been particularly strong in Eastern Europe and Asia. Between 2003 and 2006, for example, Malaysia, Philippines, Bulgaria, Chile, Turkey, Argentina, the Russian Federation, Thailand, Mexico, and Latvia more than quadrupled their share of global trial sites, and India and China's shares more than tripled.

In a regression analysis of factors driving the global allocation of clinical trials, Berndt, Cockburn, and Thiers (2006) find that changes in countries' share of

TABLE 6 Regional Share of Worldwide Clinical Trial Sites (%)

Year	2002	2003	2004	2005	2006
North America	58.2	54.1	59.8	54.3	49.5
Western Europe	30.6	26.7	24.1	27.4	27.6
Oceania	3.3	4.0	3.2	4.5	4.6
Latin America	1.7	3.5	3.7	3.3	4.3
Eastern Europe	3.8	7.4	4.9	5.9	8.1
Asia	1.1	3.1	2.5	3.2	4.0
Middle East	0.3	0.3	0.7	0.7	0.6
Africa	1.0	0.8	0.9	1.0	1.0
"Traditional" countries	92.4	85.0	87.4	86.4	82.0
"Emerging" countries	7.1	14.2	11.8	12.6	17.1
Others	0.5	0.9	0.9	1.0	0.9

SOURCE: clinicaltrials.gov.

trial sites were negatively associated with a measure of cost per patient, but positively associated with changes in the strength of patent protection for biomedical inventions. Two- or fourfold differences in the cost per patient of conducting trials in emerging countries rather than in traditional countries thus appear to be driving a substantial expansion of activity in these locations. Interestingly the share of emerging countries in this activity is highest for large, confirmatory trials (Phase III of the drug development process) and lowest for small, early-stage trials that are more closely connected with basic biomedical science, suggesting that some types of clinical research are much less strongly influenced by the clustering and proximity to leading-edge academic research effects that drive the location of drug discovery. Turning to IP issues, the positive association found here between changes in patent protection of biomedical inventions and growth in share of global clinical trials may reflect concerns of both multinational and domestic R&D performers about imitation of their products and their ability to appropriate returns from innovation. Even though sales in these countries are only a very small share of the global pharmaceutical market, and their current profitability is therefore unlikely to be a major driver of R&D decisions, it is clear that they have huge potential for future growth in sales. Changes in patent protection may therefore be important as a guarantee of future profitability in much larger markets, attracting R&D, which will realize returns far in the future. One critical connection between local R&D and local future sales may be the role that late-stage trials can play in building future demand—by familiarizing key opinion leaders in the local medical community with a product, the sponsor of the trial may realize higher volumes or better reimbursement prices once the product is launched.

One question of great interest is whether participation in global clinical trials has "spillover" effects in the sense of building local capacity to conduct independent clinical research in support of domestic drug development programs. Clearly, sustained participation in this activity will allow clinical researchers to gain experience, credibility, and skills, and to promote development of supporting infrastructure and services. But it is likely to take considerable time before emerging countries are able to design and conduct complex trials on a routine basis and develop competitive capabilities in "translational medicine"—the bench-to-bedside combination of clinical investigation with basic research that plays a critical role in drug development. This type of R&D activity relies heavily on participation by skilled physicians, who are a very scarce resource in emerging countries relative to medical needs. Substantial expansion of clinical research will require these countries to make significant investments in medical schools and training programs in order to meet increased demand both for routine care for larger and wealthier populations and for an emerging clinical research sector. It is far from clear how fast or extensive this supply response will be, or what will be the impact of rapid economic growth and consequent expansion of health care on the market for physician services.

Manufacturing and Process Innovation

While product innovation remains the major focus of the industry, it is worth noting that process innovation capabilities may play an increasingly important role in the future. Generic products account for a very large proportion of drug consumption, and this will likely grow in the future as patents expire for the current set of "blockbuster" products. Suppliers of these products compete on costs, and Indian manufacturers in particular appear to have acquired world-leading capabilities in developing low-cost manufacturing processes, which positions them to play a dominant role in supplying low-price products to both developed and developing countries. Looking further into the future, new generations of large-molecule biotech drugs will likely displace the current set of chemistry-based products, particularly for diseases such as cancer. But these drugs are notoriously expensive and difficult to produce, suggesting an important future role for manufacturing innovation that should bring costs down to the point at which these products can be profitably supplied to large lower-income markets. Clusters of activity and development of these manufacturing innovation skills are already occurring in "new" locations such as Singapore, which may become important locations for this type of manufacturing in the same way that they have for other technologies such as semiconductors.

DISCUSSION AND CONCLUSIONS

Globalization is not a new phenomenon in pharmaceuticals. At least since the 1950s, innovation in this industry has been geographically dispersed, with pharmaceutical companies conducting R&D in multiple locations around the world. Over the past 25 years, however, gains from specialization in different aspects of the drug discovery and development process, changes in IP rules, easy access to venture capital, and the refinement of collaborative business models have driven greater vertical disaggregation of the industry, along with greater geographic dispersion of R&D. Pharmaceutical companies now have more opportunities and much greater flexibility to reallocate R&D effort between internal and external projects and among different locations both within countries and around the world. While the industry continues to be dominated by vertically integrated multinational firms whose activities span the entire chain of research, from basic science to postapproval epidemiology, these firms compete, collaborate, and interact with a wide variety of new actors.

This interaction is not confined by national boundaries, and the development of high-quality, low-cost research capabilities outside the United States and Europe presents significant opportunities for multinational R&D-based pharmaceutical companies to outsource some aspects of the innovation process. The available evidence suggests that emerging countries are playing a rapidly growing role in the global research effort, albeit currently very small compared to the scale of activity in the United States and Europe. In large part, this role has been enabled by structural change in the industry creating opportunities for collaboration and markets for specialized skills and services. But it has also been facilitated by the general trends in the world economy that are driving globalization of many industries, including factors such as improvements in communication technologies, greater international mobility of labor and capital, accumulation of human capital and business infrastructure in low-cost emerging economies, and harmonization of IP rules following the TRIPS agreement.

From a U.S. perspective, this "offshoring" of pharmaceutical R&D has potential benefits to be considered as well as the costs from any loss of the stable, high-wage employment characteristic of this industry. The pharmaceutical industry currently faces what some commentators have termed a "productivity crisis." The number of new drugs approved each year appears to be stagnating, with limited progress made in recent decades in treating some major diseases, yet R&D expenditure has been growing very rapidly. One widely used indicator of research productivity is the cost per new drug approved, accounting for failed projects and the time value of money, which has been rising at an alarming rate. The most recent in a series of studies from the Tufts Center for the Study of Drug Development (Dimasi et al., 2003) estimates the present value of R&D expenditures to bring a new drug to market to be $802 million per FDA-approved new drug. In constant year 2000 dollars, this $802 million is more than double the $318 million estimated in an earlier 1991 study, and almost six times larger than the $138 million figure obtained in a 1979 analysis. Recent industry estimates are now well in excess of $1 billion per successful new drug.

There are good reasons to believe that this "crisis" may not be as severe as some media accounts suggest. For example, simply counting the number of new drugs approved is not very helpful if the "quality" of each new drug measured in terms of health impact is changing over time.[24] Nonetheless, there are serious grounds for concern. To support these continued investments in R&D in the face of these rising costs, pharmaceutical companies need to realize very substantial sales revenues, which may be difficult to sustain in the face of political pressure around the world to restrain increases in health care expenditure. To the extent that the drug development process can be made more efficient through greater flexibility in resource allocation, more experimentation in business models, and greater use of low-cost inputs, offshoring of some aspects of R&D has the welcome potential to reduce the cost of developing new drugs.

Furthermore, at least in the short term, there are limits to the types of activity that are likely to be relocated outside the industry's traditional locations. Substantial offshoring of R&D activities is most likely to occur where the research activity is relatively routinized, uses large amounts of relatively low-skilled labor, and does not need to be tightly integrated or co-located with other R&D

---

[24]See Cockburn (2007) for a fuller discussion.

activities. Large-scale, late-stage clinical trials with "low-tech" endpoints (such as measuring blood pressure) are examples of this kind of activity, and indeed the global allocation of research effort in these areas shows signs of a significant response to cost differences across countries. As suggested earlier, greater use of, for example, India's chemistry-oriented R&D capacity presents a "win-win" opportunity for U.S.-based companies to improve R&D efficiency by contracting out relatively routine work on process engineering, compound synthesis, and medicinal chemistry while focusing on other aspects of R&D.

These other aspects of the innovation process—less routinized, more science-intensive—are much less likely to relocate to low-cost locations. Decisions about where to locate science-intensive drug discovery appear to be much less sensitive to labor costs and may be driven primarily by factors such as proximity to leading-edge academic research and "cluster" externalities. The benefits of co-locating R&D labs along with competitors in locations such as Baltimore/Washington, Boston, or the San Francisco Bay area, which feature agglomerations of commercial research, university science, and academic medical centers, coupled with "thick" local markets for specialized inputs and human capital, are very substantial. Any labor cost savings from relocating R&D labs to countries such as India and China are unlikely to compensate for the negative effect of losing access to these benefits.

Substantial geographic redistribution of core R&D effort in this industry therefore seems likely to occur only if and when these offshore locations develop their own critical mass of academic biomedical science and supporting complementary infrastructure. Some countries, such as Singapore and Taiwan, have put in place major programs to create research infrastructure and attract leading academic researchers, but this will take significant amounts of time and money. Emerging countries such as India and China, which have pockets of academic excellence in biomedicine but have had historically relatively low levels of public support for biomedical research, face an uphill struggle to develop this national capacity. The United States enjoys a dominant position in the world in publicly funded biomedical research. Provided U.S. taxpayers continue to fund this level of support, and public policy sustains the institutions of Open Science that support the productivity and vitality of academic science, the United States seems likely to remain the location of choice for science-intensive pharmaceutical R&D.

Open Science, with its curiosity-driven, investigator-initiated agenda and priority- and publication-based incentives, is a distinctive and vital component of the biomedical innovation system. But, particularly in the United States, science has become increasingly "propertized" by the extension of the patent system into basic research and the enthusiastic participation of universities and individual academics in patenting and entrepreneurial activity in life sciences. While this activity has clear benefits in terms of facilitating technology transfer and attracting venture-backed investment, it also carries with it less obvious costs in terms of weakening the institutions of Open Science, limiting access to research tools and data, and forcing congruence between the agenda of academic research and commercially attractive areas of inquiry. The long-run competitiveness of the U.S. pharmaceutical industry may therefore require careful management by policy makers of conflicts between exclusion-oriented IP rights and traditional academic norms.

Patent policy may play a somewhat different role in influencing R&D location decisions through its effect, in conjunction with price regulation and government purchasing, on domestic returns to pharmaceutical R&D. As discussed earlier, policy efforts to restrain pharmaceutical spending in the United States—for example, through changes to the Medicare drug benefit or reform of the Hatch-Waxman Act—should therefore carefully consider the impact of such measures on R&D location decisions. Of course, government actions elsewhere in the world directed at lowering drug prices may work in the opposite direction. "Unenthusiastic" implementation of the TRIPS patent provisions in emerging countries and unpredictable operation of these new patent systems in practice may cause multinationals to rethink decisions to expand R&D activity in these countries. The closely watched Gleevec case in India has major implications for the future of pharmaceutical R&D in that country, whether performed by domestic competitors on their own account or in partnership with multinationals. Similarly, current efforts by the government of Thailand to force one U.S.-based company (Abbott) to lower the price of some its drugs to Thai consumers are unlikely to encourage multinational companies to engage in R&D activities in Thailand in the future. Widespread actions of this kind may have a significant negative impact on offshoring trends.

Major changes in the existing international allocation of innovative effort in the pharmaceutical industry are unlikely, particularly in the short run. Compared to other technologies, this industry moves relatively slowly: very long product development cycles, and necessarily conservative organizational structures and processes imposed by health and safety regulation, make it difficult for pharmaceutical companies to make large, transformative changes to their business as fast as some firms can in other industries. Recent rapid expansion of research capacity in low-cost emerging countries will benefit U.S.-based multinationals (and U.S. consumers of pharmaceuticals) by lowering R&D costs in some activities, but this rapid expansion is highly unlikely to cause a "tsunami" of professional job losses in pharmaceutical research. In the long run, emerging countries may succeed in developing a large enough base of local academic and publicly funded biomedical science to threaten the substantial competitive advantage that the United States currently enjoys in this area. Commercial R&D location decisions are tightly linked to publicly funded science; therefore, it will be necessary for public policy in the United States to play close attention to vitality and viability of academic- and government-supported biomedical science if the United States is to retain global leadership in the pharmaceutical industry.

ACKNOWLEDGMENTS

I thank Ernst Berndt, Jeff Furman, Jeffrey Macher, and David Mowery for helpful discussions and suggestions.

REFERENCES

Berndt, E., I. Cockburn, and F. Thiers. (2006). Intellectual Property Rights and the Globalization of Clinical Trials for New Medicines. Seminar Presentation, RAND Institution, Santa Monica, November 2006.

Berndt, E., I. Cockburn, and F. Thiers. (2007). The Globalization of Clinical Trials for New Medicines into Emerging Economies: Where Are They Going and Why? Conference Paper, UNU-MERIT Conference on Micro Evidence on Innovation in Developing Countries, Maastricht, May 31, 2007.

Chatterjee, C. (2007). Fundamental Patent Reform and the Private Returns to R&D—The Case of Indian Pharmaceuticals. Manuscript, Heinz School, Carnegie Mellon University.

Chaudhuri, S. (2005). *The WTO and India's Pharmaceuticals Industry: Patent Protection TRIPS and Developing Countries*. New Delhi: Oxford University Press.

Cockburn, I. M. (2004). The changing structure of the pharmaceutical industry. *Health Affairs* 23(1):10-22.

Cockburn, I. M. (2006). Blurred boundaries: Tensions between open scientific resources and commercial exploitation of knowledge in biomedical research. Chapter in *Advancing Knowledge and the Knowledge Economy*, B. Kahin and D. Foray, eds. Cambridge: MIT Press.

Cockburn, I. M. (2007). Is the pharmaceutical industry in a productivity crisis? Chapter in *Innovation Policy and the Economy*, Vol. 7, A. Jaffe, J. Lerner, and S. Stern, eds. Cambridge: MIT Press for the National Bureau of Economic Research.

Cockburn, I. M., and R. Henderson. (1998). Absorptive capacity, coauthoring behavior, and the organization of research in drug discovery. *Journal of Industrial Economics* 46(2):157-182.

Cockburn, I. M., J. Furman, R. Henderson, and M. Kyle. (2002). Geographic Location and the Productivity of Pharmaceutical Research. Conference paper, NBER Summer Institute, July 2002.

Danzon, P., Y. R. Wang, and L. L. Wang. (2003). The Impact of Price Regulation on the Launch Delay of New Drugs—Evidence from Twenty-Five Major Markets in the 1990s. NBER Working Paper 9874.

Dimasi, J.R. Hansen, and H. Grabowski, 2003. The price of innovation: New estimates of drug development cost. *Journal of Health Economics* 22:151-185.

Drug trials and questions. (2002). *Frontline* 19(19). September 14.

Furman, J., I. Cockburn, R. Henderson, and M. Kyle. (Forthcoming). Public & private spillovers, location, and the productivity of pharmaceutical research. *Annales d'Economie et Statistique*.

Gambardella, A. (1995). *Science and Innovation*. Cambridge University Press.

Grabowski, H., D. Ridley, and K. Schulman. (Forthcoming). Entry and competition in generic biologicals. Working Paper, Fuqua School of Business. *Managerial and Decision Economics*.

IBEF (India Brand Equity Foundation). (2006). Pharmaceuticals. Report by Ernst and Young for the India Brand Equity Foundation/Confederation of Indian Industry/Ministry of Commerce and Industry.

Kyle, M. (2006). The role of firm characteristics in pharmaceutical product launches. *RAND Journal of Economics* 37(3):602-618.

Kyle, M. (2007). Pharmaceutical price controls and entry strategies. *Review of Economics and Statistics* 89(1):88-99.

Lanjouw, J. (2005). Patents, Price Controls, and Access to New Drugs: How Policy Affects Global Market Entry. NBER Working Paper 11321.

Lanjouw, J. (2007). Patents, Price Controls and the Arrival of New Drugs: How Policy Affects International Launch Patterns. Paper presented at UC Berkeley Lanjouw Memorial Conference, March 2007.

Macher, J., and J. Nickerson. (2006). Pharmaceutical Manufacturing Research Project. Available at http://faculty.msb.edu/jtm4/PMRP%20results/. Accessed December 12, 2006.

Zhao, M. (Forthcoming). Conducting R&D in countries with weak intellectual property rights protection. Manuscript, University of Minnesota. *Management Science.*

Zhao, M., and M. Islam. (2007). Cross-Regional Ties within Firms: Promoting Knowledge Flow or Discouraging Spillover? Working Paper, Stephen M. Ross School of Business.

7

Biotechnology

RAINE HERMANS
Northwestern University and ETLA Research Institute of the Finnish Economy
ALICIA LÖFFLER
Northwestern University
SCOTT STERN
Northwestern University
National Bureau of Economic Research (NBER)

INTRODUCTION

Over the past decade, the biotechnology industry has been the focus of increasing academic and policy interest as a potential source of regional and national economic development (Cortright and Mayer, 2002; Feldman, 2003). Although the current size of the industry is quite small, particularly in terms of employment, both local and national policy makers—in the United States and abroad—have proactively encouraged local and regional investment in the biotechnology industry. In many cases, policy interest in biotechnology is grounded in the belief that, whereas traditional sectoral sources of jobs and investment are increasingly subject to erosion due to globalization, the biotechnology industry is associated with superior wages and a high level of economic prosperity and growth (Battelle and SSTI, 2006). The proliferation of biotechnology investment programs—even within regions that have little current activity in the industry—raises concerns about the effectiveness of biotechnology as a driver of regional economic development. Moreover, these policy initiatives will have a long-lived impact on patterns of regional development and on the evolution and long-term structure of the industry.

The geography of this industry, and the impact of globalization on biotechnology, will be shaped not only by policy initiatives but also, perhaps more important, by fundamental features of the economic, strategic, and institutional environment. This chapter provides an overview of the drivers, patterns, and consequences of the globalization of biotechnology and offers a preliminary assessment of historical and contemporary patterns of the geographic dispersion of biotechnology innovation. Our analysis of the distinctive nature of the globalization of biotechnology motivates policy implications aimed at ensuring continued leadership and dynamism in the American biotechnology sector.

While there has been a great deal of academic and policy interest in the biotechnology industry, the scope and extent of the industry are loosely defined, and measures of its scope, size, and patterns of geographic activity depend on the specific definitions that are used (Kenney, 1986; Orsenigo, 1989; Cockburn, et al., 1999; Cortright and Mayer, 2002; van Beuzekom and Arundel, 2006). At the broadest level, biotechnology is an industry that includes the commercialization of life science innovations in the health, agriculture, and industrial sectors, which are often referred to as the "red," "green," and "white" biotechnology sectors, respectively. While the international biotechnology industry incorporates activities in all three biotechnology spheres, the bulk of policy and academic analysis have focused on "red" (i.e., health-oriented) biotechnology. Furthermore, although the majority of privately and publicly funded biotechnology enterprises have been located in the United States, the pattern of regional and international development is quite distinct for the red, green, and white divisions. Despite ambiguities in the scope of the industry and variation across the three subsectors, "cluster-driven" growth in biotechnology has emerged as a key economic development strategy for regions and nations at all levels of economic and technological prosperity (Cortright and Mayer, 2002; Feldman, 2003). Beyond its importance for economic development policy, biotechnology is also the setting for a very active debate across several social sciences about the drivers of clustering and the impact of globalization on the importance of location in innovation.

In this chapter we examine trends related to the geographic distribution of industrial biotechnological activity, focusing on the following broad questions: What are the key drivers of innovation within biotechnology, and how do these drivers influence patterns of regional development? What are the drivers of location and clustering within the biotechnology industry, and how does globalization impact the geography of the biotechnology industry? What are the main locational patterns within the biotechnology industry, both in terms of employment and firm formation and in terms of innovation and sales? What are the main strengths and limitations of publicly available data on the biotechnology industry? Finally, how does the current geography of the biotechnology industry impact contemporary debates over the potential for biotechnology to serve as a source of regional development, innovation, and improvements in human welfare?

Overall, our analysis suggests that biotechnology remains a clustered economic activity and relies strongly on interaction with science-based university research. However, the number of active clusters in biotechnology is increasing over time. An increasing number of distinct locations in the United States are home to a significant level of biotechnology activity, and an increasing number of countries around the world support modest to significant activity within the biotechnology industry. More notably, while many countries around the world now "host" a biotechnology industry of varying importance, the activity within most countries is highly localized and often centered in a single city or metropolitan area. Although the data are inadequate to allow for a comprehensive analysis, qualitative and quantitative evidence suggests that the number of biotechnology clusters that host a significant number of viable private companies and serve as a recurrent source of innovation has increased; this increase in the number of clusters with "critical mass" is reflected in the increased dispersion of biotechnology employment, entrepreneurship, and measured innovations.

This central insight—an increase in the number of regional clusters, rather than a simple dispersion of biotechnology activity—holds a number of implications. First, the impact of globalization on biotechnology seems to be distinct from the pattern observed in traditional manufacturing sectors. While the globalization of many industries seems to reflect the increasing availability of low-cost locations for performing low-margin activities that had previously been conducted in the United States or Europe, the globalization of biotechnology reflects a "catching up" process. A few regions around the world have established infrastructure and conditions to attempt to compete "head-to-head" with leading regions in the United States. Second, the analysis highlights the small absolute size of the biotechnology industry. Using a relatively inclusive definition, total biotechnology employment in the United States accounts for less than 200,000 full-time employees, which itself accounts for well over 50 percent of global employment (van Beuzekom and Arundel, 2006). In contrast, a single company in information technology (IT) such as Hewlett-Packard employs more than 150,000 workers (Hewlett-Packard, 2006). While globalization may affect the broader economy through its impact on sectors such as IT or traditional manufacturing, the small scale of the biotechnology industry precludes it from having a significant employment impact on the U.S. economy, at least at the present time. In other words, while an increasing number of policy initiatives focus on the role of biotechnology in encouraging job creation and employment, the simple fact is that, if the biotechnology industry remains at roughly the same scale it has achieved after the past decade of rapid growth, it is unlikely to be a major driver of employment patterns and overall job growth, either in the United States or abroad.

Finally, the analysis raises several interesting questions for further study. The most important issue is one of data collection. While our understanding of the biotechnology industry is greatly facilitated by detailed public and private data-gathering efforts (including the extremely useful Organisation for Economic Co-operation and Development [OECD] Biotechnology Statistics program), there seems to be an important gap between qualitative evaluations focusing on the role of subnational clusters and the fact that most international statistics are measured only at the country level. While there have been several ambitious attempts to document the clustering of biotechnology activity among regions within the United States, there is no single source of data or unambiguous approach that allows for a comparison of biotechnology clusters on a global basis. Second, although most analyses of the industry focus on the red biotechnology sector, patterns of locational advantage and the impact of globalization are quite distinct for the green and white sectors. For example, countries such as Japan and Denmark hold leading positions in the industrial applications of biotechnology. Moreover, in contrast to the high level of academic entrepreneurship that characterizes the red sector, the green sector is largely dominated by a small number of large firms such as Monsanto and DuPont. These alternative patterns make it problematic to extrapolate from detailed studies of the health-oriented sector in analyzing the growth and geographic evolution of the industrial and agricultural sectors of the industry.

The remainder of the chapter is organized as follows. The second section provides a concise introduction to the biotechnology industry and the key drivers of innovation in this industry. Among other issues, we highlight the importance of proximity to the creation of knowledge in fostering agglomeration. We then turn to an explicit discussion of the drivers of location and clustering in the industry, extending the "diamond" framework (Porter, 1990, 1998). In adapting that framework to the biotechnology industry, we highlight the potential for catch-up by lagging regions, the potential for disagglomeration as the industry or segments of it mature, and the potential for a leading region to establish itself as a global "hub" for biotechnology research and innovation going forward. In the fourth section, we consider broad patterns and data regarding firm location, employment, and sales in the biotechnology industry. As discussed earlier, the data illustrate the small size of the industry overall and the dominance of the United States within the industry. We then turn in the fifth section to an empirical assessment of the geography of innovation, in terms of both patenting behavior and commercial sales. A concluding section discusses the key findings and implications for policy.

THE DRIVERS OF INNOVATION IN THE BIOTECHNOLOGY INDUSTRY

The Origins and Scope of the Biotechnology Industry

Biotechnology is a relatively young and still emerging sector of the economy that is focused on the application of cellular and biomolecular processes to develop or make useful products (Biotechnology Industry Organization, 2006).[1]

---

[1]There is no single definition of the industry, and different criteria are often used to define the scope of the biotechnology industry in different countries. For example, the OECD employs both a functional definition—"the application of science and technology to living organisms, as well as parts, products and models thereof, to alter living or nonliving materials for the production of knowledge, goods and services"—and list-based definitions in which firms or workers are included in biotechnology if their activities fall within the scope of a set of listed categories (van Beuzekom and Arundel, 2006). To the extent possible, we are careful to define the definition and sample by which international or intranational comparisons are made.

The origins of the biotechnology industry can be traced back to a confluence of technological, economic, and institutional shifts during the late 1970s and early 1980s: the development of recombinant DNA technology and other fundamental advances in life sciences research during the 1970s; a significant increase in funding and resources for life sciences research (both public and private, in the U.S. and abroad); and a set of policy decisions, such as the 1980 Diamond vs. Chakrabarty Supreme Court decision and the Bayh-Dole Act, that allowed the assertion of intellectual property rights over innovations based on genetic engineering, even those funded by the public sector.

The conceptual ideas underlying biotechnology date back almost 12,000 years with the domestication of plants and animals through selective breeding. However, it was not until 1973, when Stanley Cohen, Stanford University, and Herbert Boyer, University of California San Francisco, demonstrated the ability to manipulate genetic material in a practical way, that the potential for commercial applications from the science of molecular biology became apparent. Indeed, Herbert Boyer himself was one of the founders of one of the first and among the most successful biotechnology companies, Genentech. While the discoveries of the 1970s represented fundamental scientific breakthroughs and offered isolated commercial applications, such as the development of synthetic insulin and human growth hormone (McKelvey, 1996; Stern, 1995), the growth of the biotechnology industry has relied on a series of complementary technological and scientific breakthroughs of similar magnitude. These include but are not limited to the development of rapid genetic sequencing methods such as the polymerase chain reaction in the 1980s to the use of increasingly advanced IT in bioinformatics in the 1990s and the ability to integrate genomic information through initiatives such as the Human Genome Project. Biotechnology represents the confluence of many emerging disciplines and relies on discoveries from academic and government laboratories as well as commercial institutions. While the precise boundaries of the industry are admittedly fuzzy, it is useful to consider three related but distinct spheres: health-oriented, agricultural, and industry biotechnology, which are referred to as red, green, and white biotechnology, respectively.

Health-Oriented Biotechnology ("Red Biotech")

Private investment in health-oriented biotechnology has been concentrated in a small number of regional clusters, which are also home to leading universities and other research institutions. On the one hand, publicly funded life sciences research serves as an extremely important source of discoveries for health-oriented biotechnology and is dispersed broadly across universities and research institutes in the United States and abroad. However, private-sector investment in the health-oriented biotechnology industry is much more regionally concentrated. In the United States, a small number of regional clusters in areas such as San Francisco, Boston, and San Diego have served as the origin for a large share of all biotechnology innovative investment and activity (Cortright and Mayer, 2002). Although the health-oriented biotechnology sector is concentrated largely in regional clusters in the United States, there are a significant number of small- to medium-sized clusters outside of the United States, including concentrations around Cambridge (UK), the Medicon Valley (Sweden/Denmark), Singapore, Sydney, and Melbourne, among other locations. More generally, although the commercialization of health-oriented biotechnology innovation has largely involved cooperation with more established firms (many of which are pharmaceutical firms located outside of the regional clusters), health-oriented biotechnology has been closely associated with academic entrepreneurship, whereby leading university research faculty are associated with the creation of new biotechnology firms.

Agricultural Biotechnology ("Green Biotech")

The second major application segment in biotechnology is associated with the development and commercialization of "green," or agriculture-focused, biotechnology products, particularly the development of new seed traits for staples and specialized agricultural products, from corn to papayas. While cluster-driven entrepreneurship has also played a role in this sector, the bulk of investment and commercialization has been centered around a small number of large, established players, including companies such as Monsanto and DuPont. Relative to health-oriented applications, the earliest commercial applications for agricultural biotechnology were not brought to market until the mid-1990s. While diffusion of products such as pest-resistant corn and soybeans was rapid in the United Sates, there was significant opposition to the adoption of these technologies in international markets, particularly in Europe, which enacted a ban on most products until 2004. In other words, both development and initial use of agricultural biotechnology have been centered in the United States, and companies and farmers who invested in these technologies at an early stage have benefited as markets for genetically modified organisms have globalized over the past several years.

Industrial Biotechnology ("White Biotech")

Industrial biotechnology is the application of biotechnology for industrial purposes, ranging from more effective enzymes in the chemical and textile sectors to biofuels to bioremediation (i.e., environmental applications). By and large, industrial biotechnology has served as a useful source of process innovation in established industrial settings. For example, in the chemical sector, bioengineered enzymes significantly enhance yields in chemical manufacturing by lowering costs and raising productivity. Relative to the other two spheres, white (i.e., industrial) biotechnology applications appear to be far more geographically dispersed than those of red biotechnology. For example, while industrial biotechnology applications are found in the United States, leading users of these technologies are also located in Denmark, Japan, and Finland. Over the past few years, increased interest in biofuels and biotechnology solutions for the energy industry has greatly increased the level of policy interest in this third sphere of the biotechnology industry.

In the remainder of this section, we emphasize some of the distinctive features of the industry, each of which will influence the ultimate geographic dispersion of activity within the industry.

The Nature of Biotechnology Research

One of the most distinctive and pervasive characteristics of innovation in biotechnology is *duality*. Duality arises when biotechnology research makes a simultaneous contribution to both basic research and applied innovation (Rosenberg, 1974; Stokes, 1997). For example, the developments in recombinant technology and cloning in the 1970s and genomics in the 1990s allowed scientists to understand the fundamental mechanisms of gene expression and also served as the foundation for novel therapies, diagnostics, transgenic crops, biofuels, and so on. The impact of duality is extensive and undermines some of the implications of the traditional linear framework for science, technology, and innovation.[2] While the linear framework allows for a concise formulation of the relationship between the nature of knowledge and the incentives provided for its production and distribution, it fails when knowledge has both basic and applied value. Stokes (1997) reformulated the traditional linear distinction between basic and applied research by highlighting the duality of research; a discovery could simultaneously have both applied and basic characteristics (Figure 1). Stokes identified the importance of research in "Pasteur's Quadrant": Louis Pasteur's research on fermentation simultaneously offered fundamental insights that led to the germ theory of disease and was of immediate practical significance for the French beer and wine industry. Stokes argues that, rather than placing research on a single linear dimension ranging from basic to applied, it is more useful to consider two dimensions: in terms of whether research is dependent on "considerations of use" or, separately, on a "quest for fundamental understanding." Most biotechnology research takes place in Pasteur's Quadrant—individual discoveries both rely on and have influence on science and commercialization.

The production of "dual-purpose" knowledge, particularly in the disciplines

---

[2]In the traditional "linear" model, the norms and institutions supporting the production and use of basic versus applied research are separable and distinct. Under this model, applied research *exploits* publicly available basic research as an input, transforming that knowledge into innovations with valuable application. Although the linear model has been sharply criticized (Klein and Rosenberg, 1986), most formal theoretical and empirical economic research remains premised on the linear model, from assessment of the impact of university research (Jensen and Thursby, 2001; Mowery et al., 2001; Narin and Olivastro, 1992; Zucker et al., 1998a,b) to the impact of science and basic research on economic growth (Adams, 1990; Romer, 1990).

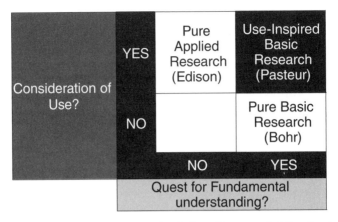

FIGURE 1 Pasteur's Quadrant. The traditional "linear" framework fails when knowledge has both basic and applied value. Since its inception, biotechnology research has been at the center of Pasteur's Quadrant, and so individual discoveries rely on and influence both science and commercialization. SOURCE: Adapted from Stokes (1997).

that underpin modern biotechnology, raises important new challenges for policy makers. For example, the past decade has seen a significant rise in the use of intellectual property rights (IPRs) over research that had traditionally been disclosed only through scientific publication. The increased role of IPR has sparked a vigorous academic and policy debate over the "anticommons effect." On the one hand, some argue that such expansions of IPRs (in the form of patents or copyrights) "privatizes" the scientific commons, reducing the benefits from scientific progress (Argyres and Liebskind, 1998; David, 2004; Heller and Eisenberg, 1998; Murray and Stern, 2007). On the other hand, a significant amount of research suggests that IPRs may also facilitate the creation of a market for ideas, encourage further investment in ideas with commercial potential, and mitigate disincentives to disclose and exchange knowledge that might otherwise remain secret (Arora et al., 2001; Gans and Stern, 2000; Merges and Nelson, 1990, 1994; Lerner and Merges, 1998). While there are many questions surrounding the use and misuse of IPRs, particularly at the interface between university and industry research, its availability may allow startup biotechnology firms to focus on the early-stage research and contract with pharmaceutical, agricultural, and chemical companies for downstream activities, including manufacturing, marketing, and distribution (Arora et al., 2001; Gans and Stern, 2003).

The Biotechnology Value Proposition and the Structure of the Value Chain

While the size of the biotechnology industry is still quite modest—relative to, say, employment or revenue in the automobile industry—the potential global demand for biotechnology products is large, mostly driven by the needs of a growing and aging world population. The promise of biotechnology to find solutions to some of the critical problems arising from population growth and demographic change, from new medical treatments to improving agricultural output and developing new sources of energy, creates a favorable environment for this sector. The world's population is not only growing, but is, in aggregate, growing older.[3] As life expectancy increases, a need to find new approaches to treat chronic diseases that affect a more elderly population will increase. At the same time, rising global trade and travel, highly porous international borders, increased urbanization, and an uneven distribution of wealth are creating optimal conditions for outbreaks of new infectious diseases with no available treatments. Similarly, the need to increase the productivity and efficiency of agricultural products to feed the rising population is becoming a critical global issue for which biotechnology may offer important solutions. The pressing need for new treatments is creating a great demand for biotechnology innovations. Likewise, global climate change, caused in part by economic development and population growth, has intensified the need for finding solutions for alternative sources of energy. Industrial biotechnology could provide some means of producing environmentally friendly biofuels.

Despite these promising opportunities, the industry faces a series of distinctive challenges in translating innovations into commercialized products and services for global markets; at least in part, these challenges are a consequence of duality. On the one hand, close interinstitutional collaborations in biotechnology contribute to the need for geographic proximity around centers of research excellence. Moreover, one manifestation of the complex networked relationship between biotechnology firms and other institutions is that many researchers in biotechnology work not only at the convergence of multiple scientific fields but also at the boundaries of multiple institutions. While these overlapping institutional affiliations are most apparent in the area of health-oriented biotechnology (Zucker et al., 1998b), agricultural and industrial biotechnology innovation also

---

[3]Demographic projections estimate world population gains from 6.5 billion in 2005 to 7.9 billion in 2025 (United Nations, 2004). The greatest growth in total population is projected in the rising nations of China and India, whose populations are expected to benefit from improved socioeconomic conditions and should drive increased needs for biotechnology innovations. The global population is also growing older. Individuals over age 60 represented 10.4 percent of the world's population in 2005; by 2050 this segment is expected to grow by 1 billion, with a total number representing 21.7 percent of a much larger total population. This trend will undoubtedly spur greater demand for new biomedical innovations and treatments worldwide. Today, the U.S. population over age 65 consumes 40 percent of the nation's biomedical output products and it is reasonable to expect similar trends worldwide. Persons aged 60 and over comprised 10.4 percent of the global population in 2005; by 2050 this component will amount to 21.7 percent of a much larger total population. By midcentury, the number of persons aged 60 and older will grow by 1 billion. The greatest advance is expected in the rising nations of China and India, whose populations will come to benefit from drug treatments and medical devices formerly available mainly to consumers in the United States and Europe (Magee, 2005).

takes place at the university-industry interface (Graff et al., 2003). Biotechnologists often need to have both scientific and commercialization acumen; they work for and with multiple organizations and institutions.

At the same time, while proximity to scientific and commercial knowledge led to the rise of concentrated geographic clusters for biotechnology innovations, the jobs created by the products of these innovations are far more dispersed. In each of the three areas of biotechnology, the value chain is highly fragmented and requires significant capital expenditures, meaning that an entrepreneurial innovator can rarely afford or find it worthwhile to commercialize an innovation independently all the way to market. As a result, the downstream users of biotechnology (e.g., physicians, farmers, or industrial managers) may have only limited if any interactions with the initial innovators or research teams. As a consequence, in each of the three segments of biotechnology, the location of innovation may be very different from the location of application and greatest use.

This pattern is most apparent in red biotechnology (Figure 2). Close connections with university and public researchers, as well as more geographically dispersed relationships with those that commercialize innovation, have contributed to a highly entrepreneurial structure in red biotechnology. This structure, combined with the presence of multiple revolutions in science and technology, has kept the industry in a state of "perpetual immaturity." The continuous flow of scientific innovations and the fragmentation of the value chain encourage the biotechnology sector to create new companies continuously. Since its inception and looking across all three industry segments, the biotechnology sector had around 1,300 companies in the United States and around 5,000 worldwide (Burrill & Company, 2004). Although successful individual biotechnology companies in the health-oriented sector have grown from startups to large firms—Genentech and Amgen being the prime examples, each with a market cap in excess of $50 billion—the sector as a whole is a study in dynamism, with new entrants appearing on the scene every year, attracting capital from both public and private sources. Once companies in the red biotechnology sector establish a proven commercial path, they often consolidate or partner with established companies for development and distribution. Consolidation, however, does not result in a gradual winnowing of companies. This trend is offset by the continuous rate of company formation that keeps the sector fragmented, particularly in health-oriented applications. The biotechnology supply chain is filled with specialized players. Firms often do not integrate vertically but instead continue to play within specific and limited stages of the value chain.

Though not as extreme as red biotechnology, green and white biotechnologies are also characterized by a reliance on the combination of university research, startup innovators, and established firms. For example, Monsanto, the leading agricultural biotechnology firm, initiated its efforts to diversify from its agrochemicals business through the establishment of research partnerships with leading universities such as Washington University in St. Louis (Culliton, 1990;

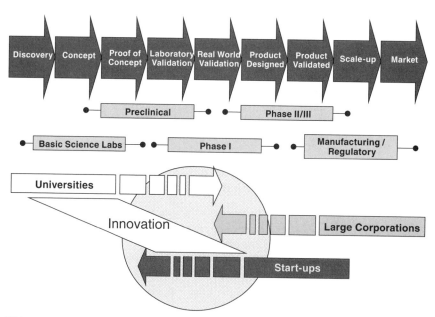

FIGURE 2 Typical value chain for a biotechnology product. Commercialization takes many steps, and, while there is geographic confluence between universites and startups, the value chain is both complex and fragmented. Biotechnological product development in biotechnology is a long and fragmented process. For example, it is estimated that an agricultural biotechnology product might take 10 years to bring to the market and an investment of $50 million to $200 million (McElroy, 2004). Similarly, a drug might take about 12 years and around $800 million (DiMasi et al., 2003). Rarely the innovator has the resources to bring the product to the market and outlicense or sell their technology to a large pharmaceutical company, which can more feasibly undertake the most expensive development (i.e., approval) phases. The value chain is fragmented with smaller companies specializing at the innovation and discovery stages and larger companies specializing in the development and distribution stages.

Nelkin et al., 1987). Since that time, Monsanto has developed significant in-house research and commercialization capabilities in agricultural biotechnology and relies on an extensive network of strategic partnerships and licensing relationships. In other words, although large established companies such as Monsanto and DuPont are ultimately responsible for the commercialization of agricultural biotechnology innovations, the origins of those innovations are divided among university research projects, startup innovators, and internal development (Pierre-Benoit, 1999). A similar pattern, but one that is less documented in the academic and business literature, is the case for industrial biotechnology, although there seems to be a smaller role for the university sector. For example, Hermans, Kulvik, and Tahvanainen (2006) document the licensing and alliance relationships between startup innovators and more established users of industrial biotechnology products in the Finnish context.

Overall, across the three distinct segments, the industry exhibits a highly dynamic structure. The dynamism of the biotechnology industry is based on its foundations in rapidly emerging scientific disciplines; its potential to address important social needs while creating significant commercial value in health, agriculture, and industry; and its orientation in terms of the commercial application of knowledge that is simultaneously of independent scientific interest.

THE DRIVERS OF LOCATION AND CLUSTERING OF BIOTECHNOLOGY INNOVATION

As mentioned earlier, the drivers of the geography of biotechnology industry and innovation are complex and changing over time. On the one hand, the geography of biotechnology reflects broad factors relating to the overall orientation of an economy to support innovative activity. The geography of innovation of the biotechnology industry is consistent with the role of location and institutions emphasized in the national innovation systems literature (Lundvall, 1992; Mowery and Nelson, 1999; Nelson, 1993). This national innovation systems literature focuses on the role that national policies and local institutions play in shaping the location and effectiveness of innovative productivity and emphasizes the important preconditions that must exist for innovative investment to be effective. Such policies and institutions include an effective intellectual property system, the availability of high-quality human resources and risk capital, and institutions (and public-private partnerships) that encourage investment and innovation in particular regions.

While the aggregate national innovation system sets the basic conditions for innovation, the development and commercialization of new technologies take place, disproportionately, in clusters—geographic concentrations of interconnected companies and institutions in a particular field. Over the past two decades, there has been an explosion of research concerning the structure and dynamics of industrial clusters and the role of location in industrial activity and innovation (see, among others, Breschi and Malerba, 2005; Krugman, 1991; Porter, 1990, 1998; Audretsch and Feldman, 1996; Saxenian, 1994).

The biotechnology industry has been of particular interest to research on industrial clusters, for several reasons.[4] First, *within* the United States, biotechnology companies and investment are clustered in a small number of regions, such as San Diego, Boston, and San Francisco (see Figure 3). Moreover, the activities and investments by companies in biotechnology clusters are focused on

---

[4]The case study and empirical literature on the regional clustering of innovation activities in biotechnology is quite large and cannot be adequately reviewed here. For a very useful recent review, see Cooke (2002).

FIGURE 3 Biotech clusters in the United States. The colored states indicate where there are both large and specialized firms in two of the three biotechnology subsectors (pharmaceuticals, research and testing, and medical devices). SOURCES: The Brookings Institution; Cortright and Mayer (2002).

research and innovation (Audretsch and Stephan, 1996). Internationally oriented case studies have also documented that, within individual countries, biotechnology innovation tends to be regionally clustered in other countries (Cooke, 2002; Hermans and Tahvanainen, 2006; Swann et al., 1998). Regional clustering of innovation-oriented activity in biotechnology is particularly striking since companies do not rely on hard-to-access natural resources, and the sciences underlying biotechnology are dispersed at universities and research institutions across the United States and abroad (Audretsch and Stephan, 1996; Feldman and Francis, 2003). A rich and nuanced literature has developed emphasizing some of the key patterns and dynamics associated with biotechnology clusters in the United States and abroad, with an emphasis on the importance of collaboration and networks among universities, startup innovators, and established firms (Koput et al., 1996; Powell et al., 2005), and the crucial role played by scientists who bridge the university-industry divide (Audretsch and Stephan, 1996; Zucker et al., 1998a,b). As well, there has been significant interest in the science policy community in the distribution of research in biotechnology, often focusing on clustering in specific locations around the world (Hoffman, 2008).

To organize our discussion of the clustering of biotechnology activity and integrate the insights of these studies of clustering in biotechnology, we build on Porter's seminal work on clusters (Porter, 1990, 1998). In Porter's "diamond" framework (Figure 4), four attributes of the microeconomic and strategic envi-

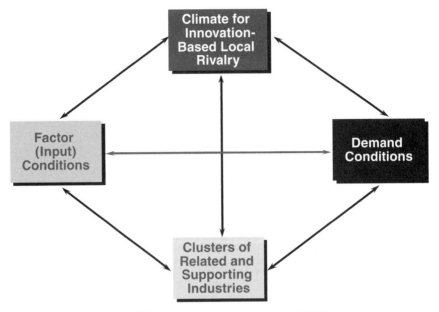

FIGURE 4 The "diamond" framework adapted from Porter (1990).

ronment surrounding a cluster support its overall competitiveness and innovative vitality: (1) the presence of high-quality and specialized inputs, (2) a local context that encourages innovative investment and intense rivalry, (3) pressure and insight emerging from sophisticated local demand, and (4) the local presence of high-quality related and supporting industries. The four elements of the diamond framework highlight the key resources and dynamics associated with the emergence and sustainability of leading clusters in all segments of the biotechnology industry. First, as mentioned earlier, the development of biotechnology innovation requires access to specialized inputs, including researchers, risk capital, biological materials, and even intellectual property. By and large, accessing these resources is most easily accomplished within a regional context, rather than across long distances or political boundaries. For example, the development of the agricultural biotechnology cluster surrounding St. Louis depended on the ability of companies such as Monsanto to draw upon and reinforce the significant expertise and research capabilities of Washington University in St. Louis.

Second, a key driver of effective clustering in the biotechnology sector seems to be competition among locally based biotechnology companies. These companies compete on the basis of attracting talent, publishing high-quality scientific research, and attracting investment and interest from venture capitalists and downstream commercial partners, many of whom are located outside the cluster. This is perhaps most apparent in some of the clusters associated with health-oriented biotechnology; for example, the Massachusetts biotechnology cluster includes more than 400 different firms, 235 of which are developing therapeutic drugs (Massachusetts Biotechnology Council, 2007).

Third, most leading biotechnology clusters are located not only near sources of high-quality basic research but also around areas with significant capacity in clinical innovation. For example, the pressures on the Massachusetts biotechnology cluster arise as much from the presence of demanding clinicians in the leading hospitals as from that of specialized genetics researchers. Similarly, the medical device cluster in Minneapolis is pushed by demanding consumers at the Mayo Clinic and related institutions, and industrial biotechnology innovation in Scandanavia depends in part on demanding customers in the chemical industry (Hermans et al., 2006).

Finally, the biotechnology cluster depends on the presence of related and supporting industries, most notably an active venture capital industry to supply managerial expertise, risk capital, and relationship experience with downstream partners as well as key pieces of infrastructure (e.g., biological resource centers, specialized seed banks and agricultural research stations, specialized equipment and tools). Each of these factors encourages the investment of sunken assets and the development of specialized capabilities that reinforce the strength and ultimately the international competitiveness of that cluster environment.

When these factors are present, geographic clustering promotes important externalities in innovation that are relevant to biotechnology. Thus, while location within a cluster enhances a firm's ability to identify opportunities for innovation, it also promotes the firm's flexibility and capacity to bring new ideas to market. Within a cluster, a company can more rapidly assemble the components, machinery, and services necessary for commercialization. Suppliers of essential inputs and "lead" buyers become crucial partners in the innovation process, and the relationships necessary for effective and efficient innovation are more easily forged among proximate firms. Reinforcing these advantages for innovation within clusters are competitive, peer, and customer pressures associated with the proximity of other, often directly competing, biotechnology firms. Clustering enables easy comparisons of performance.

As would be expected, the innovation environment of a cluster is fundamental to its competitiveness. For example, the biotechnology sector serving the needs of the Scandinavian pulp-and-paper cluster benefits from the advantages of pressures from demanding domestic consumers, intense rivalry among local competitors, and the presence of Swedish process-equipment manufacturers that are global leaders (e.g., Kamyr and Sunds, for the commercialization of innovative bleaching equipment). The Finnish pulp-and-paper industry utilizes specific biotechnological techniques in its production processes, which has also partially motivated industrial enzyme providers to construct production plants in Finland. As a consequence, enzyme applications form the largest sales within the small- and medium-sized biotechnology industry in Finland (Hermans et al., 2006).

Similar examples of cluster vitality in innovation may be observed in many fields, from pharmaceuticals in the United States to semiconductor manufacturing in Taiwan. Cluster vitality is derived from industries and resources as diverse as the fields they support. The pharmaceutical industry, of course, provides a strong source of support for biopharmaceutical innovative activity. In particular, though the natural resource requirements of the pharmaceutical industry are limited, the industry has been geographically concentrated in a small number of regional locations, including New Jersey and Switzerland, partially as a result of competition for pharmaceutical demand.

Porter's cluster framework is useful for identifying some of the key drivers of international competitiveness for an innovation-oriented biotechnology cluster at a point in time. However, the durability of cluster-driven competitiveness depends on the dynamics and evolution in the scope of clusters over time. While there is less research that focuses specifically on biotechnology clusters in this regard, recent research on clusters and in economic geography emphasizes the key factors that shape the persistence of location and the dynamics of the geography of innovation-oriented industries (e.g., Krugman, 1991; Krugman and Venables, 1995).

The Convergence Effect

Though leading clusters may stay at the forefront of innovation and activity for long periods of time, the advantages of cluster leadership are balanced against the possibility of relatively low-cost, high-growth entry by other regions (Barro and Sala-i-Martin, 1992; Dumais et al., 2002; Henderson et al., 1995). Although the effect of convergence declines as the emerging regions develop in terms of size and sophistication, the convergence effect is an important consideration when examining changing patterns of geographical investment, employment, and innovation in the biotechnology industry.

Maturity and the Dynamics of Geographic Dispersion

As industries mature and products and processes are standardized and decomposed, the benefit of localized innovation may be outweighed by cost advantages in lower-wage locations (Brezis and Krugman, 1997; Duranton and Puga, 2001). In other words, one of the key signs of industry maturity and commoditization is a significant increase in the geographic dispersion of industrial activity. For example, in recent years the *diffusion* of agricultural biotechnology products has had the consequence that the industrial activities of agricultural biotechnology are more geographically dispersed than the industry was at an earlier stage. While the development of new products continues to be centered in a small number of locations, investments in improvements in how to *use* the new seed traits and how to adapt farming practices to incorporate the new products are much more geographically dispersed.

National Versus Regional Clusters

For medium and large economies, the presence of a cluster will likely be most apparent at the subnational and even local levels (e.g., Martin and Rogers, 1995; Monfort and Nicolini, 2000). This is particularly true in the biotechnology industry, where total global employment is quite modest, and the industry is located in proximity to leading university research areas. While data limitations will often force us to examine dispersion at the national level, it is important to keep in mind that much of the clustering in the biotechnology industry takes place at the regional or even municipal level.

The Emergence of International Hubs

Finally, as mentioned earlier, the basic research upon which the biotechnology industry is built is widely dispersed across thousands of universities on a global basis. As a result, one of the more subtle dynamics of clusters—hubbing—may be particularly important in this context. Specifically, it is possible that a location with a strong cluster environment will not only benefit from the dynamics arising from local relationships but will become a "magnet" for those who are located in less-favorable cluster environments (Krugman and Venables, 1995; Venables, 1996). For example, a researcher at Johns Hopkins University might maintain a relationship with a company in Boston or Silicon Valley to take advantage of the potential for commercialization in a strong cluster environment. Going forward, there will be tension between the continuing importance of forces that have given rise to agglomeration and clustering and the increasing salience of activities aimed at bridging across clusters and even across national boundaries.

THE GEOGRAPHY OF THE BIOTECHNOLOGY INDUSTRY

We build on these insights into the geography of the biotechnology industry by undertaking a short description of the global distribution of innovation-oriented activities within the biotechnology industry. As described earlier, the industry grew out of a series of fundamental scientific breakthroughs in the 1970s and was initially concentrated among a small number of entrepreneurial firms, mostly in the Bay Area in California and around Cambridge, Massachusetts. Despite interest in the *future* of biotechnology, relatively little attention has been paid to the current state of the biotechnology industry in terms of regional patterns of employment, investment, and firm creation.

Employment

In Table 1 and Figure 5, we describe the international distribution of employment as of 2005. Drawn from Van Beuzekom and Arundel (2006), Figure 5 reports the total employment of individuals in private biotechnology firms (i.e.,

TABLE 1 Biotechnology R&D Employees by Country

Biotechnology R&D Employees (Headcounts, 2003)[a]	Total	Per Firm
United States[b]	73,520	33.5
United Kingdom[b]	9,644	
Germany (2004)	8,024	13.2
Korea (2004)	6,554	10.2
Canada[c]	6,441	13.1
Denmark[d]	4,781	17.9
France[b]	4,193	5.6
Switzerland (2004)[b]	4,143	26.4
Spain (2004)	2,884	10.4
Sweden[b]	2,359	10.9
Belgium	1,984	27.2
Israel (2002)	1,596	10.8
China (Shanghai)[d]	1,447	9.2
Finland[b]	1,146	9.3
Ireland[b]	1,053	25.7
Iceland	458	19.9
Norway[b]	283	8.8
Poland (2004)	109	8.4

[a]R&D employment: includes scientists and support staff such as technicians.
[b]Data from Critical I Report to the UK DTI (2005), based on all R&D employees in core biotechnology firms.
[c]Excludes firms with less than five employees or less than PPP $80,000 in R&D.
[d]Full-time equivalents (FTEs).
SOURCE: OECD (2006).

outside of universities and the public sector). While the definition of which firms are included in the "biotechnology" industry varies across countries (e.g., the United Kingdom reports only companies from "core" biotechnology firms, whereas Germany reports employment for any worker with biotech-related responsibilities), the most striking fact about these statistics is the small absolute size of the biotechnology industry. Using statistics collected in 2006, the global biotechnology industry directly employs less than 500,000 workers in the private sector.[5] While the industry may also support employment in related industries (e.g., pharmaceuticals, agriculture, or industrial engineering), there are no systematic survey data about the global distribution of these employment spillovers, and so we restrict our attention to the biotechnology industry per se. To put these numbers in context, the global automobile industry employs more than 8.4 million workers (International Organization of Automobile Manufacturers, 2007).

---

[5]The OECD statistics do not list aggregate employment from Japan, but, based on other (though not strictly comparable) figures, Japanese employment is likely less than that of the EU, which registered at 73,189 as of 2005.

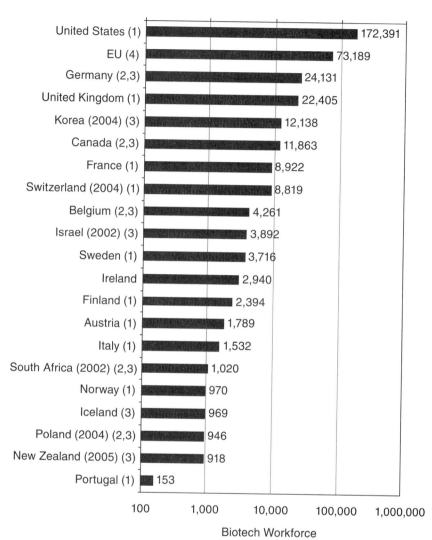

FIGURE 5 International labor distributions: (1) Data from Critical I report to the UK DTI, 2005, based on total employment in core biotechnology firms. (2) Limited to employees with biotech-related responsibilities. (3) Includes employment in both core and non-core firms active in biotechnology. (4) EU11 countries presented in this figure. SOURCE: OECD Biotechnology Statistics (Van Beuzekom and Arundel, 2006).

Figure 5 offers several striking patterns about the global distribution of biotechnology activity. First, the United States supports by far the single largest biotechnology industry. Even though the European Union (EU) has a larger total population than the United States, overall biotechnology employment is less than 50 percent that of the United States. Moreover, within the EU, the distribution of employment is highly uneven, with Germany and the United Kingdom accounting for two-thirds of total EU employment. Finally, it is useful to note that a number of small countries support significant biotechnology employment, with an employment intensity exceeding even that of the United States. For example, although Iceland has a population of only 300,000, the biotechnology industry accounts for nearly 1,000 jobs.

Table 1 provides a complementary perspective and reports biotechnology research and development (R&D) employment by country (and the number of R&D employees per firm, which we return to later). Two points are of specific interest. First, the ratio of R&D employment to total employment is extremely high; for most countries, the ratio is approximately 0.3-0.5; in other words, more than one of out every three biotechnology employees is a biotechnology researcher. This contrasts with an overall R&D employment intensity of less than 3 percent across all industries in the United States (National Science Board, 2006). With that said, it is once again important to emphasize the small absolute size of the biotechnology research workforce. Excluding the United States, no country maintains a biotechnology R&D workforce in excess of 10,000 researchers. In other words, while the absolute size of the biotechnology workforce is comparatively small, Table 1 and Figure 5 suggest that the international distribution of overall employment captures significant international differences in the distribution of labor-intensive innovation activities.

Enterprises

Figure 6 extends the analysis to the global distribution of *companies*. While the United States remains the largest single national home for biotechnology activity, it is useful to note that the EU actually accounts for a greater number of companies than the United States. Along with the earlier employment statistics, this suggests that individual EU biotechnology companies have fewer employees (on average) than their U.S. counterparts. Simply put, the scale of operations for a typical EU biotechnology firm is smaller than that of a biotechnology firm in the United States. This point is illustrated in the second column of Table 1, which records the number of R&D employees per firm by country. While the United States employs more than 33 scientists per firm on average, most EU countries employ from 5 to 15 researchers. The smaller scale of firms extends to countries in other parts of the world as well: Israel, South Korea, and China employ from 9 to 11 scientists per firm.

While some of these differences arise from the fact that the United States

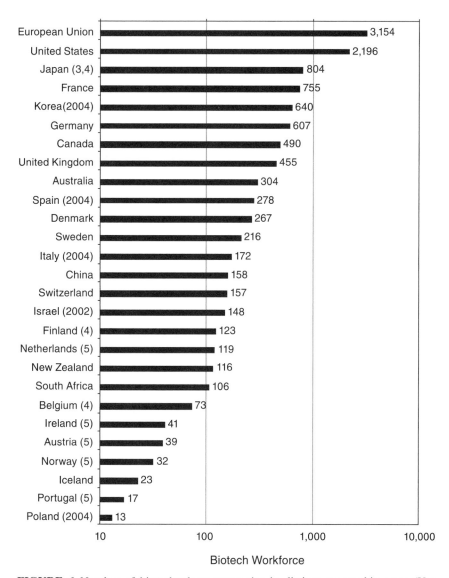

FIGURE 6 Number of biotechnology companies in distinct geographic areas (Van Beuzekom and Arundel, 2006; OECD Biotechnology Statistics).

maintains a few *very* large biotechnology companies (e.g., Amgen, Genentech), these differences in average firm size seem to reflect differences in the origin and maturity of these companies. For example, according to Europe's Biotechnology Industry Association, EuropaBio, Europe has a higher number of younger companies but fewer mature companies than the United States (Critical, 2006).

Furthermore, the European biotechnology companies seem to grow more slowly than their U.S. counterparts. By and large, young European firms are often overtaken by international competitors and even some of the oldest European biotechnology companies have been acquired by U.S. companies that have better access to financial and commercialization resources (Critical, 2006). As in the employment statistics, this concentration of small companies seems to reflect the international distribution of employment activities. Most companies reporting that consider themselves to be part of the biotechnology industry focus on innovation-oriented activities, and Figure 6 highlights the role of entrepreneurial firms in this industry.

R&D Expenditures

Finally, we examine the global distribution of R&D expenditures in Figures 7 and 8. Perhaps even more so than with the employment and enterprise activity statistics, these statistics are only partially informative data since they only measure biotechnology-related expenditures for a subset of firms that are focused in biotechnology (i.e., defined as a "core" biotechnology firm for many of the countries), and the data seem to primarily cover firms specialized in health-oriented biotechnology. While these data are flawed and should not be taken to offer a precise measure of the level of expenditures, they are still useful for highlighting broad contrasts across different countries and regions. In particular, as with the employment statistics, investment expenditures are concentrated in the United States (see Figure 7); in 2003, the U.S. investment pace is an order of magnitude higher than for any other individual country. Interestingly, there are some important differences in the investment levels of countries, relative to their employment levels. For example, while France registers a much lower level of biotechnology employment than Germany, the level of investment expenditures is similar between the two countries; moreover, this reflects real differences in investments because both countries share a common currency. As well, modest but still significant employment statistics for several countries outside the United States and Europe are also reflected in R&D investment: New Zealand and Australia support a relatively high level of investment activity, as do Asian countries such as South Korea and China (although China obviously has a very small industry on a per capita basis).

This skewed pattern of global biotechnology investment is reinforced in Figure 8, which reports the distribution within the OECD of venture capital investments. If anything, the investment bias toward the United States is even more apparent (venture capital investments are more than 12 times as large in the United States than in the second-largest target country, Germany). To the extent that venture capital funding also offers a particularly effective model for funding innovation (Kortum and Lerner, 2001), this skewed distribution of financial

FIGURE 7 Total expenditures for biotechnology R&D by biotechnology-active firms, OECD biotechnology statistics (Van Beuzekom and Arundel, 2006).

investment suggests that the United States may be extending its historical dominance in the creation and evolution of biotechnology enterprises.

Overall these empirics highlight three very important findings about the geographic distribution of innovative activity in the biotechnology industry. First, this is an industry of relatively modest size by any measure. The absolute level

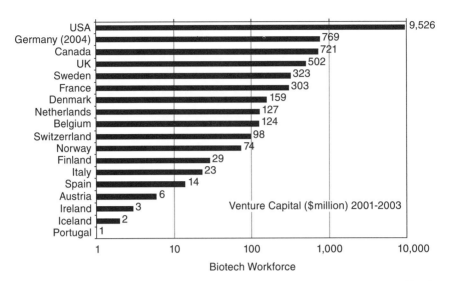

FIGURE 8 Total venture capital investments in biotechnology, 2001 to 2003 combined, OECD biotechnology statistics (Van Beuzekom and Arundel, 2006).

of employment is relatively low, and the total expenditures (and even the number of entrepreneurial firms) are relatively low in terms of absolute value. Second, relative to other industries, measures of employment and entrepreneurship activity are closely linked to innovative activity. More than one-third of all employees in the global industry are considered R&D employees (much higher than in most other industries) and most companies are focused on innovation-oriented strategies and investments. Finally, although there is an increasingly dispersed global set of locations for biotechnology (e.g., United Kingdom and Germany, South Korea, Australia), the overall level of the United States in terms of total activity is much higher.

THE GEOGRAPHIC DISTRIBUTION OF INNOVATIVE OUTPUT IN BIOTECHNOLOGY

In this section, we move beyond the general patterns of the geography of biotechnology to examine global patterns of innovative performance. Attempts to measure and benchmark innovative outputs have become common across advanced economies.[6] One approach to this activity (Furman et al., 2002; Porter and Stern, 1999) is based on a clear distinction between measures of the outputs

---

[6]A review of this process is beyond this chapter's scope. However, a good starting point is the benchmarking programs of the EU (http://trendchart.cordis.lu/).

of technological innovation (for example, international patenting) and its drivers: infrastructure, clusters, and linkages.[7] While one must be very careful in interpreting patterns based on patent data, patenting trends across countries and over time are highly likely to reflect actual changes in innovative outputs rather than spurious influences, especially in measuring innovation at the global level. Also, international patenting captures the degree to which a national economy develops and commercializes internationally important new technologies—a prerequisite for building international competitiveness on a platform of quality and innovation. In short, international patenting is "the only observable manifestation of inventive activity with a well-grounded claim for universality" (Trajtenberg, 1990).[8] With that said, our analysis of international patenting in biotechnology comes with several important caveats. In particular, the standard for patentability for many biotechnology-related innovations differs across countries (and across time within countries). To cite one example, as of 2006, the United States had granted more than 40 human embryonic stem cell patents, whereas the European Patent Office (EPO) had granted none due to an EU directive to reject human embryonic stem cells patents on "moral" grounds (Porter et al., 2006). While U.S. patent office practice has tended to allow patents that are relatively close to the arena of pure scientific "discoveries," EPO practice has tended to only allow patents when a specific industrial application has been identified. More generally, the use of patent data to identify the geography of innovation is of course limited by the fact that many innovations (even important innovations) are not patented or patentable; although this critique is particularly important in the context of a broad cross-industry study, biotechnology is an arena with a close connection between innovation and patenting (Cohen et al., 2000). With these caveats in mind, we now turn to a detailed discussion of international patterns of biotechnology patenting.

Global Biotechnology Patenting

We use several different measures reflecting the number of international biotechnology patents. In particular, we focus on the number of patents granted

---

[7]In addition to patent counts, there are some alternative measures to illustrate the distribution of biotechnology innovations. For instance, other forms of intellectual capital could also be useful to measure. On the one hand, some forms of human capital are often held as critical success factors in the science-driven business, such as outcomes of scientific research and a level of education and business experience of employees. On the other hand, the measures related to relational capital, such as collaboration networks, would be useful in assessing the significance of location of the biotechnology industry (Edvinsson and Malone, 1997).

[8]Trajtenberg (1990) provides a thorough discussion of the role of patents in understanding innovative activity, referring to their early use by Schmookler (1966) and noting their increasing use by scholars (e.g., Griliches, 1984, 1990, 1994). Our use of international patents also has often been used as a precedent in prior work comparing inventive activity across countries (see Dosi et al., 1990; Eaton and Kortum, 1996).

to inventors from a given country by the U.S. Patent and Trademark Office (USPTO), the EPO, and the Japanese Patent Office (JPO). We then combine these measures in our analysis by examining the number of triadic patent families (i.e., patents granted in each of the three major patent jurisdictions).

Figure 9 graphs the number of biotechnology patents issued by the USPTO and EPO, by the region of origin of the inventor.[9] Several striking patterns stand out. First, the United States is the dominant country of origin for biotechnology innovations, even those that are patented in Europe (i.e., where the "home bias" would favor the European inventors). Second, there was a sharp increase in U.S. biotechnology patenting by U.S. inventors during the late 1990s, a trend that is partially reflected in the EPO data and partially ameliorates from 2000 onward. USPTO patents with European inventors are associated with a much more gradual rise and achieved a 20 percent share of USPTO biotechnology patents by 2003.

Clearly, the regional patenting patterns reflected in the USPTO and EPO figures reflect a "home bias"; inventors tend to prefer domestic patent offices to foreign ones (as documented and discussed in detail by Criscuolo [2006]). This suggests at least in part that domestic biotechnology companies tend to apply for patents first in their domestic patent office and only seek foreign patents for their most significant and valuable products and processes. We attempt to address the home-bias problem by moving toward triadic patent family counts to perform more strict comparisons among biotechnology patents filed in the USPTO, the EPO, and the JPO.[10] Triadic patent families provide a more valid proxy for the economic value of patents. Patent application processes differ by country; most companies or individuals will undertake the costly process of filing a patent abroad only if the invention or process in question has significant earnings prospects.

When we turn to triadic patent families in Figure 10, a similar set of patterns emerges. The United States continues to have a dominant share, on both an absolute and a per capita basis. Furthermore, when we calculate patent per capita estimates, Japan's innovative productivity appears to be at the same level or higher than that of the EU. It is useful to note that, on a per capita basis, the

---

[9]Unfortunately, we are unable to obtain disaggregated data on assignee location using the Derwent database. The location of the inventors is often distinct from the location of the assignee of the patent. This distinction may confound an analysis of global sourcing of R&D activity. The decrease in the share of U.S. inventors described earlier does not necessarily imply a decrease in U.S. patent ownership. Therefore, we cannot rule out a geographical shift in R&D activity by U.S.-based firms. However, given the strong entrepreneurial element of the biotech industry and the findings of the OECD 2007 Compendium of Patent Statistics, biotechnology R&D work does not seem to be heavily outsourced from the United States at this point. However, constructing a database of triadic biotechnology patents by assignee country is a strong priority for future research in this area.

[10]Eurostat defines triadic patent families as follows: "A patent family is a set of patents taken in various countries for protecting a single invention. . . . Patent is a member of a triadic patent family if and only if it is filed at the European Patent Office (EPO), the Japanese Patent Office (JPO) and is granted by the US Patent & Trademark Office (USPTO)."

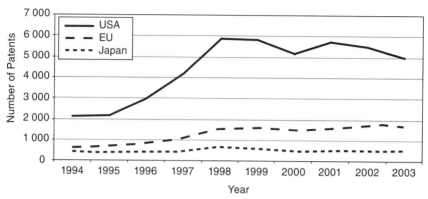

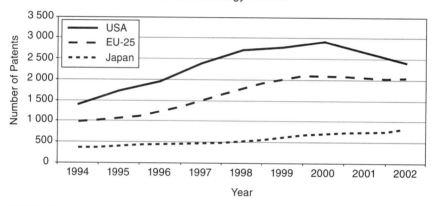

FIGURE 9 Biotechnology patent counts in USPTO and EPO by inventor's country of origin. SOURCE: OECD.

United States has only about two times the innovative capacity of Japan and the EU. Perhaps more important, these patterns provide some interesting insights into the evolution of the global biotechnology industry over the past decade or so. In particular, despite the fact that countries outside the United States started from a very low level of activity (and may benefit from the "convergence effect"), the gap between the United States and the rest of the world has persisted. While there has been a very slight convergence in the last years of our data (i.e., applications from 2000 onward), these broad patterns are consistent with the hypothesis that regional agglomeration remains an important driver of the geography of the biotechnology industry.

There are several potential explanations for this continued persistence. First,

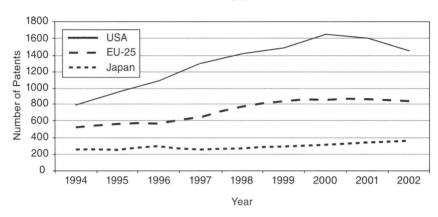

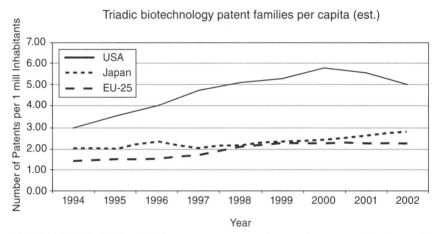

FIGURE 10 Triadic biotechnology patent counts and per capita measures by inventor's country of origin. SOURCE: OECD and authors' calculations.

and perhaps most important, the extremely rapid advances in the scientific and technological frontier in biotechnology likely reinforce the strengths of preexisting clusters, such as San Diego and Massachusetts. In contrast to environments where a single "macro innovation" diffuses first locally and then globally (resulting in convergence in incremental innovation over time), biotechnology innovation remains "perpetually immature." Second, the scale of private and public research funding in the United States continues to be very large relative to any other individual country or region. The National Institutes of Health has experienced rapid increases in its funding, and this seems to have been a complement, rather than a substitute, for venture capital and private investment. Finally, even though there are an increasing number of biotechnology innovation clusters around the world that are operating at least at "minimum scale," the United States has benefited from an environment that by and large encourages the commercialization of new biotechnology products. This is perhaps most apparent in the agricultural sector, where the strength of clusters is probably less salient than in health-oriented biotechnology, but resistance to development and commercialization efforts in Europe have allowed the United States to establish and maintain a dominant position.[11] Together, while future U.S. leadership will depend on the continued vitality of cluster environment, these patterns suggest that the United States has by and large provided a favorable environment for biotechnology innovation.

Global Biotechnology Patenting by Application Segments

We now provide a more detailed analysis of innovative output as measured by patent counts based on inventor location, which are divided into 12 patent subcategories by the same regions considered earlier. Our analysis utilizes Derwent biotechnology abstracts, the most widely utilized classification system for biotechnology patent analyses (Dalpé, 2003). Table 2 presents biotechnology patent counts and regional shares from 2000 to 2003. Note that we are no longer looking only at patent triads.

While the overall results reflect our more aggregate findings (i.e., the United States as a dominant player), Table 2 also reveals some striking differences across industrial applications. U.S. leadership in biotechnology is centered on the patent classes most closely related to red biotechnology. More than 75 percent of all U.S. patents are in the categories genetic engineering and fermentation, pharmaceuticals, and cell culture. While these classes are also active in the portfolio of the EU and Japan, an important share of patenting activity by the EU and Japan is in classes associated with green and white biotechnology. These patterns of comparative advantage can be seen most clearly when we calculate the share of patenting recorded by each region within each industrial application. We define comparative advantage as those patent classes with a higher share of domestic patenting than the country's share of the total number of biotechnology patents.[12] For example, the United States has a comparative advantage (as indicated by the boldface entries) in the classes for which it holds over 55.4 percent of all

---

[11]This pattern may be reversed in the case of stem cells, where restrictions on U.S. federal funding of early-stage embryonic stem cell research have spurred numerous international initiatives to attract key scientists and create a favorable cluster environment for stem cell commercialization efforts.

[12]The formal condition for flagging a quotient is $\frac{P_{ij}}{P_j} > \frac{P_i}{P_{total}}$, where P is the number of patents, i denotes the country, j indicates the application area, and *total* stands for the entire number of biotechnology patents within the period 2000-2003 in Derwent Biotechnology Resource.

TABLE 2 Patent Counts and Share of Patents in Biotechnology Patent Classes, 2000-2003

Code	Patent Class	U.S.	EU15	JP	Total
A	Genetic engineering and fermentation	7,125	2,671	1,655	12,138
		58.7%	22.0%	13.6%	
B	Engineering, biochemical engineering	196	166	103	479
		40.9%	34.7%	21.5%	
C	Sensors and analysis	124	77	55	245
		50.6%	31.4%	22.4%	
D	Pharmaceuticals	5,564	1,978	1,110	9,250
		60.2%	21.4%	12.0%	60.2%
E	Agriculture	1,249	391	236	2,010
		62.1%	19.5%	11.7%	
F	Food and food additives	260	286	186	712
		12.1%	7.6%	9.0%	
G	Fuels, mining, and metal recovery	44	66	45	171
		25.7%	38.6%	26.3%	
H	Other chemicals	160	204	176	504
		31.7%	40.5%	34.9%	
J	Cell culture	1,058	423	249	1,779
		59.5%	23.8%	14.0%	
K	Biocatalysis	593	548	492	1,604
		37.0%	34.2%	30.7%	
L	Purification, downstream processing	54	52	16	127
		42.5%	40.9%	12.6%	
M	Waste disposal and the environment	122	185	232	563
		21.7%	32.9%	41.2%	
	Total	16,375	6,815	4,433	29,582

granted patents. Consider, then, the areas of relative strength for the EU, such as fuels, mining, and metal recovery; other chemicals; purification, downstream processing; and waste disposal and the environment. These patterns seem to reflect historical strength by the EU in the chemical industry and related industrial applications of biotechnology. Similarly, the relative strength of Japanese inventors is apparent in areas such as waste disposal and the environment and other chemicals. Indeed, it is useful to note that the EU and Japan *both* register a higher number of patents on an absolute basis in several application categories: fuels, mining, and metal recovery; other chemicals; and waste disposal and the environment. Finally, while the bulk of U.S. patents are in classes related to red biotechnology, the United States also exhibits advantage on a relative basis in green biotechnology (the agriculture sector), reflecting in part the leading global position of Monsanto and DuPont in this application segment. Overall, these patenting patterns suggest that U.S. leadership in biotechnology is by no means monolithic. While the United States does tend to have a dominant position in red and green biotechnology, the EU and Japan exhibit innovation leadership in areas

TABLE 3 From R&D Activity to Patenting and Sales of the Biotechnology Industry

	R&D Per Capita Index	Patents Per Capita Index	Sales Per Capita Index
USA	1.00	1.00	1.00
EU	0.37	0.41	0.28
Japan	n/a	0.46	0.45

SOURCE: OECD biotechnology patents and authors' calculations (OECD, 2006, author's calculations).

related to white biotechnology. This is consistent with qualitative assessments that specific areas of biotechnology tend to be organized around clusters, with a small number of global innovation hubs.

From Innovation Activity to Sales

Of course, our analysis so far provides a limited perspective on the intensity of biotechnology activity across different regions: while evaluations of R&D employment and investment capture the intensity of R&D inputs, and patenting provides an imperfect measure of early-stage research outcomes, the impact of biotechnology ultimately depends on the ability to commercialize new technologies in the marketplace.

As such, Table 3 provides a brief examination of the relative intensity of inputs and outputs of the biotechnology industry. Biotechnology R&D expenditure, patent counts, and sales are divided by the total population within each distinctive geographic area to calculate per capita measures for each category, which are then indexed to the U.S. level (U.S. = 1.0). Both R&D investment and patenting in the EU are approximately 40 percent of the U.S. level on a per capita basis, yet sales per capita are nearly one-third lower, at 28 percent of the U.S. level. As mentioned earlier, this may reflect the earlier stage of development of many European biotechnology firms or the fact that European firms are more specialized in areas such as industrial applications, which may be associated with a lower level of sales for a given level of innovative investment (and patenting output). In contrast, although Japan is also concentrated in white biotechnology, Japanese companies exhibit a slightly higher level of patent per capita than Europe (0.46) and a much higher level of sales per capita (0.45).

Country-Specific Innovation Performance

Finally, Table 4 presents the distribution of biotechnology patent counts across a range of countries from 2000 to 2003, divided by individual application areas. These data are not strictly comparable to the official OECD triadic patent

TABLE 4 Biotechnology Patenting from 2000 to 2003 by Country

	A. Genetic Engineering and Fermentation	B. Biochemical Engineering	C. Sensors and Analysis	D. Pharmaceuticals	E. Agriculture
WPO/IB	7,979	213	139	6,488	1
USA	7,125	196	124	5,564	1,249
Canada	111	6		90	36
Mexico	4			3	2
Cuba	1			1	
Argentina	5				
Brazil					1
EPO	797	44	24	587	110
UK	653	21	16	520	93
Ireland	3	1		3	1
Germany	712	73	27	496	104
France	258	16	6	192	46
Netherlands	21	2	1	13	7
Belgium	4			3	1
Switzerland	10			7	
Austria	17	3	1	14	4
Denmark	86	2		46	4
Sweden	44		2	46	4
Finland	19	1		9	5
Norway	10			5	
Italy	31	4		28	7
Spain	21			19	5
Portugal	4			1	
Greece	1			1	
Hungary	4			3	1
Czech Republic	2	1		1	
Slovakia	1	1		1	
Poland					
Serbia and Montenegro	1				
Republic of Macedonia		1			
Russia	33	1		28	1
Turkey	1				
Israel	51	2	2	39	9
Japan	1,655	103	55	1,110	236
Republic of Korea	67	2	1	52	10
China	465	2	1	416	37
Taiwan	1				1
India	6			4	4
Singapore	6			2	4
Malaysia					
Australia	146	8	2	111	42
New Zealand	23				14
South Africa	8	7			4
Total	12,138	479	245	9,250	2,010

SOURCE: Derwent Biotechnology Resource (2006).

F. Food and Food Additives	G. Fuels, Mining and Metal Recovery	H. Other Chemicals	J. Cell Culture	K. Biocatalysis	L. Purification-Downstream Processing	M. Waste Disposal and the Environment
352	61	197	1,190	765	71	113
260	44	160	1,058	593	54	122
3	2		21	10	2	9
			1			1
					3	
102	14	87	112	160	11	32
22	6	15	99	67	9	23
	1		2			2
92	24	70	128	179	19	81
39	10	11	47	43	7	28
1	2	5	1	5	1	5
1			1	2		1
	2		2	4	1	1
2	2	3	5	9		5
	6	7	12	55		
		1	9	6	1	2
5		1		6		3
	1					2
2	1	4	7	7	3	2
2			2	7		
1				1		
				1		1
				1		
					1	
	1				1	1
		1		1		
						1
6	2		3	4	1	1
			9	4		3
186	45	176	249	492	16	232
7	1	9	9	17		5
12		11	2	33	2	12
1				1		
						1
1						1
6	5	2	22	2		5
1		2	2	1		1
				4		
712	171	504	1,779	1,604	127	563

counts presented earlier. Instead, Derwent Biotechnology Resources relies on an idiosyncratic algorithm for assigning patents (e.g., fractional patent shares) to different countries, by the country of origin of the inventors (Derwent Biotechnology Resource, 2006). With that caveat, the results are intriguing, as they deepen the broad patterns observed in our earlier U.S.-EU-Japan comparisons.

In particular, while we do not engage in a detailed application-specific examination of individual countries, there seem to be several distinct "tiers" of global activity within the biotechnology industry. First, there are several countries that exhibit a high level of overall activity, realized across several different application areas with a high number of patents in each area. These multifunctional biotechnology centers include the United States, Japan, Germany, the United Kingdom, and Australia. The presence of Australia in this category is significant; it has a strong history of basic research in the life sciences and has made significant investments in nurturing biotechnology companies and applications. Second, there is a grouping of countries that either have a broad base with only a few patents in each category (e.g., the Netherlands) or have intensive activity in a few categories (e.g., Israel). Finally, a large number of countries have only a small number of patents in biotechnology, often exhibiting only one or two patents in total. These include several European countries (e.g., Portugal, Greece), most of the Latin American and former Eastern European countries, and several of the less developed Asian economies (India, Malaysia, etc.).

Overall, these country-specific patterns reinforce several of the themes already mentioned. First, the United States exhibits persistent innovation leadership in biotechnology by a wide margin. Second, an increasing number of countries around the world seem to be displaying significant activity within biotechnology, and there is significant heterogeneity among countries in their biotechnology innovation intensity. For example, although Belgium has an advanced economy, it is a clear laggard in biotechnology innovation. Finally, as the biotechnology industry begins to spread from its origins in the life sciences sector, it will be increasingly important to distinguish the geography of innovation by individual applications; while the United States exhibits leadership in life sciences and agriculture, Denmark and Japan seem to have established leadership positions within industrial biotechnology applications.

KEY FINDINGS AND POLICY CONCLUSIONS

Key Findings

Overall our analysis suggests that both the biotechnology industry and biotechnology innovation in biotechnology remain clustered economic activities, with a strong reliance on and interaction with science-based university research. However, the number of active clusters in biotechnology is increasing over time, both in terms of the number of distinct locations in the United States that serve as the host for activity in the industry and in terms of a globalizing activity. While many countries around the world now host a biotechnology industry of varying importance, the activity within most countries seems to be highly localized. In other words, the data, though clearly inadequate to provide a complete picture, suggest that the number of biotechnology clusters that achieved "minimum scale" has increased, which is reflected in an increased dispersion in terms of employment, measures of biotechnology entrepreneurship, and measures of the geographic origins of biotechnology innovation.

This central insight—an increase in the number of regional innovation clusters, rather than a simple dispersion of biotechnology activity—holds several important implications for (1) evaluating the global biotechnology industry going forward and (2) developing effective policy to ensure continued U.S. leadership in this area.

First, our analysis suggests that the impact of globalization on biotechnology innovation seems to be different than that of traditional manufacturing sectors, such as the automobile industry or the IT sector. Specifically, the globalization of other industries reflects the increasing availability of low-cost locations to conduct activities that previously had been done in the United States. In contrast, the globalization of biotechnology reflects a "catching up" process by a small number of regions around the world that seek to compete head-to-head with leading regions in the United States.

Second, it is important to account for the range of activities now included within the biotechnology industry, including diverse applications in the life sciences, agriculture, and industry. Although most discussion focuses on life sciences—which remains the largest single segment of biotechnology in terms of employment, enterprises, investment, and patenting—the globalization of biotechnology is occurring most rapidly in industrial applications. Moreover, although the United States continues its historical advantage in agricultural applications, this may be due to political resistance in Europe and other regions rather than the presence of strong agglomeration economies within the United States. For example, the presence of extremely strong clusters with a high level of entrepreneurship that characterizes life sciences biotechnology seems to be a bit less salient for agricultural applications. The presence of multiple industrial segments—each of which is associated with distinct locational dynamics—raises the possibility that, even as individual clusters become more important within each application area, the total number of global clusters may increase with the range of applications.

Third, at least in terms of the available data, the United States maintains a very strong, even dominant, position within biotechnology. While some conceptual frameworks (e.g., the convergence effect) would suggest that early leadership by the United States would have been followed by a more even global distribution of biotechnology innovation, the "gap" between the United States and the rest of the world has remained relatively constant over the past decade or so. Indeed, it is likely that the United States has a historic opportunity to establish a long-term position as a global hub for biotechnology innovation, particularly in the life sciences and agricultural areas. In contrast to traditional debates about outsourcing, it is possible that increased global activity in biotechnology can complement rather than substitute for U.S. investment, employment, and innovation.

Finally, our analysis highlights the small size (in terms of absolute levels of employment) of the biotechnology industry. While industries such as IT may plausibly be associated with a large impact on the total workforces of individual states and regions, total employment in biotechnology is very small, although associated with very high average wages. The simple fact is that, if the biotechnology industry remains at roughly the same scale that it has achieved over the past decade or so, it is unlikely to be a major driver of employment patterns and overall job growth, either in the United States or abroad.

Policy Conclusions

The analysis holds a number of important policy implications. First, and perhaps most important, effective innovation policy concerning biotechnology must account for the broad differences between biotechnology and other sectors of the economy. The globalization of innovation in biotechnology is occurring in a much different way and for different reasons than the globalization of innovative activity in other manufacturing sectors, such as automobiles or IT. Consequently, policies that may be beneficial for these more traditional sectors (e.g., domestic R&D tax credits) may have little impact in biotechnology, where the vast majority of firms do not report positive accounting profits subject to significant taxation.

Second, there are policies that are likely to be particularly important in biotechnology, even though they may do little to stem the broader pattern of the globalization of innovation. Specifically, the biotechnology industry is extremely reliant on effective intellectual property institutions, most notably patents. U.S. leadership in biotechnology has benefited historically from a strong intellectual property environment, in many cases protecting innovations that received limited protection in other jurisdictions (e.g., transgenic mammals). Similarly, innovation in biotechnology benefits from the promotion of early-stage venture capital, including seed investments, and an effective system for technology transfer from university to industry (Mowery, 2004). While such considerations may be of modest importance for many of the sectors currently undergoing globalization, policies ensuring effective operation of the patent system, providing favorable treatment of early-stage venture capital investment, and enhancing the effectiveness of technology transfer are likely to enhance the strength of the U.S. biotechnology sector.

Recent patent reform proposals illustrate the challenge of ensuring continued U.S. leadership in biotechnology in a changing policy environment. Spurred in part by key studies emphasizing significant inefficiencies in the patent system (Cohen and Merrill [2003]; Jaffe and Lerner [2004]), numerous patent reform proposals have been advanced in the last few years, including legislation and administrative reviews. While some of these proposals seek to limit the strength of patents in areas such as business methods, biotechnology will be impacted by these reforms. Continued dynamism in the U.S. biotechnology sector requires strong and enforceable intellectual property protection, and would benefit from significant improvements in the operation of the patent system, such as reduced administrative delay and a higher level of consistency in patent grant decisio mak-ing. The danger is that reforms targeting sectors very distant from biotechnology will undermine the ability for biotechnology innovators to effectively commer-cialize their discoveries.

Third, the distinctive nature of biotechnology innovation suggests that the globalization of biotechnology innovation need not detract from U.S. strength in this area. Both the underlying science and the industry are still at a relatively early stage, and long-term American prosperity will benefit from establishing the United States as a global hub for biotechnology innovation. This can be ac-complished in several ways, most notably through investments in education and immigration policy. International leadership by American universities in the life sciences is a fundamental precondition for continued American leadership in biotechnology innovation. The biotechnology sector will benefit from policies that encourage the "best and brightest" on a global scale to study and potentially work in the United States. Significant restrictions on the ability of researchers liv-ing abroad to travel and collaborate with researchers in the United States in both public and private sectors or significant restrictions on the free flow of capital in-vestments undermines the likelihood of translating current U.S. cluster leadership into a position of durable centrality as a global biotechnology innovation hub.

Finally, an increasing number of *state* policy initiatives are focused on biotechnology in terms of encouraging job creation and employment. While providing a favorable local environment for biotechnology innovation and en-trepreneurship is important, policy makers should be careful to avoid focusing too heavily on attracting *external* investments in biotechnology. As emphasized by Feldman and Francis (2004), effective local economic development in bio-technology focuses on encouraging entrepreneurship and an effective interface with preexisting scientific institutions, rather than focusing on attracting a single large company. While there are of course cases where the "match" between an individual company and region are particularly favorable, most qualitative and quantitative evidence about the growth of biotechnology clusters emphasizes the centrality of indigenous entrepreneurship and the key role played by local university research. In addition, local policy makers must avoid excessive opti-mism about the promise of biotechnology for short-term economic development. Relative to the size and scope of other industries undergoing globalization, the absolute size of the biotechnology industry is quite modest and is likely to have only a small effect on regional employment and economic growth for the fore-seeable future.

ACKNOWLEDGMENT

We would like to thank Jeff Furman, Fiona Murray, Jeffrey T. Macher, David C. Mowery, and Stephen A. Merrill for extremely thoughtful suggestions and comments.

REFERENCES

Adams, J. D. (1990). Fundamental stocks of knowledge and productivity growth. *Journal of Political Economy* 98(4):673-702.

Argyres, N., and J. Liebeskind. (1998). Privatizing the intellectual commons: Universities and the commercialization of biotechnology. *Journal of Economic Behavior & Organization* 35(4): 427-454.

Arora, A., A. Fosfuri, and A. Gambardella. (2001). *Markets for Technology: Economics of Innovation and Corporate Strategy.* Cambridge: MIT Press.

Audretsch, D. B., and M. P. Feldman. (1996). R&D Spillovers and the geography of innovation and production. *American Economic Review* 86(3):630-640.

Audretsch, D. B., and P. E. Stephan. (1996). Company-scientist locational links: The case of biotech-nology. *American Economic Review* 86(3):641-652.

Barro, R.J., and Xavier Sala-i-Martin. 1992. Converage. *Journal of Political Economy,* 100(2): 223-251.

Battelle Technology Partnership Practice, SSTI. (2006). *Growing the Nation's Bioscience Sector: State Bioscience Initiatives.*

Biotechnology Industry Organization. (2006). *Milestones 2006-2007.*

Breschi, S., and F. Malerba. (2005). *Clusters, Networks, and Innovation.* New York: Oxford University Press.

Brezis, E., and P. Krugman. (1997). Technology and the life cycle of cities. *Journal of Economic Growth* 2(4):369-383.

Burrill & Company. (2004). Biotech 2004. Available at http://www.burrillandco.com/bio/biotech_book. Accessed August 3, 2007.

Cockburn, I., R. Henderson, L. Orsenigo, and G. P. Pisano. (1999). Pharmaceuticals and biotechnol-ogy. Pp. 363-398 in *U.S. Industry in 2000: Studies in Competitive Performance*, D. Mowery, ed. Washington, D.C.: National Academy Press.

Cohen, W., and S. Merrill. (2003). *Patents in the Knowledge-Based Economy.* National Research Council (U.S.). Committee on Intellectual Property Rights in the Knowledge-Based Economy. Washington, D.C.: The National Academies Press.

Cohen, W., R. R. Nelson, and J. P. Walsh. (2000). Protecting their Intellectual Assets: Appropriability Conditions and Why U.S. Manufacturing Firms Patent (or not). National Bureau of Economic Research Working Paper 7552.

Cooke, P. (2002). Biotechnology clusters as regional, sectoral innovation systems. *International Regional Science Review* 25(1):8-37.

Cortright, J., and H. Mayer. (2002). *Signs of Life: The Growth of Biotechnology Centers in the U.S.* Washington, D.C.: Brookings Institution Press.

Criscuolo, P. (2006). The "home advantage" effect and patent families. A comparison of OECD triadic patents, the USPTO and the EPO. *Scientometrics* 66:23-41.

Critical, I. (2006). *Biotechnology in Europe: 2006 Comparative Study.* EuropaBio, The European Association for Bioindustries.

Culliton, B. (1990). What is good for Monsanto is good for Washington University. *Science* 247: 1027.

Dalpé, R. (2002). Bibliometric analysis of biotechnology. *Scientometrics* 55:189-213.

David, P. (2004). Can "open science" be protected from the evolving regime of IPR protections? *Journal of Institutional and Theoretical Economics* 160(1):9-34.

Derwent Biotechnology Resource. (2006). Thomson, Inc Available at http://www.thomson.com/content/scientific/brand_overviews/biotech_resource.

DiMasi, J. A., R. W. Hansen, and H. G. Grabowski. (2003). The price of innovation: New estimates of drug development costs. *Journal of Health Economics* 22:151-185.

Dosi, G., K. Pavitt, and L. Soete. (1990). *The Economics of Technical Change and International Trade.* New York: Columbia University Press.

Dumais, G., G. Ellison, and E. L. Glaeser. (2002). Geographic concentration as a dynamic process. *Review of Economics and Statistics* 84(2):193-204.

Duranton, G., and D. Puga. (2001). Nursery cities: Urban diversity, process innovation, and the life cycle of products. *American Economic Review* 91(5):1454-1478.

Eaton, J., and S. Kortum. (1996). Trade in ideas: Patenting & productivity in the OECD. *Journal of International Economics* 40(3-4):251-278.

Edvinsson, L., and M. Malone. (1997). *Intellectual Capital.* New York: Harper Business.

Feldman, M. (2003). The locational dynamics of the US biotech industry: Knowledge externalities and the anchor hypothesis. *Industry and Innovation* 10(3):311-328.

Feldman, M., and J. Francis. (2003). Fortune favours the prepared region: The case of entrepreneurship and the capitol region biotechnology cluster. *European Planning Studies* 11(7):765-788.

Feldman, M., and J. Francis. (2004). Homegrown solutions: Fostering cluster formation. *Economic Development Quarterly* 18(2):127-137.

Furman, J. L., M. E. Porter, and S. Stern. (2002). The determinants of national innovative capacity. *Research Policy* 31:899-933.

Gans, J., and S. Stern. (2000). Incumbency and R&D incentives: Licensing the gale of creative destruction. *Journal of Economics and Management Strategy* 9(4):489-511.

Gans, J., and S. Stern. (2003). The product market and the market for "ideas": Commercialization strategies for technology entrepreneurs. *Research Policy* 32(2):333-350.

Graff, G., S. Cullen, K. Bradford, and D. Zilberman. (2003) The public-private structure of intellectual property ownership in agricultural biotechnology. *Nature Biotechnology* 21(9):989-995.

Griliches, Z., ed. (1984). *R&D, Patents and Productivity.* Chicago, IL: Chicago University Press.

Griliches, Z. (1990). Patent statistics as economic indicators: A survey. *Journal of Economic Literature* 92:630-653.

Griliches, Z. (1994). Productivity, R&D, and the data constraint. *American Economic Review* 84: 1-23.

Heller, M., and R. Eisenberg. (1998). Can patents deter innovation? The anticommons in biomedical research. *Science* 280:698-701.

Henderson, J. V., A. Kuncoro, and M. Turner. (1995). Industrial development in cities. *Journal of Political Economy* 103:1067-1090.

Hermans, R., and A.-J. Tahvanainen. (2006). Regional differences in patterns of collaboration, specialisation and performance. In *Sustainable Biotechnology Development*, R. Hermans and M. Kulvik, eds. ETLA, The Research Institute of the Finnish Economy, Series B 217.

Hermans, R., M. Kulvik, and A.-J. Tahvanainen. (2006). The biotechnology industry in Finland. In *Sustainable Biotechnology Development—New Insights into Finland*, R. Hermans and M. Kulvik, eds. ETLA, The Research Institute of the Finnish Economy, Series B 217.

Hewlett-Packard. (2006). Annual Report, 2006.

Hoffman, William (2008). MBBNet, http://www.mbbnet.umn.edu/scmap/biotechmap.html, University of Minnesota, MN.

International Organization of Automobile Manufacturers. (2007). Available at http://www.oica.net/htdocs/Main.htm. Accessed August 1, 2007.

Jaffe, A., and J. Lerner. (2004). *Innovation and Its Discontents: How Our Broken Patent System Is Endangering Innovation and Progress, and What to Do About It.* Princeton, NJ: Princeton University Press.

Jensen, R., and M. Thursby. (2001). Proofs and prototypes for sale: The licensing of university inventions. *American Economic Review* 91(1):240-259.

Kenney, M. (1986). *Biotechnology: The University-Industrial Complex.* New Haven, CT: Yale University Press.

Klein, S., and N. Rosenberg. (1986). An overview of innovation. In *The Positive Sum Strategy: Harnessing Technology for Economic Growth*, R. Landau and N. Rosenberg, eds. Washington, D.C.: National Academy Press.

Koput, K. W., W. W. Powell, and L. Smith-Doerr. (1996). Interorganizational collaboration and the locus of innovation: Networks of learning in biotechnology. *Administrative Science Quarterly* 41.

Kortum, S., and J. Lerner. (2001). Assessing the contribution of venture capital to innovation. *RAND Journal of Economics* 31(4):674-692.

Krugman, P. (1991). Increasing returns and economic geography. *Journal of Political Economy* 99(3):483-499.

Krugman, P., and A. J. Venables. (1995). Globalization and the inequality of nations. *Quarterly Journal of Economics* 110(4):857-881.

Lerner, J., and R. P. Merges. (1998). The control of technology alliances: An empirical analysis of the biotechnology industry. *Journal of Industrial Economics* 46(2):125-156.

Lundvall, B. (1992). *National Systems of Innovation: Towards a Theory of Innovation and Interactive Learning.* London, UK: St. Martin's Press.

Magee, M. (2005). *Health Politics.* Spencer Books.

Martin, P., and C. Rogers. (1995). Industrial location and public infrastructure. *Journal of International Economics* 39:335-351.

Massachusetts Biotechnology Council. (2007). Massachusetts Biotechnology Company Directory. Available at http://massbio.org/directory/statistics/stats_comp_yrfound.html. Accessed December 19, 2007.

McElroy, D. (2004). Valuing product development cycle in agricultural biotechnology. What is in a name? *Nature Biotechnology* 23:817-822.

McKelvey, M. D. (1996). Discontinuities in genetic engineering for pharmaceuticals? Firms jumps and lock-in in systems of innovation. *Technology Analysis & Strategic Management* 8(2):107-116.

Merges, R. P., and R. R. Nelson (1990). The complex economics of patent scope. *Columbia Law Review* 90(4): 839-916.

Merges, R. P., and R. R. Nelson. (1994). On limiting or encouraging rivalry in technical progress: The effect of patent scope decisions. *Journal of Economic Behavior and Organization* 25(1):1-24.

Monfort, P., and R. Nicolini. (2000). Regional convergence and international integration. *Journal of Urban Economics* 48:286-306.

Mowery, D. (2004). *Ivory Tower and Industrial Innovation: University-Industry Technology Transfer Before and After the Bayh-Dole Act in the United States.* Stanford, CA: Stanford Business Books.

Mowery, D., and R. Nelson. (1999). *Sources of Industrial Leadership: Studies of Seven Industries (1999).* Cambridge, UK, and New York: Cambridge University Press.

Mowery, D., R. Nelson, B. Sampat, and A. Ziedonis. (2001). The growth of patenting and licensing by U.S. universities: An assessment of the effects of the Bayh-Dole Act of 1980. *Research Policy* 30:99-119.

Murray, F., and S. Stern. (2007). Do formal intellectual property rights hinder the free flow of scientific knowledge? An empirical test of the anti-commons hypothesis. *Journal of Economic Behavior and Organization* 63(4):648-687.

Narin, F., and D. Olivastro. (1992). Status report: Linkage between technology and science. *Research Policy* 21(3):237-249.

National Science Board. 2006. *Science and Engineering Indicators 2006.* Two volumes. Arlington, VA: National Science Foundation (volume 1, NSB 06-01; volume 2, NSB 06-01A).

Nelkin, D., R. Nelson, and C. Kiernan. (1987). University-industry alliances. *Science, Technology, & Human Values* 12(1):65-74.

Nelson, R. (1993). *National Innovation Systems: A Comparative Analysis.* New York: Oxford University Press.

OECD. (2006). Statistical Definition of Biotechnology. Available at http://www.oecd.org/document/ 42/0,2340,en_2649_37437_1933994_1_1_1_37437,00.html. Accessed March 1, 2008.

Orsenigo, L. (1989). *Emergence of Biotechnology: Institutions and Markets in Industrial Innovation.* Pinter Publishers, 230 pp.

Pierre-Benoit, J. (1999). Innovating through networks: A case study in plant biotechnology. *International Journal of Biotechnology* 1(1):67-81.

Porter, G., C. Denning, A. Plomer, J. Sinden, and P. Torremans. (2006). The patentability of human embryonic stem cells in Europe. *Nature Biotechnology* 24:653-655.

Porter, M. (1990). *The Competitive Advantage of Nations.* New York: Free Press.

Porter, M. (1998). Clusters and the new economics of competition. *Harvard Business Review* 76(6): 77-90.

Porter, M., and S. Stern. (1999). *The New Challenge to America's Prosperity: Findings from the Innovation Index.* Washington, D.C.: Council on Competitiveness.

Powell, W. W., D. R. White, K. W. Koput, and J. Owen-Smith. (2005). Network dynamics and field evolution: The growth of interorganizational collaboration in the life sciences. *American Journal of Sociology* 110:1132-1205.

Romer, P. M. (1990). Endogenous technological change. *Journal of Political Economy* 98(5): 71-102.

Rosenberg, N. (1974). Science, invention and economic growth. *Economic Journal* 84(333):90-108.

Saxenian, A. (1994). *Regional Advantage: Culture and Competition in Silicon Valley and Route 128.* Cambridge, MA: Harvard University Press.

Schmookler, J. (1966). *Invention and Economic Growth.* Cambridge, Mass.: Harvard University Press.

Stern, S. (1995). Incentives and focus in university and industrial research: The case of synthetic insulin. In *The University-Industry Interface and Medical Innovation*, A. Geligns and N. Rosenberg, eds. Washington, D.C.: Institute of Medicine, National Academy Press.

Stokes, D. (1997). *Paster's Quadrant: Basic Science and Technological Innovation.* Washington, D.C.: Brookings Institute Press.

Swann, G. M., M. Prevezer, and D. K. Stout. (1998). *The Dynamics of Industrial Clustering: International Comparisons in Computing and Biotechnology.* Oxford University Press.

Trajtenberg, M. (1990). *Patents as Indicators of Innovation, Economic Analysis of Product Innovation.* Cambridge, MA: Harvard University Press.

Van Beuzekom, B., and A. Arundel. (2006). *OECD Biotechnology Statistics—2006.* OECD, Paris, France.

Venables, A. (1996). Equilibrium locations of vertically linked industries. *International Economic Review* 37:341-359.

Zucker, L., M. R. Darby, and J. Armstrong. (1998a). Geographically localized knowledge: Spillovers or markets? *Economic Inquiry* 36(1):65-86.

Zucker, L., M. R. Darby, and M. B. Brewer. (1998b). Intellectual human capital and the birth of U.S. biotechnology enterprises. *American Economic Review* 88(1):290-306.

8

Logistics

ANURADHA NAGARAJAN
University of Michigan
CHELSEA C. WHITE III
Georgia Institute of Technology

INTRODUCTION

"Alexander's army was able to achieve its brilliant successes because it managed its supply chain so well" says Michael Hugos (2006), drawing on published literature on the history of the successful campaigns of Alexander the Great. One of the critical differences between Alexander's army and the ones he defeated was the fact that his army was the fastest, lightest, and most mobile army of its time. While other armies were constrained by carts and pack animals carrying the supplies, Alexander's soldiers were trained to carry their own equipment and provisions and to live off the land as needed. Competitive advantage provided by logistics and the supply chain persists 2000 years later. In the battlefield of modern industry, globalization, demanding consumers, and efficient capital, markets have combined to create an environment that punishes mediocrity and incompetence. Firms are looking to the logistics industry to provide competitive advantage as goods and services move seamlessly worldwide. In a recent speech on global economic integration, Federal Reserve Chairman Ben Bernanke said, "Technological advances continue to play an important role in facilitating global integration. For example, dramatic improvements in supply-chain management, made possible by advances in communication and computer technologies, have significantly reduced the costs of coordinating production among globally distributed suppliers" (Bernanke, 2006). Economic integration brought about by vanishing global trade barriers has enabled firms to creatively manage their value chain, identifying sources of competitive advantage in the world stage to provide the best value for customers worldwide.

A consequence of the global dispersion of economic activity has been the

increased complexity of the supply chain. When national boundaries were imper-
meable due to geographical, technological, logistical, and political constraints,
firms could control the movement of goods through the value chain with more
certainty. Consider the movement of goods in and out of the United States. In
2004, nearly $1.5 trillion worth of goods were imported to the United States, and
$0.8 trillion of U.S. goods were exported to other countries. Nearly 16 million
20-foot equivalent units (equivalent to shipping containers that are 20 feet long,
8 feet wide, and 8 feet tall) arrive each year at U.S. ports. Roughly one-quarter
of U.S. imports and one-sixth of its exports—or about $423 billion and $139 billion worth of goods, respectively, in 2004—arrive or depart on container ships.
Containerized imports include both finished goods and intermediate inputs, some
of which are critical to maintaining U.S. manufacturers' "just-in-time" supply
chains. Supply-chain disruptions can leave manufacturers vulnerable if a necessary part does not reach an assembly plant in time. The lack of key parts could
reduce output, employment, and income for individual companies and entire
economic regions by amounts larger than the value of the delayed part—and
in areas and businesses far removed from the port where a disruption occurred.
Although concerns about disruptions in the flow of freight focus on terrorist attacks, similar economic losses could result from extreme weather, the high cost
of fuel or fuel unavailability, or labor disputes that affect freight operations or
from disruptions elsewhere in the supply chain.[1] The increasing complexity of
the global supply chain has compelled firms to look carefully at managing three
primary challenges in logistics: (1) managing visibility of information and product movement as it relates to the ability to track orders, inventory, and shipments
in real time; (2) managing costs; and (3) securing reliable service. Innovation
in the logistics industry has played a major role in helping businesses manage
these challenges. According to the U.S. State of Logistics report, although absolute logistics costs rose substantially in the United States in 2004, because of
the growing economy, the costs of logistics remained at 8.6 percent of the gross
domestic product (GDP). In fact, logistics costs have generally been declining
annually since 1995.

As shown in Figure 1, logistics costs were 10.4 percent of GDP in 1995 but
they have declined year after year annually, except for a slight increase in 2000,
reaching its lowest level of 8.6 percent in 2003 (Wilson, 2005). Innovation in logistics has been critical to offering better products and services at lower costs.

We begin this chapter with a brief definition of the logistics industry and its
role in the economy. We classify the firms involved in the industry into four categories and examine innovation practices in each of these categories in order to
understand where innovation occurs and how the locus of innovation has changed
over time among the various participants in the logistics network. We present the
results of our semistructured interviews with 20 respondents consisting of execu-

---

[1]See http://www.cbo.gov/ftpdocs/71xx/doc7106/03-29-container_shipments.pdf, p. 7.

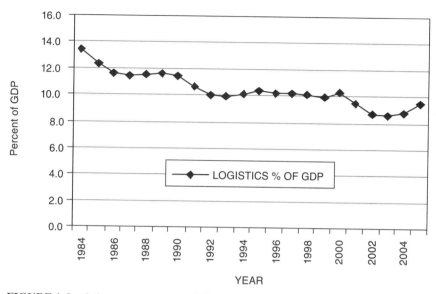

FIGURE 1 Logistics as a percentage of GDP. SOURCE: Wilson (2005).

tives from the logistics industry and industry experts. We then focus specifically on the locus of innovation, as indicated by the home country of the first inventor listed on logistics-related patents in the database of World Intellectual Property Organization (WIPO) patents we have collected, and we examine the changes in the location of innovation worldwide in the logistics industry. Our conclusions on the current state of innovation are derived from our observations from the two sources mentioned earlier: patent data and semistructured interviews with top industry executives and experts. We conclude with policy implications and our expectations for the future of innovation in the logistics industry.

INDUSTRY DEFINITION AND DESCRIPTION

Logistics management is that part of supply-chain management that plans, implements, and controls the efficient, effective, forward and reverse flow and storage of goods, services, and related information between the point of origin and the point of consumption in order to meet customers' requirements. Logistics management activities typically include inbound and outbound transportation management, fleet management, warehousing, materials handling, order fulfillment, logistics network design, inventory management, supply/demand planning, and management of third-party logistics services providers. To varying degrees, the logistics function also includes sourcing and procurement, production planning and scheduling, packaging and assembly, and customer service. It is involved in all levels of planning and execution—strategic, operational, and tactical. Logistics management is an integrating function that coordinates and optimizes all logistics activities, as well as integrating logistics activities with other functions, including marketing, sales manufacturing, finance, and information technology (Council of Supply Chain Management Professionals, n.d.). The logistics industry consists of firms offering a variety of services, including transportation and warehousing. As globalization has extended and expanded the supply chain, logistics firms have added order fulfillment, logistics network design, inventory management, and supply/demand planning to their portfolio of products.

In terms of tons transported, domestic freight transportation in the United States grew by about 20 percent annually from 1993 to 2002. In the same period inventory carrying costs as a percentage of GDP have decreased from 8.3 percent in 1981 to 2.8 percent in 2002 as shown in Figure 2. "This reduction can be at-

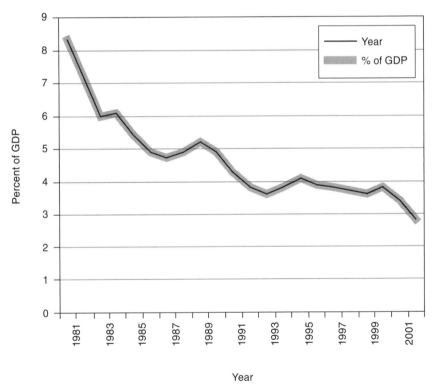

FIGURE 2 Inventory carrying costs as a percent of GDP. SOURCE: FHWA Freight Management and Operations; see http://www.ops.fhwa.dot.gov/freight/freight_analysis/ econ_methods/lcdp_rep/index.htm.

tributed to timely, reliable delivery, efficiency, and visibility in the supply chain system. These reductions are especially evident in industries such as automotive and computer hardware manufacturers[2]." Firms use logistics not only to manage cost, as Dell did in their assembly and distribution of desktops, but also to be finely tuned to customer needs, as Zara has done in fashion retailing. Dell's personal computer supply chain creates positive cash flow for the firm because it receives payment from its customers before it has to pay its suppliers by managing its supply chain efficiently.[3] Zara's supply chain allows the firm to closely observe the market's changing tastes and to respond quickly, keeping the firm in the forefront of fashion while minimizing inventory markdowns.[4] In each case, these supply chains are well-coordinated, complex global networks, and in each case the firms have focused on innovation in their supply chain in order to achieve competitive advantage (*Economist*, 2006).

Three trends have spurred innovation in logistics (Smith, 2006). The first trend is the increase in demand for high-tech goods.[5] According to the Global Insight World Industry Service database, which provides production data for 70 countries that account for more than 97 percent of global economic activity, the global market for high-technology goods is growing at a faster rate than that for other manufactured goods, and high-technology industries are driving economic growth around the world. During the 22-year period examined (1980-2001), high-technology production grew at an inflation-adjusted average annual rate of nearly 6.5 percent compared with 2.4 percent for other manufactured goods. Between 1996 and 2001, high-technology industry output grew at 8.9 percent per year, more than double the rate of growth for all other manufacturing industries. Output by the five high-technology industries represented 7.7 percent of global production of all manufactured goods in 1980; by 2001, this output had doubled

---

[2]Ginter, J. L. and La Londe, B. J. 2001. An Historical Analysis of Inventory levels: An Exploratory Study. Ohio State Working Paper and Fein, A. J. 2004. The Myth of Decline: Assessing Time Trends in US Inventory to Sales Ratios. CES 04-18. See http://www.ces.census.gov/index.php/ces/1.00/cespapers?down_key=101710#search=%22Ginter%20and%20La%20Londe%22

[3]Dell pioneered build-to-order supply-chain management in the 1990s to support its desktop business. Instead of building computers to forecasts and letting retailers sell them, Dell sells directly from its own website and call centers and then builds to order. Dell cuts retailers and distributors out of its supply chain but also gets paid up front, often before they have to pay for components. However, Dell has found it difficult to duplicate its supply-chain success in its laptop and consumer electronics businesses.

[4]Zara, a part of Spain's Inditex Group, is in the fashion apparel business. Fashion retailing is highly perishable, influenced quickly by changes in consumer and celebrity tastes. In contrast to a typical clothing company, where outsourcing manufacturing to Asia could take about 6 months to get a new design to shops, Zara completes the process in 5 weeks. Zara accomplishes this speed to market primarily by buying some garments and materials partly finished and by avoiding mass production.

[5]The five industries considered high technology by Organisation for Economic Co-operation and Deveopment (OECD) based on their R&D intensity are aerospace, pharmaceuticals, computers and office machinery, communication equipment, and scientific (medical, precision, and optical) instruments.

to 15.8 percent.[6] High-tech products typically have short product life cycles and complex value chains with design, manufacture, sales, and distribution located in different parts of the world. The role of innovation and execution of the supply chain has become paramount in this context.

The second trend influencing the logistics industry is globalization—the integration of many microeconomies into one worldwide, interdependent economy. Companies have the ability to source and sell globally and have begun to streamline their supply chains and open new markets. According to World Trade Organization statistics, world merchandise trade has grown by 70 percent since 2001.[7] According to Global Insight, trade exports are going to increase significantly in all regions of the world, with the largest increases coming from exports from China to Europe and intra-Asia trade (*Economist*, 2006, p. 5). Table 1 shows the top 25 firms in the logistics industry as identified by the Hoovers Database.[8] These companies, headquartered all over the world, are not only moving logistics services worldwide but are also creating innovation centers close to the new centers of trade. For example, Saint Gobain, a glass manufacturer headquartered in France, has a "Technology Center" for conducting multidisciplinary research and development (R&D) in Blue Bell, Pennsylvania, to support its U.S. glass manufacturing and logistics operations. It has also opened a new materials research and logistics support center in Shanghai, China, to cater to all its business divisions in Asia in 2006. In 2005, Hitachi invested in an information system company in China to help its third-party logistics (3PL) operations there.[9] According to industry experts, including the council of supply-chain professionals, reliable, timely accurate data is the keystone of the new global supply chain (Bowman, n.d.; Wilson, 2004). Information is needed from each market and manufacturing outpost for planning, to enable flexibility, and to ensure security. Software and hardware innovations that enable greater visibility of product movement to shippers and carriers have become critical to success in the industry.

Innovation in the industry has also been spurred by the growth of the Internet and e-commerce. Business-to-business commerce on the Internet has been increasing. According to *Industry Week*'s 2005 Value Chain survey, worldwide, in the 2 years between 2003 and 2005, business purchases made online increased

---

[6]U.S. Technology Marketplace. See http://www.nsf.gov/statistics/seind04/c6/c6s1.htm#c6s111p2. Accessed October 23, 2006.

[7]See http://stat.wto.org/Home/WSDBHome.aspx. Accessed August 15, 2006

[8]*Logistics services* is described in Hoover's as companies that engage in the process of planning, implementing, and controlling the movement and storage of raw materials, in-process inventory, finished goods, and related information from the point of origin to the point of consumption. We selected firms in the industry category that had logistics as one of their primary businesses. The firms are ranked by sales, not necessarily by logistics.

[9]By definition, a 3PL becomes a third party to the traditional two-party (shipper/carrier) contract for transportation. A 3PL firm is, therefore, an outsourced provider that manages all or a significant part of an organization's logistics requirements and performs transportation, locating, and sometimes product consolidation activities.

TABLE 1 Top Logistics Companies Worldwide

No.	Company Name	Location	Year Started	Sales ($ million)	Employees	Comments (source of information)
1	Hitachi, Ltd.	Tokyo, Japan	1920	2,524.80	323, 072 (Hitachi Ltd.); 7505 (Hitachi Transport System Ltd.)	Company website, Annual Report 2005, One Source Global Business Browser, Hoovers Online
2	Deutsche Post AG	Bonn, Germany	1949	MAIL: 16,486.42; EXPRESS: 23,393.09; LOGISTICS: 10,176.31	MAIL: 125,282; EXPRESS: 131,927; LOGISTIC: 148,095	One Source Global Business Browser, Hoovers Online, Annual Report 2005, Equinet Analyst Report—February 2005
3	United Parcel Service, Inc.	Atlanta, US	1907	42,581.00 (total); 5,994 (supply chain and freight operations)	407,000	One Source Global Business Browser, Annual Report 2005, Factiva
4	Compaigne de Saint-Gobain	Courbevoie, France	1665	45,058 (total revenues); 19,808.18 (building distribution)	199,630	One Source Global Business Browser, Annual Report 2005, Factiva
5	Deutsche Bahn Aktiengesellschaft	Berlin, Germany	1994	35,314	233,657	Hoovers Online, Company website
6	DHL Worldwide Network	Diegem, Belgium	1969	33,524.4	171,980	One Source Global Business Browser, Hoovers Online, Annual Report 2005
7	FedEx Corporation	Memphis, US	1971	32,294.0	184,953	One Source Global Business Browser, Factiva, Hoovers Online
8	A.P. Moller–Maersk A/S	Copenhagen, Denmark	1904	34,809	76,000	One Source Global Business Browser, Hoovers Online

continued

TABLE 1 Continued

No.	Company Name	Location	Year Started	Sales ($ million)	Employees	Comments (source of information)
9	McLane Company, Inc.	Temple, US	1894	23,373.00	14,300	One Source Global Business Browser, Hoovers Online, Factiva
10	Supervalu Inc.	Eden Prairie, US	1926	19,863.6 (total); 9,228.6 (food distribution)	52,400	Company website, Factiva, One Source Global Business Browser
11	TNT N.V.	Hoofddorp, Holland	1946 (TNT originally founded)	12,621.12 (total); 991.30 (logistics)	128,671	One Source Global Business Browser, Factiva, Hoovers Online, company website
12	Schenker AG	Essen, Germany	1872	10,912	39,000	Hoovers Online, One Source Global Business Browser, Factiva, Company website
13	Mitsui O.S.K. Lines, Ltd.	Tokyo, Japan	1964	10,920.2 (Total); 553.4 (logistics)	7,385	Annual Report 2006, One Source Global Business Browser, Hoovers Online, Company website
14	KBR, Inc.	Houston, US	The parent company, Halliburton, was founded in 1919	10,138	62,500	Factiva, Hoovers Online
15	YRC Worldwide Inc.	Overland Park, US	1924	8,741.6 (total); 447.6 (Meridian IQ)	68,000	Factiva, Hoovers Online, Annual Report 2005, Company website
16	CSX Corporation	Jacksonville, US	1980	8,618	35,000	Hoovers Online, One Source Global Business Browser
17	Norfolk Southern Corporation	Norfolk, US	1894	8,527	30,294	Annual Report 2005, Hoovers Online, One Source Global Business Browser, Company Web site
18	R.R. Donnelley & Sons Company	Chicago, US	1864	8,430.2	50,000	Company website, Factiva, One Source Global Business Browser

	Company	Location	Year	Revenue	Employees	Sources
19	Kuehne + Nagel International AG	Schindellegi, Switzerland	1890	11,276.6 (total); 1,681.2 (railroad logistics); 1,071.1 (contract logistics)	23,000	Company website, Factiva, One Source Global Business Browser, Annual Report 2005
20	Bollore Investissement	Puteaux, France	1955	7,648.7	32,808	Company Website, Factiva, One Source Global Business Browser
21	Hanjin Shipping Co., Ltd.	Seoul, South Korea	1988	5,921.1	1,641	Company website, Factiva, One Source Global Business Browser, Supplychainbrain.com
22	Ryder System, Inc.	Miami, US	1934	5,740.8 (total); 1,637.8 (supply chain solutions)	27,800	Company website, Factiva, One Source Global Business Browser, Annual Report 2005, The South Florida Business Journal—southflorida.bizjournals.com
23	C.H. Robinson Worldwide, Inc.	Eden Prairie, US	1905	5,688.9	5,776	World Trade Magazine (worldtrademag.com), company website, One Source Global Business Browser, Hoovers Online
24	Canada Post Corporation	Ottawa, Canada	1867	5,520.3	72,874	Factiva, Company Website, One Source Global Business Browser
25	Panalpina World Transport (Holding) Ltd.	Basel, Switzerland	1890	5,290.6	13,583	Company website, Factiva, One Source Global Business Browser

from 15 to 20 percent (Vinas, 2005). In addition, according to Forrester Research, of the 80 million U.S. online households, three-fourths have purchased products online (Mulpuru, 2006). Acquiring and retaining customers online necessitates providing complete satisfaction from the first click on the company's website to delivery to the door. The product has to be delivered when, where, and how the customer wants it delivered (Bhise et al., 2000). In contrast to the bricks-and-mortar economy, delivery of the product on a timely basis has become a critical part of the overall customer's purchasing experience. The Internet and computing technologies have increased the amount of information available to the buyer and manufacturer about the product and the fulfillment process. Consequently expectations relating to timely and accurate fulfillment are rising. Data from the American Customer Satisfaction Index Annual E-Commerce Report in 2005 showed a slump in scores after 2 years of progressive increases (Smith, 2005). Auction leader eBay had a 4.8 percent reduction, and Amazon.com dropped 4.5 percent in their customer satisfaction scores. Firms such as Amazon and eBay depend on the logistics function to improve and complete their customer's purchasing experience. Innovation along the supply chain, especially warehouse management software and tracking products, support firms' efforts to cope with the demands of the Internet environment.

In summary, the modern logistics industry is a complex network of firms involved in the flow and transformation of goods from the raw materials stage to the end user. Innovation in the industry has been spurred by increases in the demand for high technology products, globalization in all aspects of the value chain, and the advent of ecommerce.

PATTERNS OF INNOVATION IN THE LOGISTICS INDUSTRY

One of the most challenging aspects of understanding innovation in logistics management lies in the accepted wisdom that every product has its own unique value chain. Thus, innovation is primarily a pull phenomenon[10] for firms in the logistics industry, with new products and services being developed in response to specific customer needs. The competing pressures of managing global supply chains cost-effectively while increasing visibility, which allow a firm to monitor events and exceptions in real time, have created an environment ripe with opportunity for innovation. We adopt a broad definition of innovation that is not limited

---

[10]John Seely Brown and John Hagel III define pull and push systems in the context of innovation as follows: "Push systems contrast starkly with pull ones, particularly in their view of demand: the former treat it as foreseeable, the latter as highly uncertain. This difference in a basic premise leads to fundamentally different design principles. For instance, instead of dealing with uncertainty by tightening controls, as push systems would, pull models address immediate needs by expanding opportunities for local participants—employees and customers alike—to use their creativity. To exploit the opportunities that uncertainty presents, pull models help people come together and innovate by drawing on a growing array of specialized and distributed resources" (Brown and Hagel, 2005).

to technological breakthroughs or new products. As Rogers notes, "Innovation is an idea, practice, or object that is perceived as new by an individual or other unit of adoption" (Rogers, 1995). Consequently logistics innovation could improve internal efficiency within a logistics firm or could help serve shippers better. For the purposes of this study, logistics innovations could be new to the world or new to the particular context of the firm and its stakeholders.

To understand the locus of innovation in the logistics industry, we begin with a classification of firms in the industry. Our analysis of innovation patterns in the industry is presented in the context of this taxonomy. We present the results of our semistructured interviews with top management personnel and experts in the industry on innovation practices of prominent firms in the industry. These semistructured interviews, though few in number, allow us a fine-grained look at innovation practices in the industry. Then we present an analysis of patent data from the WIPO to provide further insight into the innovation practices in the logistics industry and how they have changed over time.

Classification of Firms

To identify meaningful patterns in innovation behavior, we needed to classify firms according to the role they play in the industry before we discuss the interview data and patent data. Figure 3 depicts our classification of firms into

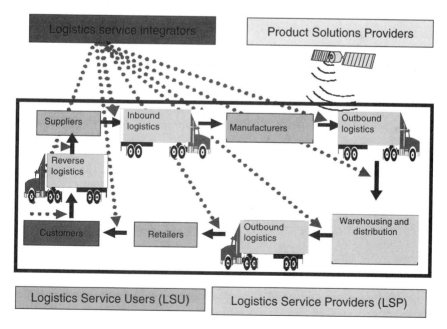

FIGURE 3 Classification of firms in the logistics industry.

four categories based on a simple logistics value chain.[11] They are logistics services users, logistics service providers, logistics service integrators, and product
solution providers. In this section, we focus on describing the classification since
our subsequent sections discuss innovative activity along the categories detailed
in the following paragraphs.

Logistics service users (LSUs) are shippers, such as Dell and General Motors (GM), who manufacture products and use logistic services. There are many
variations among the firms in this category. Some LSU firms manage important
parts of the logistics function while outsourcing parts of the supply chain such
as transportation planning, execution, and warehousing to logistics service providers and logistics integrators. In other firms, the purchasing and transportation
departments have combined to form supply chain departments. Beginning in the
mid- to late 1990s these departments were given the responsibility for logistics
management as defined earlier. As LSU firms began to focus more on core operations in the late 1990s, they began outsourcing many aspects of the supply
chain, giving rise to 3PL providers and fourth-party logistics (4PL) providers.
3PL and 4PL providers enable logistics outsourcing and provide logistics value-
added services.[12] A recent survey of 381 LSUs revealed that 69 percent of them
outsourced their logistics operations to 3PLs (Stoffel, 2006). However, as innovation in the supply chain becomes critical for business success and the competitive
advantage it confers becomes more evident, there has been a move to bring parts
of the activity within the organization again. For example, Vector SCM was a
joint venture created in December 2000 by Con-way and GM to manage GM's
global supply chain. However, in June 2006, GM declared its intent to enter
into negotiations to purchase Vector from Con-way. GM noted that, by integrating Vector into the firm, it had made a strategic decision to resume more direct
control over its logistics functions.[13] Most LSUs are not interested in acquiring
the physical assets involved in the physical transportation of their goods. Rather,
they look to increase their capability in supply-chain management to effectively
partner with 3PLs and 4PLs to leverage their combined capabilities. According
to Razat Gaurav, vice president of global transportation and distribution for i2,
companies "want to let 3PLs execute against their plans and then rely on their
specialized local knowledge. If you are highly reliant on international sources of
supply, you also need some internal competence" (Bartels, 2006).

Specialized industry and firm-specific supply-chain challenges are best understood by LSUs. These firms use their specialized knowledge to innovate
products and processes as solutions to solve their particular business challenges.
A survey of LSUs by McKinsey and the Institute for Supply Chain Management

---

[11]We thank Tushar Dave of Satyam for helping us develop these categories.
[12]Please see section on Logistics Service Providers for 3PL and 4PL definitions; see also footnote 9.
[13]http://www.menloworldwide.com/mww/en/newsroom/prarchives/29_Jun_2006.shtml. Accessed
July 20, 2006.

at the University of Munster found that the best supply-chain performers create innovations to increase efficiency in logistics by borrowing from efficiency programs in other parts of their business process, such as lean manufacturing (Grosspietsch and Kupper, 2004). Consequently, LSUs generate efficiency-based supply-chain innovations that draw on business practices outside the supply-chain function. The United States, with more than one-fourth of the world's GDP, depends heavily on logistics for movement of goods within the country and across its borders. It provides a fertile environment for innovation among LSU firms with U.S. operations regardless of where it is headquartered.

Logistics service providers (LSPs) are firms, such as Fedex, UPS, Con-way, and Ryder, that offer partial or complete logistics solutions. These firms could be 3PL or 4PL providers. The difference between 3PL and 4PL, to be explained in more detail later, lies primarily in the value provided to customers and the basis of firms' competence, although the boundary defining the scope of activity is hazy.

3PLs are firms that enable logistics outsourcing. Drawing on their core business, whether it be forwarding, trucking, or warehousing, they have moved into providing other services for customers. 3PLs, in general, are commodity transportation service providers who have moved into higher-margin, bundled services such as warehousing and inventory management. Exel Plc and Penske Logistics are among leading 3PL firms headquartered in the United States with expanding international operations.

In the past decade, competitive pressures have created more complex supply chains as LSUs deal with multiple customers, suppliers, transportation providers, and government organizations worldwide. 4PLs, such as UPS Supply Chain Solutions, emerged in the late 1990s in order to manage the flow of information and to coordinate the movement of goods. The 4PL firm wants to position itself as an extension and part of its customer's business environment. For example, UPS operates a computer repair service center in Louisville, Kentucky. The Digital Products Division of Toshiba America Information Systems, Irvine, California, has used the UPS repair service since 2004 for repair of it laptop computers. UPS has decreased the turnaround time by two and a half days since they took over repairs for Toshiba (Violini, 2006). A key differentiator between 3PLs and 4PLs is that while many 3PLs focus on providing value-added services to their customer with a view to maximizing the utilization of the assets of the parent LSP, 4PLs develop solutions tailored to meet the unique and special needs of each customer, without regard to a parent company's service offerings and operations. The 4PL's parent's assets are given due consideration; however, the overarching goal of the 4PL is to serve the customer's specific needs and provide visibility and efficiency to the logistics process (Craig, 2003).

3PLs and 4PLs are the backbone of the logistics system. In the past the primary task of an LSP was to transport goods from one location to another. Today, as we have noted earlier, all the firms along the supply chain have to work in concert in order to achieve cost-effective and flexible solutions. LSPs turn to innovation to improve supply-chain efficiency as demanded by LSUs, their customers. Innovation among the LSPs has focused on creating new supply-chain solutions such as "dock-to-stock" delivery systems,[14] increased information visibility through implementation of innovative information software and hardware systems, and better asset utilization through load-and-back haul management.

LSPs, working closely with their customers, have expanded their operations as their customers have globalized. As the locus of operations expand, so too has the locus of innovation for these firms. Innovation among LSP firms draws on local environments, and solutions created depend on local challenges.

Logistics service integrators (LSIs) such as Accenture, IBM, and Satyam are firms that are platform-neutral[15] and non-asset-based in that they do not own means of cargo transportation such as trucks, ships, or planes. The key difference, as per our definition, between 4PLs and LSIs is asset ownership. The value proposition that LSIs bring to their customers is knowledge of specific business processes going beyond a firm's logistics needs. LSIs work with a variety of product solution providers, including vendors of products based on radiofrequency identification technologies (RFIDs), enterprise resource planning (ERP) software vendors, software developers, and transportation providers to provide customized solutions for their customers, the LSUs. In the past decade, firms that have traditionally focused on strategy consulting have set up supply-chain divisions to exploit the need for unbiased and process-capable integrators. On the other end of the spectrum, firms specializing in the provision of information technology services have recognized the vital role that information plays in supply-chain management and have created divisions specializing in supply-chain services. LSIs recognize that innovation in the supply chain lies in the management of shipment visibility and requires analysis of critical supply-chain information (Pande et al., 2006).

Innovation among LSI firms is often collaborative in nature. These firms work closely with LSUs and LSPs to enhance supply-chain operations. LSI innovation often arises in the United States, where many management consulting firms have begun to offer supply-chain integration services. Their innovative efforts are likely to draw upon their knowledge of products, processes, and practices in contexts outside the supply chain. Supply-chain innovation among LSIs may be expected to increase their innovation efforts, especially in business processes and in information integration.

Product solution providers (PSPs) such as SAP, Oracle, Manugistics, i2,

---

[14]Robert Handfield and Ernest Nichols (2002) define "dock-to-stock" delivery as those supplier deliveries of component parts that are made directly to the plant floor and end up in finished goods by the end of the same day.

[15]LSIs are platform-neutral in this context in the sense that they do not work necessarily with one brand of hardware or software. They examine the business process to determine the best solution.

and SAVI Technologies are firms that create products that are used in logistics management. They often work with firms in each of the other three categories to enable efficiency, visibility, and integration.[16] Some examples of products developed by these firms include asset tracking technologies such as RFID hardware and software, ERP systems, and warehouse management software. Figure 4 shows how different asset tracking technologies are deployed from the truck level to the product level. Organizationally, these firms begin as vendors of a specific product that fulfills a specific niche in the supply chain. In time, they grow to offer deeper and broader product and service solutions but are generally not complete LSPs with transportation assets.

PSPs are the most prolific innovators in the logistics industry, consistent with the "pull" characteristic of the logistics industry. They are specialists and create new products and solutions that address specific challenges in parts of the logistics network. For example, RFID manufacturers offer innovative passive RFID products that help LSUs track inventory at the product level. The same manufacturer may offer innovative active RFID products that assist 3PL firms with fleet monitoring. As the demands on the supply-chain network increase, PSPs can be expected to continue to offer innovations to enable LSUs and LSPs to manage the supply chain more effectively.

Innovation among PSPs, unlike firms in the categories discussed earlier, draws upon basic and applied research and development. According to National Science Foundation (NSF) reports, the United States ranked sixth among countries with reported R&D-to-GDP ratios with a ratio of 2.9 percent in 2003. The business sector performed nearly 70 percent of the U.S. R&D in 2004. Besides performing the majority of U.S. R&D, the business sector was also the largest source of R&D funding in the United States, providing 64 percent ($199 billion) of total R&D funding in 2004.[17] U.S.-based PSP firms draw on this rich tradition of R&D to provide most of the innovation in the logistics industry.

The preceding taxonomy has been developed so that data gathered through interviews and patent analysis may shed light on where, when, and why innovation is occurring in the logistics industry. Depending on their primary business, firms are classified as LSU, LSP, LSI, or PSP. In the sections that follow, we ex-

---

[16]Efficiency in supply-chain management often relates to minimizing the time that elapses between procurement of raw material and the delivery of an order to a customer while minimizing the total cost of procurement, transportation, inventory management, and warehousing and reducing variability in execution. Visibility in the supply chain relates to the ability to track orders, inventory, and shipments in real time. Visibility allows a firm to monitor events and exceptions in real time so that it may proactively manage supply-chain activities. Supply-chain integration involves sharing and combining information flows relating to critical business processes including customer service management, procurement, product development and commercialization, manufacturing flow management and support, physical distribution, outsourcing and partnerships, and performance measurement.

[17]US R&D continues to rebound in 2004; see http://www.nsf.gov/statistics/infbrief/nsf06306/. Accessed December 22, 2006.

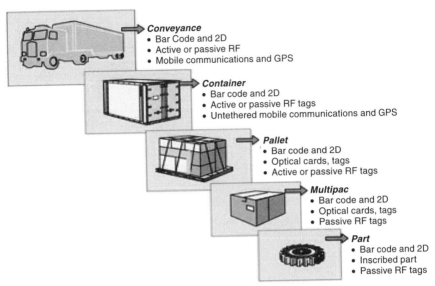

FIGURE 4 Asset tracking technologies. SOURCE: Available at http://www.ops.fhwa. dot.gov/freight/intermodal/freight_tech_story/freight_tech_story.htm#toc1. Accessed July 27, 2006.

plore patterns of change in innovative activity by examining interview and patent data for the industry as a whole and for each category separately. We begin by presenting the results of our interviews with executives in the industry.

Semistructured Interviews with Executives in the Industry and Industry Experts

To delve into the innovation process of the logistics industry, we conducted semistructured interviews with top executives in the industry and with industry experts. A copy of the questionnaire used for discussions with industry executives is available from the authors on request. We interviewed a total of 16 firms and 4 industry experts who were faculty at leading universities and representatives of industry associations. The firms include 4 LSUs (2 non-U.S.), 2 LSIs (1 non-U.S.), 4 LSPs (2 non-U.S.), and 6 PSPs (all U.S.). Since the respondent pool is small, there was concern raised by many respondents about the nature of information revealed. In order to obtain the most comprehensive information, we agreed that all information would be kept anonymous and that the names of responding firms would not be revealed.

The four firms classified as logistics users are large multinational firms with global operations in different manufacturing industries. Two of them are headquartered in the United States, one is in Europe, and one is in Asia. The two LSIs are among the world's largest consulting organizations. One of these firms is headquartered in the United States and one is in Asia; however, both have operations worldwide. One of the four LSPs is a 3PL organization located in Asia while three of them provide the complete range of logistics services as 4PLs. Two of the three 4PL firms are headquartered in the United States and one of them is headquartered in Europe. All three of the 4PL firms have worldwide operations and are among the largest providers of logistics services in the world. The six PSPs represent a range of products and provide hardware, software, or both for specific use in logistics. These firms are among the global leaders in their product categories, are headquartered in the United States, and have worldwide operations. Two of these PSPs have their primary supply-chain business functions in Europe, and we conducted our interview with their supply-chain personnel in Europe. Finally, we talked to two industry experts in the United States and two in Asia.

During the course of our conversations with the two LSI firms, we realized that, although they fill an important role in the industry, these two firms did not actually create new products or services. Rather, they reduce the friction in the supply-chain system, using products and services developed by the other three categories. Since our interest is focused on the changing locus of innovation in the industry, we have classified the responses of the LSI firms along with the responses provided by the industry experts for this section since LSI firms are prominent players in the industry and have valuable insight into the innovation activity in the industry as a whole.

The respondents were asked about the evolution of the supply-chain organization within their own firm. The mid-1990s appear to have been watershed years for supply-chain services for logistics users and providers. The user organizations we talked to began their logistics and supply-chain division, usually combining their purchasing and transportation departments of the past, around 1995. In subsequent periods, our LSU respondents noted that there has been partial outsourcing of parts of the logistics process such as transportation of raw materials and finished goods, inventory management, and warehousing. Some of them reversed their decision to outsource some logistics services but LSU firms differed in terms of logistics activities conducted within the firm and activities outsourced. Around the same time, the 3PL and 4PL firms we interviewed began offering more value-added services by adding warehousing and inventory management to providing transportation. Separate divisions offering supply-chain services such as transportation, warehousing, cross-docking, inventory management, packaging, and freight forwarding began in the 3PL and 4PL companies between 1995 and 1997. These divisions worked closely with customers to provide complete transportation solutions and used the parent company's fleet only if it was the best solution for the customer. The supply-chain services organizations among the users and the transportation providers of our respondents are usually headed up by a vice president who reports directly to the CEO or COO. This organizational arrangement shows the importance placed on the supply-chain activity by the LSU and LSP firms. The PSP companies in our sample were usually started when the innovator applied technology solutions to the logistics domain. All the PSP firms we talked to were created during or after 1990. Once again, most of their activity picked up around 1997. It is clear that the Internet, computing, and communications technologies created the right environment for firms to develop new products and services for the logistics industry just as the LSUs and LSPs were recognizing the need to effectively manage supply chains as the cornerstone to success in a global context of competition.

All respondents to the survey identified globalization—defined in the survey instrument as the increased mobility of goods, services, labor, technology, and capital throughout the world—as the primary driver of innovation in the industry. They noted the consequences of globalization for their business. First, globalization has increased competition in the LSU home countries. Consequently, firms can no longer trade cost for quality of product or service. Rather, they need to offer cost-competitive products of high quality to increasingly demanding consumers. LSUs are responding by looking to supply-chain solutions within their own firms and to partnering with LSPs and LSIs to create maximum visibility of information relating to the flow of product and money through the value chain. According to our respondents, globalization has also led LSUs to locate various value chain activities, such as product development and manufacturing, in different parts of the world. LSUs look for logistics partners who can provide global logistics solutions as they expand their supplier base and their markets. As LSUs and LSPs extend their product development, manufacturing, marketing, sales, and service activities around the world, they are confronted with unique country-specific challenges. For example, while labor costs in the emerging economies are attractive, the infrastructure for transportation and the regulations of local, regional, and national authorities create barriers to effective transportation solutions. LSUs and LSPs, sometimes with the help of local LSIs, have developed unique solutions in response to these challenges using their local subsidiaries and partners.

All the respondents concurred that most often users drove innovation in the logistics industry. An innovation typically began by identification of a business problem by the LSU. Sometimes, LSUs with their own supply-chain organizations developed unique solutions based on their particular needs. The LSUs we interviewed estimated that about one-fourth of the innovations they implemented in the past decade came from within their firms. Firms in the industry also worked collaboratively with subsidiaries and suppliers to innovate. For example, a global glass manufacturer we interviewed developed and deployed a simple supply-chain solution from within its own local operations in India, created a solution in partnership with another division within the same company in South Korea, and worked with an LSP to deploy a supply-chain innovation. The glass manufacturer looked first to internal R&D capabilities before seeking help from outside the firm. In general, industry-specific solutions were more likely to be created with internal R&D resources. In other situations, the LSUs communicate the supply-chain challenge to LSPs, and these firms then create innovative solutions, often working closely with LSUs.

According to our respondents, innovation at a PSP begins by working with a customer on a solution to a specific business problem. A leading RFID innovator and manufacturer mentioned that the firm also looks for feedback on its products from industry analysts, academia, and other external stakeholders. Internally, the customer service division, along with the sales, research and development, and professional services divisions, looks for new ways to enhance the customer experience. Our PSP respondents noted that they share the cost of development with the LSU when the solution is customized and later scale the product for wider market adoption. One of the PSPs we interviewed, a leading software firm, acquires smaller firms with the requisite innovation. Sometimes it just acquires the rights to the technology, leaving the smaller firm alone to continue its innovative trajectory. At other times, the entire firm is acquired. After acquisition, the firm may be completely integrated into the larger PSP business or may be left alone for product-development purposes while the customer interface functions—marketing, sales, and service—are handled by the larger firm. The same software firm noted that incremental innovations arise by continuously tracking customer complaints and wish lists. The next new release of the product addresses the issues most often raised by customers. PSPs and LSPs that have operations in many countries release products on a global basis but permit the local entities to have great freedom to adapt the product to local conditions. According to our U.S.-based respondents, products are released first in the United States and are subsequently scaled for global release.

We asked our respondents about their R&D process. The LSU and LSP respondents commented that, over the past decade, innovation has become more dispersed, both organizationally and geographically, and solutions involve multiple firms in locations all over the world. In the past, many supply-chain solutions were developed in-house. Today's best in class order management, warehouse/inventory management, and transportation management systems from PSPs provide the majority of functionality necessary to support new product development. Economies of specialization encourage PSPs to innovate, as can be expected from the "pull" system of innovation prevalent in the logistics industry where firms address immediate needs by expanding opportunities for local participants to use their creativity and innovate by drawing on a growing array of specialized and distributed resources. According to our LSU and LSP respondents, the active participation of PSPs in supply-chain innovation began about a decade ago as globalization combined with advances in communication and information technologies. The PSP respondents supported this view, stating that they focused on specialized niches and worked with LSPs and LSUs to customize products if necessary. The respondents felt that the consequence of the modular R&D process increased innovative output and allowed for greater specialization in all parts of the logistics network. Interestingly, LSIs have arisen as a response to the challenge of integrating modular inputs from the PSPs. LSUs in particular noted that LSI firms fill an important role in encouraging innovation among PSPs, since they facilitate the widespread adoption of innovative products with their integration expertise.

We then asked the respondents about the organizational structure of the supply-chain-related R&D activity. Among the LSUs and LSPs, there are significant structural differences in the way innovation is managed within the organization, with the trend shifting from large unwieldy product development groups to more focused, small development teams. For example, one firm we interviewed uses a dedicated core team, leveraging Six Sigma methodology[18] to drive product development; another firm, using techniques based on the systems development life cycle (SDLC) process,[19] uses a champion along with three leads (business, IT, and customer organization) to design, develop, and deploy the new products. The PSPs have a more traditional approach to R&D and have separate R&D departments to focus on both basic and applied research. For example, one of the PSPs we interviewed had a separate division focusing on RFID and related technologies. They worked closely with universities to improve electronic circuitry in tags or communication between tags and readers. A separate division worked with the customer service department to monitor and understand customer requests and complaints and then worked on creating customer- or industry-specific solutions. In general, PSP respondents noted that they detected the need in recent years to work more closely with their customers while developing new products.

---

[18]Six Sigma (6σ) is a business-driven, multifaceted approach to process improvement, reduced costs, and increased profits. With a fundamental principle to improve customer satisfaction by reducing defects, its ultimate performance target is virtually defect-free processes and products (3.4 or fewer defective parts per million [ppm]). The Six Sigma methodology, consisting of the steps "Define, Measure, Analyze, Improve, Control," is the roadmap to achieving this goal. Within this improvement framework, it is the responsibility of the improvement team to identify the process, the definition of the defect, and the corresponding measurements. The primary goal of Six Sigma is to improve customer satisfaction, and thereby profitability, by reducing and eliminating defects. Defects may be related to any aspect of customer satisfaction including high product quality, schedule adherence, and cost minimization.

[19]SDLC is a systematic approach to problem solving and is composed of several phases, each comprised of multiple steps including the software concept stage, which identifies and defines a need for the new system; the requirements analysis phase defining the information needs of the end users; the architectural design step, which creates a blueprint for the design with the necessary specifications for the hardware, software, people, and data resources; the coding and debugging phase to create and to program the final system; and finally the system testing phase, which evaluates the system's actual functionality in relation to expected or intended functionality.

Another interesting phenomenon we observed from our respondents is a shift in the geographic locus of innovation. Our respondents felt, based on the number of new products and services introduced, that the United States has historically been the hub of innovation for logistics for LSPs, PSPs, and LSUs headquartered there. According to our respondents, most of the innovation activity for these firms—idea generation, product development, and deployment—was conducted in the United States between 1995 and 2005. More than half of our U.S. respondents noted that the United States continues to be the primary source of innovation; however, about one-third of our respondents noted that innovation activities have shifted either to multiple locations around the world or to specific research centers in Europe and or Asia. These companies cited the need to increase operational efficiency, and for better information and shipment visibility in import and regulatory operations as reasons for geographic dispersion of innovation. However, one of our LSPs noted that it is in the process of creating regional centers of innovation to identify and develop products and services relevant to its geographic areas. The regions assigned to the centers were broadly continental in scope, defined as Asia-Pacific (including Australia and New Zealand), Europe, the Middle East, Africa, and the Americas. Larger firms are able to leverage global resources to develop efficient solutions on a cost-effective and industry-wide basis. However, local conditions may not permit these solutions due to regulations and lack of appropriate infrastructure. Understanding and adapting to local conditions is critical to successful innovation for firms in the industry. According to one of our respondents, the creation of regional innovation centers is expected to confer a competitive advantage. The firm believed that innovative output for the firm as a whole would be enhanced by the decentralization of innovation resources.

In dollar terms, the shift is still quite insignificant; less than 10 percent of total logistics-related development dollars are spent outside the United States, according to our U.S. respondents. For firms headquartered in Europe and Asia, a similar pattern has been observed by our respondents: more firms are going beyond country borders to develop innovations in response to customer needs. According to one of our respondents, "the move is not for cost reasons but for logistics reasons," meaning that increases in trade among the nations of the Asia-Pacific region and Europe and the Americas increase the need to be sensitive to local conditions and to develop country-specific solutions where necessary. These strategies are consistent with two of the five motives put forth by Gerybadze and Reger (1999) through their survey of 21 firms on the R&D internationalization process. In the logistics industry, based on the information provided by our respondents as to their reasons for creating R&D facilities in markets outside their home country, we find that firms tend to locate R&D centers in markets close to their lead customers. In addition, they exploit unique logistics-related capabilities available in developed and emerging economies to which their lead customers have expanded.

Zander (1999) offers a taxonomy of international innovation networks based on the extent of international duplication of technological capabilities and international diversification of technological capabilities. Early evidence on the capabilities of the innovation networks being developed in the logistics industry indicates that these networks belong to the dispersed category, in that capabilities at each of the centers combines core logistics R&D capabilities and diversified capabilities developed with local know-how and expertise in response to local lead customers.[20] R&D centers located outside the home country have been created by the LSUs and LSPs with a mandate to use and develop local expertise to create new products in developed and emerging-economy markets.

Respondents noted that increased collaboration among LSUs and LSPs is a significant shift in innovative patterns, especially in the past 5 years. LSU respondents indicated that they work closely with LSPs to develop custom solutions for their business needs. According to one LSU respondent, collaboration is necessitated, especially in a global context; this respondent provided an illustration based on the firm's experience in China. He noted that unique solutions are required in each market due to differences in infrastructure and resource availability. Variations in the physical network in each market can be expected because physical infrastructure in many emerging economies is inadequate and communication infrastructure is insufficient. However, the LSU respondent could ill afford to decrease information visibility for any reason and expected continuous monitoring and timely shipment delivery. The firm worked closely with a U.S.-based LSP's division in Hong Kong to ensure that the communications infrastructure was sufficiently robust and seamlessly integrated with the LSP's fleet and package tracking system, with the LSU's ERP systems in the United States, and with the supplier's ERP system in China. Similar incidents were shared with us by other respondents. The respondents acknowledged that close collaboration among the firms in the logistics network is necessary in order to create innovation that is effective system-wide.

Comparing responses between firms headquartered inside and those headquartered outside the United States, we found that U.S. firms had a global view of innovation. Innovative activity for these firms was more decentralized, both organizationally and geographically. In contrast, innovation occurred primarily in the home country for respondents from Europe and Asia. Although all the respondents stressed the importance of collaborative innovation, LSUs and LSPs in the United States have been more active in collaborative innovation. The global market was the focus of innovation of U.S.-based respondents, whereas respondents in Asia were focused on innovations for the local markets.

---

[20]Zander (1999) differentiates between international duplication and international diversification of technological capabilities. According to Zander, firms belonging to the *internationally duplicated* category include those where foreign units are typically involved in the same kind of technologies that are represented at home. Firms in the *dispersed* category have significant duplication or overlap of technologies across locations but also harness a number of technologies for which units in foreign locations have developed world product mandates.

Looking to the future, respondents believed that there would be greater emphasis on the intercontinental supply chain due to increased product flows between Asian countries and the United States and Europe. In addition there was expectation among the LSP respondents that there would be more mergers and acquisitions by LSPs leading to consolidation in that category. This activity will be driven by increased competition for larger shippers and the need for an increased global footprint. These trends will spur greater development of logistics R&D centers in the lead markets, especially in Asia. The PSPs in our survey primarily had marketing outposts in markets outside the United States. We expect that they will join the LSUs and LSPs to extend their R&D presence in their international markets.

In summary, our interviews with executives in the logistics industry revealed interesting insights into the innovation process. The dynamic management of the supply-chain process—going beyond transportation—as a distinct value-enhancing activity began in most organizations in the mid-1990s. Innovation is driven primarily by user needs and firms collaborate closely to develop solutions. Logistics innovation follows the global dispersion of markets and production capacity in most industries, although the extent of offshore migration of innovation-related activities in logistics remains modest. Our respondents noted that they are just beginning to create R&D centers in countries other than their home country, often to remain close to lead markets and to respond quickly to changing user needs. Increasing global trade has accentuated the need for LSUs and LSPs to create local solutions that take into account differences in physical infrastructure, availability of reliable transportation providers, and government and export rules and regulations. However, the PSPs' innovation-related activities appear to remain more "homebound" in the United States. Innovation has necessitated increased and closer ties among the LSUs, LSPs, and sometimes PSPs in the last 5 years in order to create products and services that are creative, efficient, and cost-effective throughout the supply chain.

Analysis of Patent Data

There are many organizations that collect and organize patent data, including country patent offices and proprietary databases such as those created by Thomson's Delpion. Each database has valuable information. We used the patent database created by the WIPO, a specialized agency of the United Nations. The WIPO website (http://www.wipo.int) has a patent search option for all patents filed worldwide. This option was used to do a search on patent filing data. The patent search uses a Boolean search option and hence enabled us to restrict patents to only those related to innovation in the logistics industry. The search was restricted to following four phrases in the description of the patent: supply chain, logistics, inventory, and freight. For the period between January 1990 and December 2004, we identified 2,268 unique successful patents filed worldwide that were relevant to the logistics industry after checking for incomplete data and multiple patents issued for the same innovation in different jurisdictions. We acknowledge that the cutoff year of 2004 may create some data truncation problems and so we interpret the data for the year with caution. Additionally, patenting may not be the best measure of innovation in an industry that is primarily focused on service to its customers. The logistics industry is quite unlike industries such as pharmaceuticals and biotechnology, where patenting is ubiquitous. However, the data on patenting reveal important patterns and trends that allow us to cautiously interpret the changing locus of innovation in the industry. A database was developed with the following fields from the WIPO database augmented by firm-level information from company websites: the patent name; date of filing; date of publication; international classification; application number; country and continent in which the patent was first filed; the inventor's name, home country, and home continent; and the name, home country, and home continent of the assignee. Each assignee was further designated as an LSU, LSP, LSI, or PSP using the description of the company's business from the company website and other published sources.

Figure 5 shows patent filing by continent of the first inventor between 1990 and 2004. The figure indicates that worldwide patent filing for supply-chain products climbed very slowly between 1990 and 1996, increasing from 17 in 1990 to 27 in 1996. The number of patents filed almost doubled in 1997 and then climbed dramatically until 2001. The economic downturn caused by the crash of Internet stocks and the 9/11 terrorist attacks may be reason for the slight decrease in patents filed between 2001 and 2004. While the patent filing levels have not yet reached the peak of 480 patents filed worldwide in 2001, the trend appears to be increasing. The number of patents filed in North America are nearly double the number filed in any other continent in any year and contributed about 60 percent of the patents filed each year between 1990 and 2004. The number of patents filed in Asia and Europe is steadily increasing: Europe's share of total patents hovered between 20 and 30 percent of total patents filed between 2000 and 2004 and Asia contributed around 7 percent in the same time frame. Patents filed in Australia[21] have shown an impressive increase, moving from 1 patent filed in 1996 to between 14 and 17 each year between 2001 and 2004. Logistics innovation is beginning to occur in Africa, where 11 patents were filed between 2001 and 2004. There were only 3 patents filed from inventors in South America during the entire period from 1990 to 2004. Australia, Africa, and South America together have contributed less than 1 percent of the total patents filed in any year.

Figure 6 shows the patents filed between 1990 and 2004 as indicated by the country of the first inventor for the top five countries with patents. Consistent with the perspective of the survey respondents discussed earlier, most of the innovation has taken place in the United States in this period. U.S.-based first inventors filed 1,340 patents. The number of patents filed by the U.S.-based inventors

---

[21]We included New Zealand in the numbers for the continent of Australia.

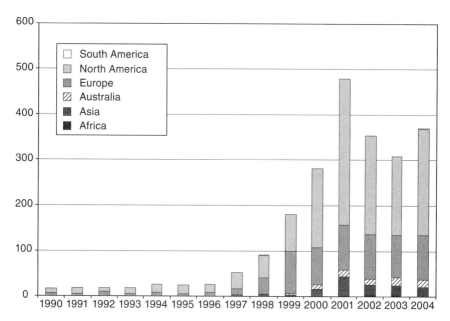

FIGURE 5 Patents filed by continent of first inventor between 1990 and 2004.

almost doubled from 49 patents in 1998 to 81 in 1999, doubled to 167 in 2000, and almost doubled again in 2001 to 308. The number declined to 214 in 2002 and even further to 169 in 2003 but shows signs of increasing again in 2004. In a distant second place is Germany, with 250 patents filed between 1990 and 2004. The number of patents filed by Germany-based inventors increased dramatically from 3 in 1997 to 19 in 1998 and then nearly tripled to 55 in 1999 and remained between 25 and 37 between 2000 and 2004. Following Germany, in third place, is the United Kingdom with 150 patents filed between 1990 and 2004. Japan and Australia filed 63 and 62 patents, respectively, in the same period, rounding out the top five countries when considering patents filed by country of first innovator.

Considering the locus of innovation over time, inventors from nearly 40 countries have been involved in innovation in the logistics industry since 1990, although Figure 6 shows only the top five countries. The increase in the number of inventor countries from 6 in 1990 to 26, 26, 25, and 27 in 2001, 2002, 2003, and 2004, respectively, is an indication of the increasing dispersion of innovative activity. Another indication of dispersion of innovative activity is the fact that five countries that filed in 2002 did not have an inventor file a patent in 2003 and four countries that filed in 2003 did not have a patent originate in their country in 2002. Eleven countries filed over 20 patents between 1990 and 2004 including

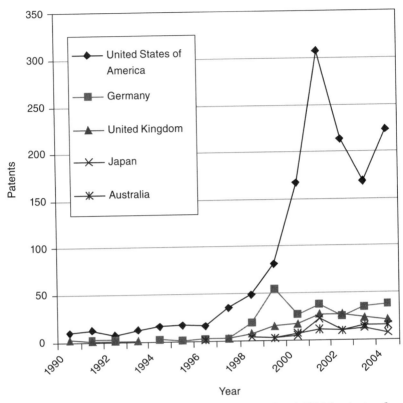

FIGURE 6 Patents by country of first inventor between 1990 and 2004 for the top five countries.

Australia, Canada, Denmark, Finland, France, Japan, the Netherlands, Singapore, and Sweden since 2000. It appears that China, India, Israel, and Korea are just beginning to become involved in logistics innovation.

A total of 149 patents had at least one inventor from a country different from the first inventor.[22] We examine those patents to understand patterns of cross-country collaboration. There is one instance of inventors from multiple countries in 1990 and then one in 1996 with no patent activity between 1991 and 1995 with inventors from multiple countries. After climbing to 29 instances of collabora-

---

[22]There were four patents that had inventors from three countries. They were not separated since they all involve the United States as the first inventor and a clear second inventor. Therefore, we counted each patent only once whether they had inventors from two countries or three focusing on the second inventor. We do provide details about the instances where inventors from the United States collaborated with inventors from two other countries in the discussion below and in Figure 8.

tion among inventors from different countries in 2001 and 2002, there is a sharp drop in collaborative activity to 20 instances in 2003, which may be an artifact of the beginning of the war in Iraq. These data indicate that cooperation in innovation in the logistics industry began only in the late 1990s and confirms the data provided by our survey respondents. Although U.S. inventors have more patents that involve inventors from other countries, the share of U.S. patents with a U.S. inventor as the first inventor and one or more foreign inventors (60 out of 1340, or 4.5 percent) is in fact somewhat smaller than the share of patents with German first inventors and foreign co-inventors (23 out of 250, or 9.2 percent), or logistics patents with British first inventors (16 out of 250, or 10.7 percent).

In 1999, inventors from the United States collaborated with inventors from China and Great Britain. In 2000, inventors from the United States collaborated with inventors from the Netherlands and Italy in one of the two instances of collaboration with the Netherlands. In 2001, inventors from the United States collaborated with inventors from Canada and the United Kingdom in one of three instances of collaboration with Canada and inventors from the United States partnered with inventors from the United Kingdom and Germany in one of four instances of collaboration with United Kingdom. AU, Australia; BE, Belgium; BM, Bermuda; BR, Brazil; CA, Canada; CN, China; DE, Germany; FR, France; GB, United Kingdom; IE, Ireland; IL, Israel; IN, India; JP, Japan; MX, Mexico; NL, Netherlands; SG, Singapore; TW, Taiwan.

Table 2 shows the countries with whom first inventors from the United States have partnered over time. Inventors from the United States have partnered with inventors in over 17 countries between 1990 and 2004 for a total of 60 patents.

TABLE 2 Patents with Inventors from Multiple Countries with United States as First Inventor

	US																	Grand Total
	AU	BE	BM	BR	CA	CN	DE	FR	GB	IE	IL	IN	JP	MX	NL	SG	TW	
1997			1				1											2
1998					1													1
1999		1			1*	2												4
2000					1		1						1		2*			5
2001		2			3*	1	1		4*					1				12
2002	1	1			2		3		6		1		1					15
2003					1		2		1	1		1		1				7
2004	2		1				4	1	1		1	1				2	1	14
Grand Total	3	4	1	1	7	2	12	3	13	1	2	2	1	1	4	2	1	60

NOTES: There are four instances when there were inventors from three countries and these are noted with an asterisk. We have counted these instances against the second inventor country.

While inventors in the United States partnered with inventors from only three countries in 1999, they partnered with inventors from nine countries in 2004, indicating an increasing scope of cooperation. U.S. inventors partnered with inventors from the United Kingdom and Germany most often during this period. However, 2004 data indicate that inventors from the United States have one patent with an inventor in India, one with an inventor in Taiwan, and two with inventors in Singapore, showing an increasing trend of cooperation of U.S. firms with the Asia-Pacific region. The share of all U.S. logistics patents accounted for by patents with inventors from multiple countries for which U.S. inventors appear as first inventors varies from a low of 2 percent in 1998 to a high of 7 percent in 2002. After a drop to 4.1 percent in 2003, in 2004 the number of U.S. co-invented patents rose to 6.2 percent of total U.S. logistics patents.

Table 3 provides the details about the 37 patents between 1990 and 2004 by first inventor's country, where the second inventor is from the United States. Once again, the collaboration shows a sharp increase in 2000 and is flat after that point. In 2003 and 2004 the inventors from the United States supported inventors in five other countries—the most in the 15 years studied. Most of the collaboration, where the second inventor is from the United States, was with Germany and the United Kingdom, which is similar to the pattern when the inventor from the United States is the first inventor. These data must be interpreted with caution, however, since instances of collaboration account for a small share of overall logistics patents. For instance, the share of all German logistics patents with U.S. inventors in second position fell from 6 percent in 2003 to 3 percent in 2004; the share of all British logistics patents similarly classified was 8 percent in 2004 and 5 percent in 2005. Data in Tables 2 and 3 combine to show that U.S. inventors in the logistics industry are increasingly collaborating with inventors worldwide. However, given the growth in patents over the period studied, the proportion

TABLE 3 Patents with First Inventors from Different Countries and with Second Inventor from the United States

	1990	1997	1999	2000	2001	2002	2003	2004	Grand Total
Belgium						1			1
Canada			5	4				1	10
Germany				2	4	2		1	9
Denmark								1	1
Spain						1			1
France				1			1		2
United Kingdom	1	1	1		1	2	2	1	9
Israel								1	1
Japan				1	1		1		3
Grand Total	1	1	1	7	8	7	7	5	37

of co-inventions in logistics with U.S. inventors listed as the first inventors or second or lower appears to be limited.

Figures 7 through 10 examine patterns in assignee patent data. While assignee data are not meaningful when considering geographic location of innovative activity, they do indicate the kinds of firms that own the patents. Figure 7 compares patents filed by category of assignee over time. Only those countries that had more than 50 patents between 1990 and 2004 are shown in Figure 10. A brief discussion on assignee data provides some insight into patterns relating to resource allocation. Examination of patents filed in the past 15 years indicates that PSPs have been the dominant assignee worldwide with 1,095 patents, followed by 678 for the LSUs. Of the 302 patents that have been assigned to individuals designated as experts in our study, 13 patents were assigned to universities and 36 patents were assigned to individuals associated with universities. The number of patents filed by PSPs worldwide shows a steady increase from 1997 through 2001 in absolute terms. The share of logistics patents assigned to PSP almost doubled from 30.8 percent in 1997 to 60.4 percent in 1998. After climbing to nearly 68

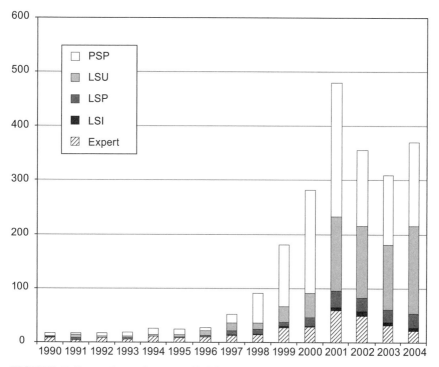

FIGURE 7 Comparison of patents filed by category of assignee between 1990 and 2004.

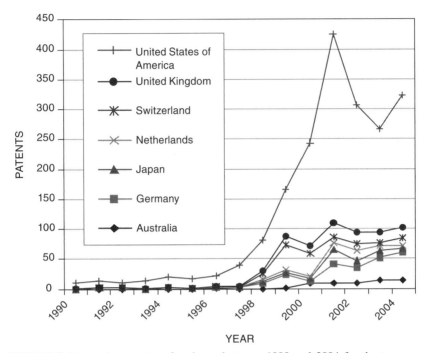

FIGURE 8 Patents by country of assignee between 1990 and 2004 for the top seven countries.

percent of total patents in 2000, the PSP share of overall logistics patents declined to 41 percent in 2004. The patents assigned to LSPs doubled between 1999 and 2000 and doubled again in 2001. Additionally, and consistent with our earlier discussion on the recent participation of LSIs in the logistics network, a few patents have been assigned to LSIs since 1997. Our discussions with LSIs and industry experts indicate that the role of the LSI in the industry is often focused more on process improvement and less on patentable innovation. A large number of patents have also been assigned to LSUs, suggesting that user firms are investing in developing logistics capabilities as a critical business advantage. In 2004, more patents were assigned to LSUs than to PSPs, with the LSU share of total patents rising from 15.7 percent in 2000 to 44.1 percent in 2004.

Figure 8 shows the distribution of patents by assignee home country or country where the assignee is headquartered. It is worth noting from the figure that, although the firms headquartered in the United States have nearly seven times as many patents as the next country—Germany—the countries of Switzerland and the Netherlands have 129 and 53 patents assigned to them. Further analysis shows

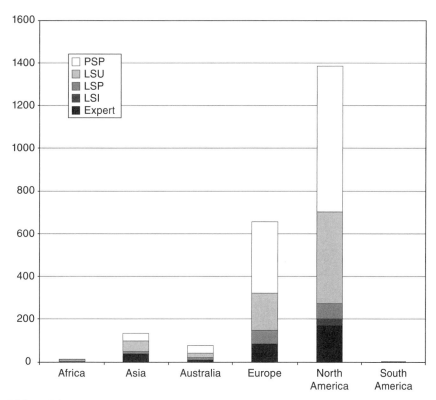

FIGURE 9 Patents by category of assignee by continent.

that the individuals listed as first inventor for the patents assigned to Dutch and Swiss firms were mostly from Germany, the United States, or Belgium.

In the geographic distribution of patents shown in Figure 9, firms assigned the most patents were from the two developed-economy regions represented broadly by North America and Europe followed by the emerging economies of the Asia-Pacific region. Three times as many patents were generally assigned to firms in North America as were assigned to firms in Europe, and the order of magnitude was even larger when compared to the Asia-Pacific region. A closer examination of the data reveals that PSPs in the United States were assigned the most patents between 1990 and 2004, with more patents than all the patents worldwide for experts, LSIs, and LSPs combined and almost as many patents as patents assigned to LSUs worldwide. In contrast to the prolific innovation of PSPs in the United States, LSUs were assigned the most patents in most of the European countries, with the exception of Germany and Sweden. Among the countries in Asia no particular pattern could be discerned. Experts were assigned the most patents in China, LSUs in Japan, and PSPs in India and Singapore.

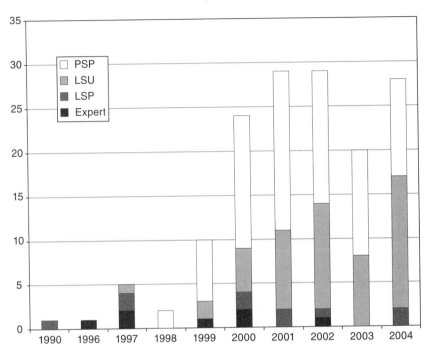

FIGURE 10 Patents filed by category of assignee between 1990 and 2004 when inventors are from different countries.

Figure 10 presents the number of patents with inventors from different countries for each assignee category over time. Although more than one-half of co-invented logistics patents were assigned to PSPs (80 out of 149), in fact the share of PSP-assigned patents that were co-invented (7.3 percent) is similar to the share of co-invented LSU patents (7.7 percent) or co-invented LSP patents (6.3 percent). After peaking in 2002 at 10.7 percent, collaborative PSP patents decreased to 9.3 percent in 2003 and 7.1 percent in 2004. One explanation for this trend may be that firms are creating local R&D centers to create innovative products and solutions for local consumption.

Our study of patent data indicates that logistics innovation gained importance in the mid- to late 1990s. In addition, we found that most of the innovation activity remains in the United States. However, logistics innovation has become increasingly dispersed. For each of the past 5 years, inventors from more than 20 countries had filed logistics-related patents with nearly 60 percent of the patents coming from inventors based in the United States. More patents have been assigned to PSPs—firms who provide software and hardware to enhance communications, tracking, and visibility along the supply chain—than to users of logistics services and providers of logistics services until 2004, when users were assigned nearly 44.1 percent of all patents. Since 1996, the number of patents with inventors from different countries has increased but the trend has been flat since 2000. Inventors from Germany have the highest proportion of collaborative patents, averaging more than 13 percent since 2000, and 11 percent of the United Kingdom's logistics patents during the same period reflect collaborative invention. Inventors in the United States have partnered with inventors in over 17 countries, primarily in Europe. However, when considering the share of patents created through collaboration (with the first inventor located in the United States), only about 5 percent of the patents were co-invented between 2000 and 2004. There is evidence of recent partnerships with inventors in the Asia-Pacific region.

DISCUSSION

An analysis of innovation in the logistics industry is handicapped by the lack of readily available data. Most of the innovation activity is confidential information, and the level of investment in logistics innovation is hard to track specifically. Financial and operational data about this aspect of their business is not publicly disclosed. To the extent possible we have relied on public sources of data to augment our understanding of the phenomena learned through private conversations. Where public data are lacking, we have relied heavily on the information obtained from our interviews, although our conclusions are not based on any single source of information.

Our study examined multiple aspects of innovation in the logistics industry through interviews with industry insiders and experts and through a study of patent data. While the logistics industry's innovation may not be perfectly reflected in patent data, we believe that, with cautious interpretation, they provide valuable information about important trends in innovation in logistics.

In addition we also examined the innovation activity of the top logistics firms, as listed in the Hoovers database, in order to understand innovation in logistics. For each of the top firms we looked at company websites, articles in industry periodicals, and databases such as Factiva and One Source Global Business.

As noted earlier, we find that innovation is gaining importance in the logistics industry. The advent of new technologies and globalization has inspired firms to look for new solutions for the challenge of business in today's competitive landscape. Innovation is also becoming more global than it has been in the past, although the United States continues to lead the innovative effort in the industry. Inventors in the United States have been the most productive throughout the past 15 years and we do not observe any threat to this dominance. Inventors from Japan and Australia have become more productive in the past 5 years, and many European countries have become more active in logistics innovation as trade increases within the European Union. Since 2000, there has been increased collaboration in the industry. Inventors from different countries collaborated on about 6 percent of the patents filed, 88 percent of which were filed after 2000. Nearly half of the intercountry collaboration involved a U.S. first inventor but these patents were a very small proportion of the total U.S. patents. U.S inventors collaborated with inventors in 17 other countries when they were listed as the first inventor, and inventors from 9 countries when they were the second inventor. PSPs were the most frequent collaborators but more recently LSUs and LSPs have been more actively collaborating for innovation.

Our data lead us to the conclusion that logistics innovation is most often a pull phenomenon. The users of logistics services identify a business challenge. As the supply-chain process has increased in complexity due to the increasing scope of business activity, solutions to these challenges require specialized and often distributed resources. The LSU "pulls" the innovation from the PSPs and LSPs with requisite specialized capabilities. More recently, LSI firms have entered the logistics value space. They are software and hardware platform- and product-neutral and focus on providing the best solution to the customer. We tracked few innovations from these firms. However, if their role increases in the logistics landscape, we may see an increase in innovative activity among the integrators.

Product innovations, especially those related to new technologies, are enabled through collaboration. For example, DHL Worldwide, an LSP, has recently unveiled its RFID pilot project, which it is developing in partnership with IBM, an LSI. Savi Technologies, a PSP and recently acquired by Lockheed Martin, leverages its leading position in RFID solutions through alliances with leading firms in data collection, services, software, and solutions for the logistics industry.[23]

We see a growing trend of innovation in different parts of the world as the outcome of globalization. We have noted that a few LSUs and LSPs are beginning to locate logistics R&D centers in different parts of the world in order to be close to the customer in lead markets. We believe that these centers are used to create capabilities that allow greatest responsiveness to local customer needs while exploiting local expertise and relationships. Nearly two-thirds of the executives responding to a survey by *The McKinsey Quarterly* in September 2006 stated that they are concerned with increasing risks from disruptions to their supply chain. The environment is ripe for firms in the logistics industry to continue to innovate and address users' concerns about their supply chain.

POLICY IMPLICATIONS

Archibugi and Iammarino (1999) offer a taxonomy of globalization of innovation with a view to understanding the implications for national policy. Based on the locus of innovation and the nature of exploitation, the three categories they suggest are the international exploitation of nationally produced innovations, the

---

[23]See http://www.savi.com/partners/program.shtml.

global generation of innovation, and global techno-scientific collaborations. Our study of the logistics industry indicates that innovation in the logistics industry has mostly belonged to the first category—international exploitation of nationally produced innovations. Logistics innovations have generally been developed for the home market, predominantly the United States, and then scaled for the international market. However, the U.S. logistics industry is slowly moving toward the second and third categories—global generation of innovation and global techno-scientific collaborations. Accordingly, one of the most relevant policy issues for the logistics industry is the role of governments in creating and enforcing patent law. Patent law enforcement is disparate and unequal in many parts of the world. Our study shows that U.S. logistics firms have been prolific in innovation compared to their peers in other parts of the world. With the globalization of logistics users and suppliers, innovations developed in the United States should be adopted in every country without concern for misappropriation of intellectual property (IP). Inadequate IP protection is a concern for many industries, and the logistics industry is no exception. As seen from the data, innovation in logistics is in the early stages of its life cycle. While there is no evidence that IP protection is hindering innovation now, the role of formal IP protection is likely to assume greater importance as innovation increases in the industry.

Reducing trade barriers and standardizing import rules would also help innovation in the U.S. logistics industry. Market liberalization in many of the world's fastest-growing economies would increase opportunities for U.S. logistics firms. Data indicate that the U.S. logistics industry is the most innovative and the most advanced in the world. Liberalization would provide the opportunity to extend the market reach of U.S. logistics firms and add to their capabilities by being exposed to different contexts and conditions. For example, Brunei, Singapore, and Thailand have recently signed the Multilateral Agreement on the Full Liberalization of All-Cargo Services. This agreement allows airlines to operate unlimited cargo on any route within the three countries (Bower, n.d.). This type of agreement enables logistics firms to create innovative solutions to better respond to customer needs since they have fewer constraints imposed on the flow of goods, information, and funds.

A skilled workforce is an essential component for successful innovation in any context. However, as discussed earlier in this chapter, the logistics industry has become a high-tech complex network with information flow becoming the critical source of competitive advantage. Graduate enrollments in science and engineering fields in U.S. universities reached a record high of 566,800 in the fall of 2003, according to data collected by NSF through 2003.[24] Encouraging higher education, especially in the fields of mathematics, computer science, and

---

[24]S&E Indicators 2006 Available at http://www.nsf.gov/statistics/seind06/c2/c2h.htm. Accessed January 22, 2007.

engineering, is essential so that the requisite skilled workforce is available for the United States to continue its dominance in logistics innovation.

Since the events of September 11, 2001, supply-chain security has been receiving much publicity. According to a recent report by the Council of Supply Chain Management Professionals, none of the programs instituted by the U.S. government have been successful in eliminating or significantly reducing supply-chain vulnerability (Wilson, 2005b). While most of the attention and funding has been directed toward enhancing airport and passenger security, little attention has been paid to the rest of the transportation system. The shipment of containers represents one of the greatest risks in the cargo supply chain. In 2004, over 10 million loaded cargo containers were imported into the United States, representing about $1.43 billion of containerized goods moving through U.S. ports each day.[25] Despite the security programs enacted by Congress, such as the Maritime Transportation Security Act and the Customs-Trade Partnership Against Terrorism, experts agree that security along the supply chain remains woefully inadequate (Wilson, 2005b). It is critical that the U.S. government devote greater resources to address this security risk by increasing funds and manpower for adequate implementation of security programs. There is no evidence to suggest that firms in the U.S. logistics industry have received federal funding for security-related innovation. The innovative power of the U.S. logistics industry could be harnessed for creating new products for increasing supply-chain security. The logistics industry has been up to the challenge when faced with many changes in the environment; with the right incentives from the government, the industry could become a partner in creating a safer, more secure, supply chain through innovation.

CONCLUSION

The Council of Supply Chain Management Professionals has been publishing an annual report on the state of the logistics industry for the past 17 years. According to the latest report, after declining for most of the last 7 years, logistics costs as a percentage of the nation's GDP are pushing upward. After hovering around 10 percent of GDP during the 1990s, logistics costs as a percent of GDP decreased as interest rates declined and business cost pressures mounted. However, the most recent estimate is for 2005, when logistics costs were about 9.5 percent of GDP, as shown in Figure 1. Estimated logistics costs in 2005 totaled $1.183 trillion—an increase of $156 billion over 2004 and the largest year-on-year change in the 17-year history of the report. Shippers are feeling pressure from two directions. One factor is the steady climb in interest rates, which has pushed up inventory-carrying costs. But the biggest cost driver has been rising

---

[25]Liner Shipping Facts and Figures. World Shipping Council. Available at http://www.worldshipping. org/liner_shipping-facts&figures.pdf. Accessed December 26, 2007.

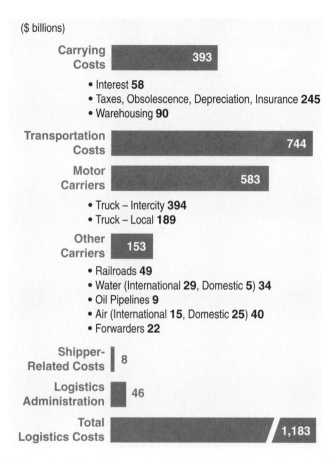

FIGURE 11 Costs in the logistics market, 2005. SOURCE: http://www.logisticsmgmt. com/article/CA6352889.html. Accessed August 28, 2006.

transportation expenses, which reached $744 billion in 2005, up from $636 billion in 2004. Soaring fuel prices, a driver shortage in segments of the trucking industry, and diminished competition have all come together to raise rates across all modes, and for trucking in particular. Figure 11 shows the breakdown of the costs in 2005.[26]

Managing these increasing costs is going to become even more critical in the future. Domestic freight transport is expected to increase by another 65 to 70 percent by 2020. International shipments are expected to increase even more

---

[26]See http://www.logisticsmgmt.com/article/CA6352889.html. Accessed August 28, 2006.

over this period (by about 85 percent).[27] As the need for logistics management grows, businesses will need to be adaptable and flexible to changing environmental conditions. In a world of synchronized trade, collaboration among all parts of the logistics network becomes critical. Innovation in logistics is essential for the flow of goods, information, and funds to be seamlessly choreographed. Some factors such as interest rates and fuel prices are beyond the control of the firm, but investment in innovation, especially that directed at reducing variability in quality of service and uncertainty in the supply chain, shows the greatest promise for the future of the industry.

ACKNOWLEDGMENT

We are grateful to the Sloan Foundation for generous support of this research.

REFERENCES

Archibugi, D., and S. Iammarino. (1999). The policy implication of the globalization of innovation. *Research Policy* 28(2,3):317-336.

Bartels, N. (2006). 21st century logistics. *Manufacturing Business Technology*. March.

Bernanke, B. (2006). Global economic integration: What's new and what's not? Available at http://www.federalreserve.gov/boarddocs/speeches/2006/20060825/default.htm. Accessed August 28, 2006.

Bhise, H., D. Farrell, H. Miller, A. Vanier, and A. Zainulbhai. (2000). The duel for the doorstep. *The McKinsey Quarterly* 2:32-41.

Bower, E. Z. (n.d.). Off shoring operations by United States firms in logistics and distribution services: Implication for ASEAN. Available at http://www.us-asean.org/ASEAN/Svcs_FDI_Comp_Paper.doc. Accessed August 15, 2006.

Bowman, R. J. (n.d.) Supply chain management: The perils of going global. Available at www.supplychainbrain.com, category SCM Technology. Accessed August 15, 2006.

Brown, J. S., and J. Hagel III. (2005). From push to pull: The next frontier of innovation. *McKinsey Quarterly* 3. See www.Mckinseyquarterly.com Accessed July 26, 2006.

Council of Supply Chain Management Professionals. (n.d.). Supply chain management/logistics management definitions. Available at http://www.cscmp.org/Website/AboutCSCMP/Definitions/Definitions.asp. Accessed July 26, 2006.

Craig, T. (2003). 4PL versus 3PL—A business process outsourcing option for international supply chain management. *World Wide Shipping*. December/January. See http://www.ltdmgmt.com/mag/4pl.htm.

Economist. (2006). The physical Internet: A survey of logistics. June 17.

Gerybadze, A., and G. Reger. 1999. Globalization of R&D: Recent changes in the management of innovation in transnational corporations. *Research Policy* 28(2,3): 251-275.

Grosspietsch, J., and J. Kupper. (2004). Supply chain champs. *The McKinsey Quarterly* 1:2-4.

Handfield, R., and E. Nichols. (2002). *Supply Chain Redesign*. Englewood Cliffs, NJ: Prentice Hall: p. 13.

---

[27]See http://www.ops.fhwa.dot.gov/freight/freight_analysis/nat_freight_stats/docs/05factsfigures/table2_1.htm. Accessed August 27, 2006.

Hugos, M. (2006). *Essentials of Supply Chain Management.* Hoboken, NJ: Wiley and Sons. 2nd edition, pages 7-9.

McKinsey Quarterly. (2006). Understanding supply chain risk: A McKinsey Global Survey. September.

Mulpuru, S. (2006). US eCommerce: Five year forecast and data overview. October 12. Available at www.Forrester.com. Accessed November 8, 2006.

Pande, A., R. Raman, and V. Srivatsan. (2006). Recapturing supply chain data. *McKinsey on IT.* Spring 2006, pages 16-21.

Rogers, E. M. (1995). *Diffusion of Innovations.* 4th Ed. New York: Free Press, p. 11.

Singh, M. (2004). A review of leading opinions on the future of supply chains. Supply Chain 2020. Working paper. Available at http://ctl.mit.edu/public/opinions_future_supply_chains.pdf. Accessed July 26, 2006.

Smith, B. (2005). Customer satisfaction is the wrong measure. *Gallup Management Journal.* April 14. Available at http://gmj.gallup.com/content/15850/Customer-Satisfaction-Is-the-Wrong-Measure.aspx. Accessed December 22, 2006.

Smith, F. (2006). Going with the Flow: How to Succeed in a Macro-trend Environment. Speech given at George Washington University, February 8, 2006. See http://www.fedex.com/us/about/news/speeches/gwu.html?link=4. Accessed July 26, 2006.

Stoffel, B. (2006, July). Navigating the world of outsourcing. *LQ Magazine* 12(3).

Vinas, T. (2005). IW value-chain survey: A map of the world. *Industry Week.* September 1. Available at http://www.industryweek.com/ReadArticle.aspx?ArticleID=10629. Accessed August 9, 2006.

Violini, B. (2006). What can logistics do for you. *Global Services.* June. Available at www.globalservicesmedia.com. Accessed July 15, 2006.

Wilson, R. (2004). 15th Annual State of Logistics Report. Presented at the National Press Club, Washington, D.C., June 7.

Wilson, R. (2005a). U.S. state of logistics. Available at http://www.logistics.or.jp/member/pdf(mazazineinformation)/20050809.pdf. Accessed August 26, 2006.

Wilson, R. (2005b). 16th Annual State of Logistics Report. Presented at the Ronald Reagan Building and International Trade Center, Washington, D.C., June 27.

Zander, I. (1999). How do you mean "global"? An empirical investigation of innovation networks in the multinational corporation. *Research Policy* 28(2,3):195-213.

9

Venture Capital

MARTIN KENNEY
University of California, Davis
MARTIN HAEMMIG
UniBW Munich, Germany and Leiden University, Netherlands
W. RICHARD GOE
Kansas State University

INTRODUCTION

In 1946, the first venture capital (VC) firms were established in the United States with the objective of providing financial backing and business assistance to entrepreneurs in exchange for repayment in capital gains. These pioneering VC firms soon discovered that technology-based innovations most consistently yielded the greatest returns. Today, the ideal/typical VC firm is the U.S.-style limited partnership that is embedded in a local entrepreneurial ecosystem and invests in technology-related deals. The pioneering VC firms were also motivated to diffuse venture capitalism nationally and internationally. In retrospect, it would appear that they were successful. In the past two decades venture capitalism has spread globally. There are now domestically owned VC firms operating in at least 40 nations. Also, there is an increasing number of VC firms that have established offices or begun investing in multiple nations, or both (i.e., VC firms operating across national borders). The international diffusion of VC investing suggests that there are entrepreneurial ventures that merit funding in many nations.

In 2006, there can be little doubt that VC financing of entrepreneurial high-technology ventures plays a significant role in the U.S. national innovation system. Venture capitalists have backed nearly all of the significant U.S. information technology (IT) firms established during the past four decades. These include 3Com, AMD, Apple, Applied Materials, Cadence, Cisco, Google, Intel, Oracle, Netscape, Seagate, Silicon Graphics, Solectron, Sun Microsystems, Yahoo!, and many others. In biotechnology, the VC-financed firms include Amgen, Biogen, Cetus, Centocor, Chiron, Genentech, and many others. In the nontechnology fields, important VC-funded firms include Federal Express, jetBlue, eBay, Home

Depot, and Office Depot. A number of these firms have changed the way human beings work and interact or, to borrow a phrase from Steven Jobs, "they have changed the world." Recently, new VC-financed firms, such as Facetime, MySpace, and YouTube, are driving yet further change.

If one agrees that the VC-financed firms are critical to the U.S. position in the global technology economy, then the globalization of the VC industry is an important topic. The VC industry is not significant in terms of either direct employment or the total capital under management. Rather its significance lies in the role of venture capitalists in finding, funding, and assisting entrepreneurs whose firms will be instruments of Schumpeter's (1939) "creative destruction" or successful in creating "new economic spaces." Previous studies have attempted to measure the employment contribution of firms funded by VC and found it to be extensive (see, e.g., Global Insight, 2004).

Our examination of the globalization of the VC industry proceeds as follows. The first section introduces VC as an organizational form. The second section examines the academic research on international VC investing. The following section describes the history of international VC investing, which began in the 1960s and has since grown enormously. The fourth section examines the reasons for the globalization of the VC industry and is followed by a quantitative section describing the international flows of VC. In the sixth section the growth of the VC investment in China is discussed. The concluding section discusses the implications of the findings for our understanding of the globalization of VC. We find that there is little evidence at this time to indicate that the globalization of VC is having a negative effect on the U.S. innovation system. The existing evidence does suggest that U.S. VC firms are finding viable investments in other nations and foreign VC firms continue to find viable investments in the United States.

VENTURE CAPITAL AS AN ORGANIZATIONAL FORM

Professional VC firms are the subject of this chapter and, as far as is practicable, buyout (BO) and angel investors are omitted from our analysis. Private equity (PE) firms are organizations that invest in firms with the aim of later selling this equity at a higher price to capture the capital gains. VC is a subset of PE firms. We do not include BO firms because they acquire existing firms and thus are involved not in supporting emerging firms but rather in acquiring and reorganizing existing firms. Angel investing refers to equity investment in young firms by individuals or groups of individuals using their own funds and is a practice that is hundreds of years old. The difference between angels and venture capitalists is that the venture capitalists are professionals operating an organization that has raised institutional money.

This chapter treats private VC as the ideal type and does not examine corporate VC. The reasons for omitting corporate VC are threefold: First, as a whole, corporate VC is much less important than private VC. Second, many corporate

VC operations have very different organizational structures, degrees of indepen-dence, compensation, and reporting relationships to the parent corporation than do the private VC firms. Finally, most corporate venture capitalists do not perform the lead-investor role, preferring instead to invest after, or in conjunction with, the private VC firms.

The narrow definitions of VC are slippery in the real world. For example, "venture capital" investing in Japan and Korea traditionally has been in the form of loans (Clark, 1987; Kenney et al., 2004; Kuemmerle, 2001). The data collected on "venture" investment often commingles VC and BO—and even angel invest-ments if they are sufficiently large.[1] In Europe, the acronym EVCA stands for the European Private Equity and Venture Capital Association, and BO continues to be the dominant investment pattern in Europe (Gompers and Lerner, 2001), al-though the relative shares may be changing. Despite the disparate definitions, the term "venture capital" in this chapter refers to equity investors in young firms.

The VC firm is a small financial services professional organization (usually employing a total of less than 30 persons) that functions primarily to (1) assess business opportunities; (2) provide capital; and (3) actively engage, monitor, advise, and assist the firms in its portfolio (i.e., those firms in which VC has been invested). By investing, the venture capitalist accepts a substantial tranche of illiquid equity that converts its status to something like a "partner" to the en-trepreneur. The goal of the venture capitalist is not only to increase the value of that equity, but to eventually monetize the investment through a liquidity event such as an initial public stock offering or sale to another investor so it can reap the results of its investment. The final way of "reaping the reward" is firm failure and bankruptcy. In all of these scenarios, the venture capitalist "exits" the invest-ment (i.e., ends its ownership role in the firm). This is necessary to complete the process because the VC firm's investors must be paid by liquidating the holdings. In environments in which exit is impossible, venture capitalists cannot invest.

The economics of VC are characterized by high risk and high returns. Invest-ing in young firms is risky—many fail and become total losses. The compensation for the failures comes from investments that yield 10, 20, or even 100 times the initial capital invested by the venture capitalists. This asymmetric return profile means that venture capitalists only invest in firms offering the opportunity for ex-tremely large returns. To be clear, venture capitalists are industry-sector agnostic, but as a generalization, during the past five decades, the sectors that most often generate such opportunities are the information and communication technolo-gies. The biomedical fields are the only other ones with a long history of good returns. Of course, many other investment fields, such as energy in the 1970s, superconductivity, and now, possibly, nanotechnology, have come and gone with minimal returns.

Operationally, venture capitalists invest only after rigorous reference check-

---

[1]All statistics used in this chapter are for VC only unless otherwise noted.

ing (due diligence) and, in return for capital, the venture capitalist receives equity and a seat on the board of directors from which to actively monitor and assist the firm's growth. After investment, the ideal-typical VC firm provides assistance ranging from practical needs such as providing advice on issues that a fledgling firm might encounter, introducing contacts, and assisting in securing necessary executive talent, to more abstract needs such as providing "legitimacy" (Aldrich and Fiol, 1994) to help overcome "liabilities of newness" (Stinchecombe, 1965).

PREVIOUS RESEARCH ON VENTURE CAPITAL GLOBALIZATION

In many respects, the globalization of VC firms is puzzling. The academic literature suggests that VC investing is strongly localized (Sorenson and Stuart, 2001), although Florida and Kenney (1988) found that venture capitalists in financial centers such as New York and Chicago exported capital to technology centers such as Silicon Valley, where the entrepreneurs and deals were clustered. More recently, Kogut et al. (2007) showed that even the early U.S. VC firms co-invested with distant firms. Also, by the mid-1980s, many East Coast VC firms established branch offices in other regions, particularly in Silicon Valley (Kenney and Florida, 2000). Perhaps this indicates that the emergence of the global VC firm should have been expected given the increasing globalization affecting nearly every industry, particularly those industries funded by venture capitalists.

The international dimensions of VC are receiving increased attention from scholars (Gompers and Lerner, 2001).[2] In the 1980s and 1990s, studies examined the establishment of VC firms in other countries (Clark, 1987; Green, 1991; Kuemmerle, 2001; Manigart, 1994). There also have been a number of cross-country comparative studies, both at the institutional level (Kenney et al., 2004; Manigart et al., 1996) and in comparisons of practice (Pruthi et al., 2003).

The literature offers three explanations for the uncanny success of VC as an institution in the United States and its relative slow diffusion to most other nations. The first explanation is inspired by the work by La Porta et al. (1998), which suggests that English common law-based nations have had the most successful VC industries. This success is ascribed to the proposition that non-common law legal regimes offer less protection to the owners of capital (Bottazzi et al., 2005; Hege et al., 2004; Lerner and Schoar, 2005). The legal/governance explanations may explain part of the cross-national variation, but these studies do not control for technological capability or entrepreneurial environment, which the extant literature suggests is important for understanding the subnational regional success in VC investing.

The second explanation is a variant on the governance explanation that focuses on the nexus between ownership and control. Here, the existence of a local

---

[2]See Zalan (2004) and Wright et al. (2005) for comprehensive reviews of the academic literature on VC globalization.

stock market as the main source of capital for growing firms is singled out as a critical variable for vibrant VC industry growth. For example, Black and Gilson (1998) argue that economies with stock market-based financial systems have stronger VC industries than economies with bank-based systems, presumably because new firms can raise capital in these markets. In a cross-national comparison, Jeng and Wells (2000) found that the numbers of inital public stock offerings (IPOs) in the domestic market were the strongest driver of VC investing.

Since IPOs are highly visible markers of wealth creation, one would expect that they would have a demonstration effect. However, Stuart and Sorenson's (2003) research suggests that in the United States this effect is very local. The stock market-based explanation has a powerful appeal because it resonates with the obvious need for there to be exits for the reproduction of VC. This conclusion is drawn from the success of VC firms in the United States and the United Kingdom (unfortunately, the European data used for empirical studies during this period came from the EVCA and do not distinguish between BO and VC). More recently, this perspective has been brought into question because IPOs on foreign markets have been the exit strategy of choice for Israeli and Chinese firms (Rock, 2001; Zero2IPO.com, 2005).

These corporate governance/financial system arguments have been questioned. For example, a study by Kaplan et al. (2003) of funding agreements from 23 nations (not including the United States) found that U.S.-style contracts could be written in a wide range of legal regimes and were used by more experienced venture capitalists. The implication is that legal systems might not be as significant an obstacle as some believe. In a study to be discussed further later in the section, Guler and Guillen (2005, p. 31) found that legal effects vanished after controlling for national strength in science and technology. Using national-level variables, Allen and Song (2005) confirmed the dubiousness of the "rule-of-law" position. They found that law and order were negatively related to VC investing. These studies suggest that governance and financial system explanations for the success of VC may be overemphasized. Given that venture capitalists have experienced some recent success investing in China, it may be that an English common law-based legal system may not be required.

There is other evidence to confirm that the traditional arguments about legal environments may not be as important as previously thought. Through an analysis of contracts written by U.S. venture capitalists in foreign markets, Kaplan et al. (2003) conclude that in nearly any environment it should be possible for VC firms to write contracts or develop various mechanisms to ameliorate legal, regulatory, fiscal, and structural obstacles to investment and exit, although the contracts may initially appear cumbersome.[3] We would extend this to suggest that if successful exits (through either acquisition or a public offering) occur

---

[3]Lerner and Schoar (2005), studying private equity, come to different conclusions. However, this may be due to the difference in operation of PE and VC firms.

in these environments, then the contracts and other organizational features can coalesce into understood routines and be taken for granted. After routinization, the contracts would appear to the participants as unproblematic (e.g., offshore investment vehicles in China).

The finance literature research on the global diffusion of venture capitalism is remarkable because it ignores the fact that the predominant deal flow for venture capitalists has come from the information and communication technology (IT) field and the biomedical field. For example, in the United States from 2002 to 2005, investments in IT as a share of total VC investments ranged between 58 and 60 percent. Between 18 and 23 percent of total investments were in the life sciences, while the remainder were scattered across other industries. In Europe between 51 and 57 percent of the total investment was in IT, while the life sciences received a further 21 to 28 percent. In Israel, IT has been far more dominant, receiving 70-76 percent of total VC investments. In contrast, the life sciences received 16-19 percent (original data from VentureOne, Ernst & Young, and Martin Haemmig, 2006). These data indicate that, in each major Western VC market, more than 80 percent of all VC investments have been in the IT and biomedical fields. In the hotbeds of investment activity (e.g., Silicon Valley, Boston, Israel, Stockholm, Cambridge [England], Austin), the percentages of VC invested in these technology areas are likely to be even greater. This strongly suggests that VC investments in any nation cannot be explained without considering the nation's technological base in general, and its IT and biomedical innovatory capabilities in particular.

Few studies of VC globalization have controlled for the VC recipient nation's technological base. Using a sample of Organisation for Economic Co-operation and Development (OECD) nations, Astrid Romain and Bruno van Pottlesberghe (2004) found that measures of technological strength such as patenting and research and development (R&D) investment were significant predictors of an increase in VC investment in that nation. In examining the overseas investments by U.S. venture capitalists, Guler and Guillen (2005, p. 30) found that a one-standard-deviation increase in a nation's U.S. patents led to a 77.5 percent increase in the number of ventures receiving investment from U.S. VC firms. Further, a one-standard-deviation increase in scientific publications led to a 113.4 percent increase in the number of ventures receiving investment from U.S. VC firms. No other measures, including stock market capitalization, political constraints, or number of students studying in the United States, were as important.

Although Guler and Guillen (2005) do not address the success of domestic VC firms, their study does provide evidence that suggests support for Romain and van Pottlesberghe's (2004) conclusion that technological capabilities are necessary to attract VC investment. As Avnimelech et al. (2005) argue, the VC industry in the United States and other nations has co-evolved with the technology industries that venture capitalists fund. In short, scientific and technological advance are the fuel for creating firms capable of generating the returns necessary to support a VC industry. This would suggest that models purporting to explain national experience in attracting VC investment (from either national or external sources) that do not control for technological capability are fundamentally misspecified.[4]

With a few exceptions, previous studies have examined national VC industries and thus measure the diffusion of VC as a social function. The phenomenon of interest in this chapter is the globalization of VC (i.e., VC firms operating across national boundaries). In a recent study of global VC investment patterns, Megginson (2004, p. 25) found that there was some evidence of "significant [international] convergence in funding levels, investment patterns, and realized return." And yet, he concluded that, because national capital markets have remained relatively segregated and legal systems remain different, "it appears that no truly integrated global VC market will likely emerge in the foreseeable future." There is evidence to qualify this conclusion.

Today, an increasing number of venture capitalists are investing successfully across borders. One important method of investing internationally is syndication. Recently, there have been a number of academic studies on the international syndication of VC investments. For example, Mäkelä and Maula (2006) found that the presence of foreign venture capitalists and top managers with foreign experience increases the probability that a portfolio firm will list on foreign markets. Pagano et al. (2002) found that R&D-intensive firms were more likely to undertake a foreign IPO, and this was supported by Hursti and Maula (2007). The apparent growing tendency for international syndication is creating a global network of VC firms. This has occurred in conjunction with the increasing number of foreign listings on stock exchanges such as the NASDAQ and the London AIM. With these changes, the grounds for Megginson's (2004) conclusion may be weakening.

THE HISTORY OF VC GLOBALIZATION

During the past half-century, venture capitalism has diffused internationally in the sense that numerous nations have indigenous VC industries and increasing numbers of venture capitalists are investing across national borders. Although this chapter is most concerned with cross-national VC investing, the presence of local venture capitalists is often important because they are usually more tightly linked to local entrepreneurs and can function as intermediaries for larger foreign VC firms. Additionally, when foreign VC firms decide to enter a new market,

---

[4]Perversely, the omission of technology variables in academic research is mirrored in "VC" investing in certain nations. For example, in a survey of British venture capitalists, Murray and Lott (1995) found that "technology projects had to meet more rigorous selection criteria than non-technology projects" and "investors imposed higher investment return 'hurdle rates' at each stage of investment other than seed capital." Though it is difficult to establish the direction of causality, this resonates with the perception that the U.K. VC industry is, in large measure, a PE industry.

they will often form a partnership or even acquire a local VC firm. This suggests that the growth of local venture capitalists and the entry of foreign venture capitalists into a market are intimately related.

A rough way of tracking the diffusion of indigenous VC activity is through counting the number of national VC organizations. In 1973, the National Venture Capital Association was established in the United States as the first national VC organization. Since then, at least, VC associations in 36 other nations have been formed. In addition to the United States, other early VC associations were established in Canada. These were followed by the establishment of a number of different national associations in Europe (largely contemporaneously) in the early 1980s. VC associations gradually spread to southern Europe and then later to eastern Europe. Because of the small size of the local VC industry in many of these nations, they had very few members. Over time, venture capitalists in Asia formed associations. The most recent major nation to form an association is China. Brazil has the only VC association in Latin America while South Africa has the only one in Africa.

More recently, regional VC/PE organizations have been formed. For example, the EVCA was formed in 1983 and is the best organized of the regional VC associations. In 2001 the Asia Pacific Venture Capital Alliance was formed by the national organizations of Hong Kong, Korea, Malaysia, Singapore, and Taiwan. This was followed in 2002 by the formation of an organization called the Emerging Market Private Equity Association. This suggests that supranational structures are emerging, possibly leading to a global umbrella organization in the future.

The fact that a nation has an association does not prove, in and of itself, that it contains a vibrant VC industry. The existence of an association does provide a certain visibility, and the association can lobby the government to improve the environment for both VC and entrepreneurship. In the United States, Israel, and Taiwan, the national associations have had policy impacts. The EVCA has played an important role in lobbying the European Union (EU). At a minimum, the existence of national and regional VC associations demonstrates the broad diffusion of VC investing.

Cross-national VC investing has evolved gradually. The first important period was in the 1970s when a number of European financial institutions established U.S. subsidiaries or invested in U.S. VC firms. Corresponding roughly to the 1980s, a second period occurred when U.S. East Coast VC firms opened offices in Europe (especially in London) in search of European investments. In the mid-1980s, several U.S. West Coast VC firms were formed to invest in Taiwan and other parts of Sinophone Asia. Contemporaneously, a few Japanese VC firms began investing abroad. At the end of the decade, Taiwanese VC firms began investing in the United States. In addition, there was also an effort to establish VC firms in various developing nations during the 1980s. A third period of international VC development occurred in the 1990s and is discussed in the next section on current cross-national linkages.

Pioneering venture capitalists were convinced of the importance of VC and the benefits it could provide to society. Early VC firms tended to invest locally (Hsu and Kenney, 2005). However, Kogut et al. (2006) showed that cross-regional co-investment occurred very early in the development of the U.S. VC industry. The initial efforts to globalize the VC model were missionary-like initiatives by the U.S. pioneers. The first effort was in 1960 when the Rockefeller VC operation opened an office in Brussels. Unfortunately, it had few successes and soon was abandoned (Wilson, 1985, p. 220). In 1962, American Research and Development (ARD) helped organize the Canadian Enterprise Development Corporation and European Enterprise Development Corporation. None of these initial efforts were sustainable and they were eventually discontinued.

Several U.S. technology startups that were backed by U.S. VC firms began to provide large capital gains in the late 1960s. For example, ARD's $70,000 investment in DEC in 1957 appreciated to more than $350 million in 1969. This provided incentives for both foreign and domestic investors to invest in VC funds. Because of the informal nature of the VC industry during this early period, little is known about its global interconnections. In the early 1970s, pioneering VC firms were established in the United Kingdom and then in the Netherlands and France (Manigart, 1994, p. 535). These marked the acceptance of the VC concept in Europe. Despite these new firms in Europe, there were only a few new entrants per year and very few investment opportunities.

The first sustained global operations were by European financial institutions that either invested in U.S. partnerships or established VC investment operations in the United States. For example, in 1970, New Court Securities, which was an arm of the European Rothschild family, opened in New York City. Also, both the large French bank Indo-Suez and Guardian Ventures Limited of Canada established U.S. VC branches. Contemporaneously, Genstar Corporation of Canada became the sole investor in Sutter Hill Ventures in the San Francisco Bay Area. These pioneers invested in the United States, not in their home nation.

Peter Brooke, the founder of TA Associates and one of the leading U.S. VCs, played a pivotal role in early international diffusion efforts. In 1971, he was appointed as a founding director of Sofinnova, which was capitalized by various French banks with the mandate to invest in both France and the United States (Advent International Corporation, 1986). In 1974, Sofinnova opened an office in San Francisco and made a number of successful investments in the nascent field of biotechnology. In 1975, the Dutch investment company Orange Nassau started a U.S. fund managed by TA Associates.

In the 1970s, some U.S. VCs believed Europe would provide significant investment opportunities. For example, in 1971, U.S. venture capitalist Philip Greer (1971) opined that Europe was attractive because the lack of competition meant that firm valuations were lower. He recognized that exits would be difficult because the only stock market having the requisite liquidity was the London Stock Exchange.

For this reason the prime exit strategy would be selling portfolio firms to U.S. corporations. Greer also identified a shortage of entrepreneurs in Europe. He stated, "The entrepreneur concept is typically American and such activities have been discouraged in Europe." These obstacles were compounded by a paucity of well-rounded managers, longer investment periods, a shortage of second- and third-stage financial sources, a European financial community that did not welcome outsiders, and higher costs of operation. Although Greer was optimistic about Europe, the reasons he identified were exactly those that stymied the development of the European VC industry up until the 1990s (PE investing was more successful).

After a severe downturn in the mid-1970s, the U.S. industry reawakened and a few more U.S. VC firms entered Europe. In 1980, Peter Brooke established an international VC firm, Advent International, with a London office. From there, it soon expanded to Belgium in 1982 and Singapore and Malaysia (SEAVIC) in 1983. In addition to these offices, TA/Advent established linkages with Four Seasons Venture Capital in Sweden in 1982, TVM Techno Venture Management in 1983 in Munich, Germany, Advent Techno-Venture in 1984 in Tokyo, Alpha Associes in 1985 in Paris, and Horizonte Ventures in 1985 in Austria. The Advent network was the first VC firm with a global presence.

Another pioneer was Apax, which was the result of a merger of a boutique international investment banking firm established in 1972 by Ronald Cohen in the United Kingdom, Maurice Tchénio in Paris, and Alan Patricof in the United States. In 1981, Cohen raised their first U.K. £10 million fund with the assistance of Patricof, who also invested in the fund. In 1983, Tchénio raised a fund in France in which Patricof also invested. From this union, Apax grew to be one of the leading international VC/PE firms, with offices in New York, Menlo Park, London, Paris, Milan, Munich, Madrid, Tel Aviv, Stockholm, and Zurich.

As U.S. firms entered Europe, European firms continued to enter the U.S. market. For example, in 1981, the British firm 3i opened an office in the United States, although it withdrew in the early 1990s (Coopey and Clark, 1995, p. 179). Atlas Venture, a division of the Dutch ING Bank, opened an office in Boston in 1986. In 1982, Vincent Worms and Thomas McKinley at Partech International launched a Global Venture Fund for Banque Paribas in San Francisco and later in Paris. Yet another firm, Alta Berkeley, was formed in 1982 in London in cooperation with the U.S. firm Burr, Egan, and Deleage. Thus a cadre of transatlantic VCs came into being.

Despite the growth in VC operations, there was a continuing lack of high-quality deals in Europe (Murray and Lott, 1995). As the Chairman of Apax Partners, Ronald Cohen, put it: "Toward the middle of the decade (1980s) there was a general shift away from business risk. This was partly the result of burnt

TABLE 1 Net Returns in European and U.S. Venture Capital and Buyout Investments by Stages as of December 31, 2005[a]

	1 yr EU	1 yr USA	3 yr EU	3 yr USA	5 yr EU	5 yr USA	10 yr EU	10 yr USA
Balanced	32.7	24.3	2.8	11.7	−2.7	−3.5	7.6	18.9
All VC	25.4	15.6	0.6	7.5	−4.0	−6.8	5.3	23.7
Buyouts	20.9	31.3	7.9	16.3	5.0	5.2	12.6	9.2

[a]Net internal rates of returns to investors in EU funds formed during the period 1986-2005 and U.S. funds formed during the period 1986-2005. These rates were calculated from Thomson Financial, the National Venture Capital Association, and European Venture Capital Association data.

fingers from start-up investments in the early 1980s, but also because of a move towards the quicker returns to be made from backing MBOs and exiting in a rising market" (quoted in Coopey and Clarke, 1995, p. 171). The theme of low returns for VC investing in Europe has been a constant refrain (see, e.g., Murray and Marriott, 1998) and is substantiated in Table 1, which compares returns from different investment stages in Europe and the United States. These data demonstrate that the long-run returns for early-stage investing in Europe have been far lower than in the United States. Even for the balanced funds category, U.S. funds outperformed European funds. It has only been since the stock market bubble's collapse in 2000 that European VC funds have performed roughly as well as U.S. funds. The disappointing European VC returns contrast with the superior performance of European PE funds.

A full explanation for the low returns to VC investing in Europe is outside the scope of this paper. However, it is possible to list some of the salient elements of such an explanation. Important elements include the relative weakness of European universities, corporations, and nations in the information technologies, with the possible exception of a few large corporate laboratories and a few universities. A similar but not quite as powerful advantage would be true in the biomedical sciences. Also, the enormous and very discerning U.S. IT market was a significant advantage. This meant Europe had fewer entrepreneurs, smaller concentrations of entrepreneurs, and slower growth in its entrepreneurial firms. As a result, a path-dependent logic was set in motion, building upon significant first-mover advantage in the United States, particularly in Silicon Valley and Boston. These factors may partially explain why the United States developed both VC and BO investing, whereas European investors emphasized BO investing.

Asia

Japanese corporate venturers and large Japanese VC firms, nearly all of which were subsidiaries of large Japanese financial institutions, began globalizing slightly later in time than the initial European VC firms. In 1983 the largest Japanese VC firm, JAFCO (a Nomura Securities affiliate), established its international office in Hong Kong. In July 1984, it established a Menlo Park, California, office and, in 1986, opened a representative office in London (Kuemmerle, 2001). Facilitated by the easy money available during the Japanese economic bubble of the 1980s, other Japanese VC firms and industrial corporations invested in Silicon Valley and other locations at extremely high valuations. Unfortunately, in too many cases, these investments failed. As a result, many but not all of these Japanese venturers retreated from these global ventures. Some firms (e.g., JAFCO) continued to invest globally. During the dot-com boom, Softbank (a Japanese technology conglomerate) began investing globally. These global investments included the purchase of a large stake in Yahoo!. Today, Softbank is an active investor in China.

From the 1980s onward, international development organizations, particularly the International Finance Corporation (IFC), which is a member of the World Bank Group, made a concerted effort to implant venture investing in the developing nations of Latin America, East and Southeast Asia, and Africa (Aylward, 1998; Fox, 1996). For example, the IFC, in concert with the U.S. Agency for International Development, British development agencies, and other European aid organizations, provided the initial capital for the first VC funds in India, Korea, and Southeast Asia (Dossani and Kenney, 2002; Kenney et al., 2004). The modus operandi differed by nation. In some cases (e.g., India), the funds were invested in domestic VC firms. In other cases, the funds were invested in an international VC firm willing to create a country fund. This is exemplified by Southeast Asia Venture Inc. (SEAVI), which was operated by Advent International and headquartered in Singapore. At this time, the financial returns were mixed, even in nations where VC firms would later succeed. Despite the failures, these initiatives helped promote other benefits such as the training of VC personnel, changes in the legal system, and an increasing awareness of entrepreneurship (Dossani and Kenney, 2002; Kenney et al., 2004). The Taiwanese VC industry started in 1982 and grew rapidly. In the late 1980s and early 1990s, the largest Taiwanese VC firms established Silicon Valley operations. In many developing nations, however, VC funds failed due to inadequate investment opportunities (Aylward, 1998; Fox, 1996).

More recently, India has been attracting attention from international VC firms. As was the case with many developing nations, India received a spate of investment from U.S. VC firms during the Internet bubble that ended in 2000 (Dossani and Kenney, 2002). The ensuing downturn dried up most of the VC investment in India. From 2003 onward, the growing practice of international outsourcing of business services encouraged some global VC firms to invest in Indian service delivery firms. Thus far, there have been only a few exits in the form of mergers (e.g., Daksh was purchased by IBM for $170 million in 2004, Spectramind was purchased by Wipro for a total of approximately $150 million in a process in 2003, and 52 percent of MphasiS was purchased by EDS in 2006 for $380 million).

These were good exits, although they did not provide the high multiples that have recently been obtained from investments in particular Chinese firms. In part, this is because the Indian market is not as large as the Chinese market. Furthermore, there are no legal or language barriers to foreign firms, and Indian firms do not yet have global-class technology. At this point, India is mainly confined to labor arbitrage opportunities (Dossani and Kenney, 2007). Given the types of recent R&D investment by multinational corporations in India, the skilled managerial and technical personnel in the Indian labor force, and the growing expertise of Indian firms in software and software services, it is not unreasonable to expect the emergence of technology-based Indian startups that are globally competitive within the next 3 years (Dossani and Kenney, 2007).

In the 1990s, the environment changed for VC investing, even for the elite Silicon Valley venture capitalists. Initially, the changes were subtle. Historically, it was necessary for VC firms from other nations to quickly establish an international office in Silicon Valley if they wanted to be considered a global player and take advantage of the investment opportunities found there. In contrast, because of the lucrative deals available locally, U.S. VC firms in Silicon Valley responded to global opportunities at a much more gradual pace. Their awareness of international investment opportunities was heightened by the fact that the business plans of an increasing number of the U.S. firms in which they invested (e.g., fabless semiconductor firms) were predicated upon using offshore assets, particularly for manufacturing. Furthermore, the startup teams for these firms frequently had at least one member born overseas (Wong, 2005). In addition, a steady flow of successful IPOs in the United States by venture-backed Israeli firms achieved sufficiently large returns to draw the attention of Silicon Valley VC firms. Finally, it became apparent that in certain promising technologies (e.g., wireless and software security), the United States was not the clear technological or market leader. Taken together, the development of VC industries in new nations, the emergence of a greater number of international investment opportunities, and strategic issues, such as ensuring that the firms in which they had invested were properly positioned in the correct markets, encouraged all VC firms, including the elite Silicon Valley firms, to make international investments and develop global strategies.

VENTURE CAPITAL GLOBALIZATION PATTERNS

There are few empirical studies of the reasons for VC globalization (for exceptions, see Cumming, 2002; Haemmig, 2003). The reasons for establishing overseas operations differ by firm and home market circumstances. First, until recently, only the largest national markets, such as the United States, offered a sufficient number of high-quality investments to support a large VC firm.[5] VC firms established in smaller markets and seeking critical mass, such as Atlas

---

[5]The EU is probably a sufficiently large market.

Ventures of Holland or Sofinnova and Partech of France, necessarily must invest internationally. In many cases, this occurs through co-investment with a local firm (e.g., a Dutch VC firm invests in a German deal or a German VC firm invests in a U.K. deal). Often, the foreign firm offers only financial support and is not an active investor providing other services to the portfolio firm. The nonlocal VC firm is normally a passive investor.

A deeper commitment is to establish an office overseas. For example, 3i opened its U.S. operation because it was believed essential to be present in the most dynamic VC market (Coopey and Clarke, 1995, p. 357). The branch office can provide a variety of services to its headquarters. For example, if it is located in a leading entrepreneurial cluster it can provide market information to the home office to prevent it from investing in "me-too" startups. Alternatively, in technologies where the skills may be available in multiple locations, the branch office may notice initiation of investment by leading venture capitalists in the United States and pass this information to its parent office so that it can fund domestic startups—receiving such a signal early could prove a significant advantage. In this case, the foreign branch creates a window into market developments.

Foreign branch offices can provide services to parent firms, such as providing introductions to potential suppliers, customers, or strategic partners. Foreign offices may begin as a listening post or contact point for later-stage investing while also serving as a marketing differentiator in the home market if domestic entrepreneurs seek a VC firm capable of providing introductions in the foreign market. For non-U.S. firms, a U.S. office might be useful for their portfolio company in securing follow-on investments from U.S. VC firms that can help "certify" the foreign firm. This may increase the value of the non-U.S. firm, particularly if it is to have a public offering on a U.S. stock exchange, or if the firm is being considered as an acquisition target by a U.S. corporation.

Depending on subjective decisions about an entrepreneur's trustworthiness and excellence, and facing a need to monitor the managers of firms in which they invest, the international investments of VC firms should pose considerably greater difficulties compared to their domestic investments. In one of the few studies of foreign VC investors operating in another country, Mäkelä (2004) found that foreign VC firms in Finland tended to have lower commitment to the Finnish firms in which they were invested compared to domestic investors. On the positive side, the foreign VC firms provided an important legitimating function to the Finnish firms in which they were invested. In comparing foreign and domestic VC firms operating in India, Pruthi et al. (2003) found that domestic firms were more apt to provide advice and monitoring for their portfolio firms while foreign firms placed greater emphasis on strategic positioning. Despite these benefits, Lara Baracel (2004) found that U.S. VC firms have enjoyed significantly less success (measured by exits) with foreign investments compared to domestic investments.

The motivation for VC globalization differs by nation. In a survey of VC firms, Haemmig (2003) found significant differences by home region (see Table 2).[6] The most striking aspect regarding motives was how different the U.S. firms were from those in Europe and Asia. U.S. firms reported two primary motives: first, a desire to invest in technologies that were superior abroad or in industrial sectors that had promising startups in other nations (e.g., mobile telephony and security software), and second, some believed that foreign deals were less expensive. In addition, some wanted to assist their portfolio firms in globalizing their markets. U.S. VC firms were not motivated to invest overseas by pressures or difficulties within the U.S. market. The reasons Asian firms gave for globalization were an insufficient domestic deal flow and the potential of higher returns from abroad—nearly the polar opposite of the U.S. firms. For the Asian firms, the only other significant issue was a need to operate in larger markets. European VCs agreed with Asian firms on the insufficient domestic deal flow, but they also felt a need to assist their portfolio firms in globalizing. Higher potential returns and larger markets were also mentioned. The most remarkable difference between the Europeans and the Asians was that the Asian VCs did not believe that they needed to assist their domestic firms in internationalizing their investments. In summary, U.S. VCs responded to business and technology opportunities abroad, while European and Asian VCs were reacting to difficulties in their domestic markets in terms of either deal flow or returns.

In contrast to most academic work that models globalization as a conscious strategy from its inception (e.g., Guler and Guillen, 2005), two-thirds of Haemmig's (2003) respondents answered that their first foreign investment was due to an opportunity presenting itself rather than a conscious strategy. At the time of the interview, 89.5 percent of the firms claimed that they had developed a written strategy for their international activities after the fact to rationalize their globalization efforts. This suggests that the initial investments were sufficiently successful to encourage a greater commitment to overseas investing.

These results are reinforced by the responses to the question of whether their international investments and operations were "add-on" businesses or were of strategic importance. Only 32 percent of the U.S. firms saw their international investments as being of strategic importance. In contrast, 68 percent of the European firms and 91 percent of the Asian firms saw their international investments as strategic. These responses suggest that many U.S. firms are motivated by an "opportunity pull," whereas Asian and European firms are motivated by the push of being located in inadequate markets. In turn, U.S. firms may be less than committed to their foreign operations due to the voluntary nature of their decision to globalize. Of course, the U.S. firms are not monolithic. Some were founded with an international mandate and, thus, are entirely committed to international investing.

---

[6]These results are from Haemmig (2003), who in 2001 conducted interviews at 95 VC firms in 12 nations (25 in the United States, 38 in Europe, and 32 in Asia). The definition of a globalized firm in Table 5 was whether it invested more than 10 percent of its total capital outside its geographic region.

TABLE 2 Reasons for International Investments by VC Firms (in percent)[a]

	U.S. (n = 28)	Europe (n = 34)	Asia (n = 33)
Insufficient domestic deal flow in sectors	0	24	34
Higher return potential outside home country	0	16	31
Need to bring portfolio firms international	12	24	3
No other choice but going to the main markets	0	16	16
Less competitive in foreign countries	24	3	0
New emerging technology that is superior to the U.S.	28	0	0
Industry sector is global (telecommunication/wireless, biotech, IT)	24	13	9
Other	12	5	6

[a]May not sum to 100 percent due to rounding error.
SOURCE: Haemmig (2003).

In summary, there does not appear to be a single evolutionary logic for globalization at the firm level, despite the unmistakable tendency toward internationalization of investment at the industry level. Nevertheless, the relational aspects of VC investing and the concentration of good deals in relatively few locations mean that VC investing is not "global" in the sense that VC capital flows with equal ease to all parts of the world. The United States in general, and Silicon Valley in particular, continues to be the center of the global VC industry. This is predicated on a number of advantages: the most experienced VCs, the most venture capital, the most sophisticated markets, the most experienced pool of managers, the most sophisticated entrepreneurial support network, and, in most fields, the best technologists. These advantages will erode slowly if at all. Current globalization does not threaten the U.S. innovation system. One current hotbed of investment—Taiwan—has very few startups that compete with U.S. firms. The other major hotbed—Israel—does generate firms that compete with U.S. firms. However, Israel is so small that it poses little significant threat and, moreover, the amount of VC invested in Israel is growing only gradually. The current pattern of globalization seems to be reinforcing the centrality of the U.S. VC industry, even as other nations are experiencing a growth in VC investment.

A QUANTITATIVE OVERVIEW OF THE GLOBAL
VENTURE CAPITAL INDUSTRY

Multinational flows of VC have become significant. Using data from the National Venture Capital Association, EVCA, Israeli Venture Capital Association, and the Asian Venture Capital Journal (2001), we aggregated data on the investment flows between and within four regions—the United States, Europe, Asia, and Israel. These data indicate that cross-regional VC investing (i.e., a VC firm based in one region investing in a firm in another region) is significant for all four regions. The data in Table 3 indicate that approximately 79 percent of the total VC invested in 2005 came from U.S. VC firms. In comparison, 15 percent came from European VC firms, 4 percent came from Asian VC firms, and 1 percent came from Israeli VC firms. These data indicate that the United States was closest to self-sufficiency in its VC investment flows compared to the other three regions. Approximately 93 percent (21,914 of 23,447) of the total VC invested in U.S. companies in 2005 came from U.S. VC firms. Furthermore, of the $24.9 billion invested by U.S. VC firms in 2005, approximately 88 percent (21,914 of 24,925) was invested in U.S. companies.

The United States exported roughly $3 billion in VC. This is approximately twice the $1.5 billion it received from the other regions.[7] European firms were the largest external investors in the United States, contributing $840 million in VC funds. Europe received $2 billion in inflows from other nations with the United States providing the preponderance ($1.8 billion). This represented 33 percent of the total VC investment in Europe. In Israel, U.S. VC firms invested almost as much capital as did the Israelis ($158 million vs. $208 million). Israeli investments outside the country were insignificant. In Asia, the United States was also the largest investor, providing $798 million in VC funds. This was a greater sum that the total intraregional investments by Asian VC firms. European investments in Asia were only one-third that of U.S. firms.[8] The bulk of the Asian outward investment was directed toward the United States (80 percent or $502 million). Asian investments in Israel or Europe were relatively negligible. In 2005, the total VC invested interregionally was approximately $5 billion, or 18 percent of the VC invested internationally. Of this, the United States provided nearly 60 percent. These aggregate statistics confirm that the United States remains the global center for VC investing, both as an investor and recipient of investment.

Although Megginson (2004) may be correct that a global market for VC has not yet emerged, these data demonstrate that there are now large international flows of VC. These international flows are even larger if one considers that intra-European and intra-Asian flows also involve investment across national boundaries, albeit within each region. Although we do not present the data on the flow of investments in VC funds, U.S. VC firms invested a greater percentage abroad than U.S. investors provided to foreign-based VC firms. This suggests that U.S. investors may have a bias toward U.S. firms and, perhaps, trust them to make the investments abroad. In summary, the quantitative data strongly suggest that, with the possible exception of the United States, national VC markets are not autarchic and a global deal market is emerging.

---

[7]The $3 billion allocated to Europe by U.S. venture capitalists includes capital that was raised by U.S. venture capitalists for Europe in Europe (e.g., Accel Europe).

[8]The capital allocated to Asia by U.S. venture capitalists includes capital that was raised by U.S. venture capitalists for Asia in Asia (e.g., Walden International has many Asian investors, but it is based in San Francisco).

TABLE 3 Interregional Flows of Venture Capital Investment (in millions of U.S. dollars) by Location of Firm and Location of Investment, 2005

Location of VC Firm (origin of investor)	2005 Location of Investments by VC Firms					
	North America ($M)	Europe ($M)	Israel ($M)	Asia ($M)	Rest of World ($M)	Total ($M)
North America	21,914	1,837	158	798	218	24,925
Europe	840	3,486	35	163	59	4,583
Israel	139	30	208	0	0	377
Asia	502	118	4	502	3	1,129
RoW	52	42	7	59	231	391
TOTAL ($M)	23,447	5,513	412	1,522	511	31,405

SOURCE: Compiled by Martin Haemmig, www.martinhaemmig.com, from data provided by National Venture Capital Association/Venture Economics, European Venture Capital Association, Asian Venture Capital Journal, and Israeli Venture Capital Association.

THE CHINA SYNDROME

In the past 5 years China has become a fertile environment for VC investing, despite the fact that it does not have an adequate legal or financial system. What China does have is a booming economy with large numbers of increasingly wealthy consumers and trained engineers. Also, it is experiencing rapidly increasing R&D expenditures in both industry and government (Jefferson and Gian, 2007). This vortex of opportunity has attracted business persons of all types and venture capitalists are no exception. Foreign venture capitalists have flocked to invest in China in both technological and nontechnological fields. Because of the topic of this chapter, this section is confined to VC investing in China and does not examine the state of Chinese high technology, except in relationship to the firms that have funded by VC. With about $1.9 billion invested in 2006, China ranked second after the U.S. in total VC investment (Ernst & Young 2007). In 2007, venture capitalists raised $5.8 billion in funds dedicated to investment in China, which was nearly 50 percent more than in 2006 (Zero-2-IPO, 2008).[9] This increase was, in large measure, motivated by the successful earlier exits by VC-backed Chinese firms on international stock markets. The data in Table 4 indicate that the most active VC firms are foreign, particularly those from the United States. Firms from other nations, such as Softbank from Japan, also are present. However, from 2005 to 2006 more domestic Chinese VC firms were entering the Top Twenty. Whether this is a trend or anomaly is not certain, but is an interesting phenomenon to watch.

There are currently four avenues for venture capitalists to exit their investments in Chinese firms. The first avenue is acquisition by foreign multinationals.

---

[9]Some or possibly a large portion of these funds may be used in PE transactions.

TABLE 4 Top Venture Capital Firms in China by US$ Invested, 2005 and 2006

Rank	Firm Rank 2006	Nationality	Firm Rank 2005	Nationality
1	IDG Technology Venture Investment	U.S.	IDG Technology Ventures	U.S.
2	SAIF Partners (Softbank)	Japan	SAIF Partners (Softbank)	Japan
3	Sequoia Capital China	U.S.	Venture TDF	China
4	Legend Capital (corporate)	China	CDH Investments	Singapore/U.S.
5	Granite Global Ventures	U.S.	DFJ ePlanet	U.S.
6	Softbank China VC	Japan	Softbank China VC	Japan
7	Walden International	U.S.	Granite Global Ventures	U.S.
8	JAFCO Asia (corporate)	Japan	Intel Capital China	U.S. (corporate)
9	Intel Capital (corporate)	U.S.	3i	U.K.
10	CDH Ventures	Singapore/U.S.	NewMargin Ventures	China
11	iD TechVentures Ltd.	China	Warburg Pincus Asia LLC	U.S.
12	WI Harper	U.S.	Doll Capital Management	U.S.
13	Doll Capital Management	U.S.	Actis China Limited	U.K.
14	Qiming Venture Partners	China	Sequoia Capital China	U.S.
15	DT Capital Partners	China	Shandong High Technology Investment	China (government)
16	Venture TDF China LP	China	Pacific Venture Partners	Taiwan
17	Capital Today Group	China	Shenzhen Capital Group	China (government)
18	Orchid Asia Group	Hong Kong	Legend Capital	China (corporate)
19	CEYUAN Ventures	China	WI Harper Group	U.S.
20	GSR Ventures	China	DragonTech Ventures	Hong Kong

SOURCE: Zero2IPO (various years).

There have been a number of successes, including Yahoo's $1 billion purchase of 40 percent of the Chinese e-commerce firm Alibaba, the purchase of Longshine in 2005 for $30 million by the U.S. firm Amdocs, and TDK's purchase of ATL for $100 million in 2005. The second avenue is acquisition by Chinese firms. A number of these have provided high returns to their investors.

The third exit window is listing on foreign markets, since Chinese firms backed by foreign VC cannot list on Chinese markets. Foreign VC firms play a vital role in advising and preparing Chinese firms for listing on the U.S. NAS-DAQ. Listing in the United States is important, because it has the largest and most liquid markets. Conversely, Chinese listings are becoming important for U.S. exchanges. In 2004 there were 21 VC-backed IPOs of Chinese firms, 10 of which were in the United States. These constituted more than 10 percent of the 93 VC-backed IPOs in the United States that year. In 2005, 20 VC-backed Chinese firms listed internationally. Of these, 8 exited in the United States. That year, there were only 57 VC-backed IPOs in the United States (Zero2IPO, 2005). Moreover, many Chinese listings performed well in the aftermarket. The final exit is in the increasingly active Chinese stock markets.

U.S. VC firms have been intimately involved in the Chinese startups listing on U.S. markets. Using data from U.S. Securities and Exchange Commission filings for IPOs, we extracted the names of the venture capitalists serving on these firms' boards of directors. The preponderance of the firms had foreign venture capitalists on their board of directors (this includes non-Hong Kong venture capitalists located in Hong Kong). The representation of Silicon Valley venture capitalists is particularly striking. This suggests that these individuals may be transferring Silicon Valley-like practices and routines for managing high-technology startups to China.

Do these Chinese startups have unique or global-class technology that might threaten U.S. dominance? Table 5 lists Chinese firms with IPO in the United States from 1999 to 2005. The activities of these firms suggest that the Chinese startups exiting in the United States are "me-too" emulators of overseas business models, semiconductor design firms operating at the lower-technology end of the marketplace (e.g., Vimicro and Actions), semiconductor fabrication firms that are direct competitors with Taiwanese firms (e.g., SMIC), or firms that are not based on technology but instead rely on business models that may not be adaptable outside of China. An example of such business models is Focus Media, which rents space in elevators on which it installs flat panel screens displaying advertisements. In summary, nearly all of these firms serve the rapidly growing and underserved Chinese consumer market or are part of a global division of labor that does not directly affect U.S. firms.

This relatively negative assessment of the technology base of these VC-backed Chinese firms does not imply that China will not rapidly improve its technological capacity. Given the current trajectory, it is possible that some firms with unique global or near-global class technology may appear within 3 to 5 years. Certainly, the success of Huawei and ZTE,[10] which were not VC financed,

---

[10]Huawei and ZTE are the largest Chinese telecommunications equipment firms and have become significant competitors in not only the Chinese market but also global markets. Examples of their developed nation competitors include Cisco, Lucent, Nortel, and Alcatel. In 2006, Huawei's global sales were $8.2 billion, of which 58 percent were outside of China (Huawei, 2007). In 2005, ZTE's global sales were $2.9 billion (*Byte and Switch*, 2006), and it competed particularly fiercely in the wireless infrastructure space.

TABLE 5 Chinese Entrepreneurial Ventures' IPOs on the U.S. Markets, 1999-2005

Firm Name	Year of IPO	Activity	Analogous Offshore Firms
AsiaInfo Holdings	2000	Software and online services for China	Many firms
Netease.com Inc	2000	Online games	Many firms
Sohu.com Inc	2000	Web portal	Yahoo!
China Finance Online	2004	Chinese financial information	Hoovers
KongZhong Corp	2004	Wireless downloads (e.g., ring tones)	Many firms
Ninetowns Digital	2004	Software for import/export from China	Many firms
Elong Inc	2004	Travel site	Expedia
TOM Online Inc	2004	Wireless downloads (e.g., ring tones)	Many firms
Baidu.com Inc	2005	Web search	Google
Hurray! Holding	2005	Wireless downloads (e.g., ring tones)	Many firms
Vimicro International	2005	IC design	Many firms
Watchdata Technologies	2005	Smart card operating system for China	Many firms
China Medical Tech.	2005	Ultrasound cancer treatment equipment	Many firms
China Techfaith Wireless	2005	Handset design	Many firms
Ctrip.com International	2003	Travel site	Expedia
51job Inc	2004	Job website	Monster
Linktone Ltd	2004	Wireless downloads (e.g., ring tones)	Many firms
Semiconductor Mfg. Int.	2004	Semiconductor foundry	TSMC (Taiwan)
Shanda Interactive	2004	Multiplayer games	Many firms
The9 Limited	2004	Multiplayer games	Many firms
Focus Media Holding	2005	Advertising in elevators	Not technology
China.com Corp	1999	Wireless downloads (e.g., ring tones)	Many firms
Asiacontent.com	2000	Internet solutions and online advertising	Many firms
Sina.com	2000	Web portal	Yahoo!
Wherever.net	2000	Discounted mobile telephony (defunct)	Many firms
Actions Semiconductor	2005	IC design	Many firms
Suntech Power	2005	Photovoltaics	Many firms

SOURCE: Compiled by authors.

suggests such an outcome. The enormous investments by the Chinese government, the return of well-trained Chinese workers from overseas, and the growth of the Chinese economy make negative predictions about the level of Chinese technology and innovation unlikely to hold true in the long term.

The evidence suggests that international VCs are more active than domestic VCs in China. For example, approximately 75 percent of the VC funds invested in China recently have come from foreign sources (Zero2IPO, 2005). Although it is difficult to be certain, there is evidence to suggest that clusters of entrepreneurial firms, VC firms, and other startup service providers may be emerging in Beijing and Shanghai (Kenney et al., 2004). All Chinese firms listed on the NASDAQ were focused on the Chinese internal market with the exception of the Semiconductor Manufacturing International Corporation, which was a semiconductor fabrication firm competing directly with Taiwanese firms. The two Chinese semi-conductor design firms listed on the NASDAQ—Actions and Vimicro—appear to be focused on the domestic market. At this point, these firms are not competing directly with U.S. design firms, which focus on cutting-edge technologies.

The firms that have listed thus far are not direct competition for U.S. high-technology firms. However, since U.S. VC firms are employing Chinese professionals, there will be a transfer of skills from the United States to China in the craft of venture investing. Furthermore, the substantial numbers of startups that have grown and managed successful exits imply that a class of experienced entrepreneurs is emerging in China. Already, a number of these entrepreneurs are being employed by U.S. VC firms or are establishing their own VC firms with significant investment from U.S. VC firms. The final missing ingredient is world-class technology. Despite the fact that China does not appear to have global-class technological opportunities, the rapidly growing domestic market is creating numerous opportunities for substantial capital gains attracting venture capitalists from around the world.

DISCUSSION

One question motivating this book is the availability of reputable quantitative information on R&D globalization. Only 5 years ago, it was nearly impossible to measure VC industry globalization because of the lack of comparable statistics and collection standards. The globalization of the VC industry has increased the demand for standardized information on industry trends throughout the global economy. At the behest of the EVCA and National Venture Capital Association, the comparability among North American, European, and Israeli data has improved. Thomson and Ernst & Young have initiated a global standardization initiative that is now being extended to Asia; this will further improve comparability. Within 5 years, a robust global reporting system should be in place.

While the globalization of VC has diffused this institution to many other nations, Silicon Valley in the United States unambiguously continues to be the center of the VC industry. One issue of concern is whether investment in foreign companies by U.S. VC firms contributes to eroding the competitiveness of the United States in high-technology industries. U.S. VC firms have invested in European companies for at least two decades. They soon learned to invest primarily in BO deals because there were too few startup deals. Recently, there have been better early-stage deals in Europe, which has attracted investment by U.S. VC firms. For example, the recent acquisition of Skype by eBay for $4 billion is an exit that will likely attract more U.S. VC investments in European firms. In addition, Israel continues to attract U.S. VC investment. An investigation of VC investments in Taiwanese companies found that Taiwanese startups do not appear to be in direct competition with U.S. startups, with the exception of the area of semiconductor design (U.S. Government Accountability Office, 2006). In sum, at this juncture it would appear that the investments of U.S. VC firms in Europe, Israel, and Taiwan have had a negligible effect on U.S. technological competitiveness.

The cross-national investments by U.S. VC firms that have caused the greatest concern are those in China. This is due to the belief that these investments are assisting in the development of foreign competition for the U.S. lead in high technology. This may be a valid concern for the future. Chinese firms such as Huawei, ZTE, and Lenovo (none of which were VC-funded) are already serious competitors in certain markets. However, as we have discussed, Chinese firms funded by VC do not yet appear to pose competition at a global level. The other chapters in this book will answer the question of whether global-class technology is currently being developed, either in Chinese firms and universities, or in the R&D operations of multinational firms in China. This chapter did not examine VC investing in India, which up until the last 18 months has been quite limited, but in 2007 began to grow rapidly.

There is little reason to believe that investment by foreign VC firms in the United States has had a negative effect on U.S. technological competitiveness; European VC firms have a long history of investing in the United States. It is more likely that the inflow of European VC has provided a net benefit to the U.S. economy—European governments believe this to be the case. One benefit for the United States is that, at a minimum, European VC firms, as well as Israeli VC firms, nearly always pressure their portfolio firms to open an office in the United States. Their objective has been to access U.S. knowledge and markets. In return, this validates and reinforces the United States as the center of global technology. It is difficult to interpret this as negative.

There are recent developments in international stock markets that may have implications for U.S. technological competitiveness. A debate in the industry and exemplified by a 2005 article in the *Venture Capital Journal* suggests that the willingness of firms to list on U.S. markets may be decreasing due to the high cost of complying with Sarbanes-Oxley (SOX) legislation (Sheahan, 2005). The argument is that U.S. markets may be losing their centrality in VC investing, particularly for smaller IPOs. For example, of a total of 78 IPOs in 2005 by 30 top-quartile U.S. VC firms, only 45 were listed in the United States. The remaining 33 were listed in non-U.S. markets including London (LSE and AIM), Taiwan, Hong Kong (main board and GEM), Japan (JASDAQ), Korea (Kosdaq), Singapore, and Malaysia. Whether this recent trend will endure is uncertain as it is also possible that these offshore IPOs involve lower-quality firms. SOX may be shifting inferior listings to markets with looser disclosure and listing standards. Alternatively, shifting listings away from U.S. markets could impact the health of U.S. equity markets by encouraging listing abroad. Despite this shift in listing, the larger exits for VC-financed high-technology firms continue to be on the

NASDAQ. So, the suspicion that SOX legislation is convincing firms to list in other markets has not been definitively proven.[11]

CONCLUSION

The VC industry has experimented with globalization for the past 40 years. In the last 10 years, and particularly since the 2000 U.S. stock market meltdown, globalization has advanced rapidly as the largest VC firms have established cross-national partnerships and overseas offices and have co-invested abroad. We expect this trend to continue to grow as other nations develop significant clusters of expertise and firms worthy of VC investment, global corporations accelerate their practice of acquiring foreign startups, and exit markets are increasingly global.

The reasons for globalization vary not only by VC firm, but also by nation. Until recently, U.S. VC firms, particularly in Silicon Valley, experienced little pressure to globalize. For European and Asian firms, better returns and more investment opportunities were prime motivators for investing and operating in the United States. For the European firms, the ability to assist their portfolio firms abroad (in the U.S. market) was also an important globalization motivator. In contrast, Asian VC firms did not attach as much importance to these reasons, suggesting that the Asian firms in which they were investing had minimal global operational ambitions (e.g., nearly all of the Chinese firms that went public on NASDAQ did not target external markets). Although U.S. VC firms have been privileged by being located in the world's largest and most lucrative investment environment, many major U.S. VC firms are globalizing.

In 2007 the VC industry was well into the process of globalization. The IT industries have always been the core business field for VC investing and U.S. firms have dominated this field. At this time, there is no indication that any other business field, such as biotechnology, nanotechnology, superconductivity, or green energy technologies, will displace the IT sector in the near future in terms of the size and speed of capital gains generation. Should the U.S. ability to generate IT and biomedical innovations capable of being commercialized for large capital gains decline, VC firms will shift their investment to more promising locations and technologies (if any emerge).

The continuing globalization of R&D described in the other chapters of this volume suggests that VC firms will also continue to globalize as investment opportunities proliferate. At this point, VC industry globalization does not threaten U.S. leadership in technological innovation. Extrapolating from the historical record, what would be most likely to threaten innovation leadership would be U.S. decisions that weaken the flow of innovative opportunities and the supply of technically proficient, well-trained entrepreneurs (e.g., changes in immigra-

---

[11]Doidge et al. (2007) dispute the hypothesis that SOX has caused listings to be moved to London markets.

tion policy, research funding policy, tax policies, or secrecy that stifles the free flow of engineers, scientists, and information). There is little evidence to suggest that innovation is a zero-sum game (i.e., if Country A innovates, then Country B cannot). Having said this, it is undeniable that technological expertise is not as concentrated in the United States as it was even one decade ago. Moreover, technological creativity is based on the talent of human beings, and they are more mobile than ever.

Venture capitalists are opportunistic and their firms' investments will flow to opportunities wherever they are located. There can be little doubt that the addition of VC to a region that already has the technical capabilities and a fledgling entrepreneurial environment can accelerate and feed the development of a virtuous circle of further entrepreneurship and a concomitant increase in VC investment. A supply of VC without a large number of high-technology entrepreneurs will not ignite new firm formation. The globalization of VC will not result in a lack of available capital in the United States. Policies that reduce the number of global-class technologists and entrepreneurs in the United States will directly affect decisions by venture capitalists to invest in the United States.

ACKNOWLEDGMENTS

Martin Haemmig thanks Jesse Reyes (formerly Vice President at Venture Economics) and John Gabbert (until recently at VentureOne) for providing data. Martin Kenney thanks the Alfred P. Sloan Foundation for supporting his research on the globalization of the venture capital industry. The authors thank Rafiq Dossani and John Taylor of the National Venture Capital Association for comments. We owe a debt of gratitude to Jeffrey Macher, David Mowery, and two anonymous reviewers for their invaluable critiques and comments that sharpened our thinking and improved the chapter.

REFERENCES

Advent International Corporation. (1986). *Advent International Corporation.* Boston, MA.
Aldrich, H. E., and M. C. Fiol. (1994). Fools rush in? The institutional context of industry creation. *Academy of Management Review* 19(4):645-670.
Allen, F., and W. Song. (2005). Venture Capital and Corporate Governance. Wharton Financial Center Working Paper 03-05.
Asian Venture Capital Journal. (2001). *The 2000 Guide to Venture Capital in Asia, Volume 1.* Hong Kong: Asian Venture Capital Journal.
Avnimelech, G., M. Kenney, and M. Teubal. (2004). Building Venture Capital Industries: Understanding the U.S. and Israeli Experience. Berkeley Roundtable on the International Economy, Working Paper 160.
Aylward, A. (1998). *Trends in Venture Capital Finance in Developing Countries.* International Finance Corporation Discussion Paper Number 36. Washington, D.C.: The World Bank.

Baracel, A. L. (2004). Performance of Cross-Border Venture Capital Investment. Paper presented at the Financial Management Association Annual Meeting, New Orleans, La., October 7-10, 2004.

Black, B., and R. Gilson. (1998). Venture capital and the structure of capital markets: Banks versus stock markets. *Journal of Financial Economics* 47(3):243-277.

Bottazzi, L., M. D. Rin, and T. Hellman. (2005). What Role of Legal Systems in Financial Intermediation? Theory and Evidence. University of Bocconi Working Paper 233.

Byte and Switch. 2006. ZTE Key Financials, 2002-2005. Available at http://www.byteandswitch.com/document.asp?doc_id=99265&page_number=2&table_number=2. Accessed February 1, 2007.

Clark, R. (1987). *Venture Capital in Britain, America and Japan.* London and Sydney: Croom Helm.

Coopey, R., and D. Clark. (1995). *3i: Fifty Years Investing in Industry.* Oxford: Oxford University Press.

Cumming, D. J. (2002). Contracts and Exits in Venture Capital Finance. Working Paper, University of Alberta School of Business.

Doidge, C. A., G. A. Karolyi, and R. M. Stulz. (2007). Has New York Become Less Competitive in Global Markets? Evaluating Foreign Listing Choices over Time. Fisher College of Business Working Paper No. 2007-03-012. Available at http://ssrn.com/abstract=982193. Accessed July 2007.

Dossani, R., and M. Kenney. (2002). Creating an environment for venture capital in India. *World Development* 30(2):227-253.

Dossani, R., and M. Kenney. (2007). The next wave of globalization: Relocating service provision to India. *World Development.*

Ernst & Young. 2007. Acceleration: Global Venture Capital Insights Report 2007. Available at http://www.ey.com/Global/assets.nsf/International/SGM_Global_VC_Insights_2007/$file/Global_VC_Insight_Report_2007.pdf.

Florida, R. L., and M. Kenney. (1988). Venture capital, high technology and regional development. *Regional Studies* 22:33-48.

Fox, J. W. (1996). The Venture Capital Mirage: An Assessment of USAID Experience with Equity Investment. Center for Development Information and Evaluation Working Paper, U.S. Agency for International Development, Washington, D.C.

Global Insight, Inc. 2004. Venture Impact 2004: Venture Capital Benefits to the U.S. Economy. Available at http://www.nvca.org/pdf/VentureImpact2004.pdf#search=%22venture%20capital%20total%20employment%22.

Gompers, P., and J. Lerner. (2001). The venture capital revolution. *Journal of Economic Perspectives* 15(2):145-168.

Green, M., ed. (1991). *Venture Capital: International Comparisons.* London and New York: Routledge.

Greer, P. (1971). Venturing in Europe. *SBIC/Venture Capital* (December):1-3.

Guler, I., and M. F. Guillen. (2005, November). Institutions, Networks, and Organizational Growth: The Internationalization of the U.S. Venture Capital Firms.

Haemmig, M. (2003). *The Globalization of Venture Capital: A Management Study of International Venture Capital Firms.* Zurich: Swiss Private Equity and Corporate Finance Association.

Hege, U., A. Schwienbacher, and F. Palomino. (2004, February). Determinants of Venture Capital Performance: Europe and the United States. RICAFE Working Paper 001. Available at http://ssrn.com/abstract=482322.

Hsu, D., and M. Kenney. (2005). Organizing venture capital: The rise and demise of American Research & Development Corporation, 1946-1973. *Industrial and Corporate Change* 14: 579-616.

Huawei. (2007). Financial Highlights. Available at http://www.huawei.com/about/info.do?cid=-1002 &id=64. Accessed February 1, 2007.

Hursti, J., and M. V. J. Maula. (2007). Acquiring financial resources from foreign equity capital markets: An examination of factors influencing foreign initial public offerings. *Journal of Business Venturing* 22(6):833-851.

Jefferson, G. H., and G. Gian. (2007). Science and technology takeoff in China? Sources of rising R&D intensity. *Asia Pacific Business Review* 13(3):357-371.

Jeng, L., and P. Wells. (2000). The determinants of venture capital funding: Evidence across countries. *Journal of Corporate Finance* 6:241-289.

Kaplan, S. N., F. Martel, and P. Strömberg. (2003). How Do Legal Differences and Learning Affect Financial Contracts? NBER Working Paper 10097.

Kenney, M., and R. Florida. (2000). Venture capital in Silicon Valley: Fueling new firm formation. In *Understanding Silicon Valley: The Anatomy of an Entrepreneurial Region*, M. Kenney, ed. Stanford, CA: Stanford University Press.

Kenney, M., K. Han, and S. Tanaka. (2004). Venture capital industries in East Asia. Pp. 391-427 in *Global Change and East Asian Policy Initiatives*, S. Yusuf, M. Altaf, and K. Nabeshima, eds. Oxford: Oxford University Press.

Kogut, B., P. Urso, and G. Walker. (2007). The emergent properties of a new financial market: American venture capital syndication from 1960 to 2004. *Management Science* 53(7):1181-1198.

Kuemmerle, W. (2001). Comparing catalysts of change: Evolution and institutional differences in the venture capital industries in the U.S., Japan and Germany. Pp. 227-261 in *Comparative Studies of Technological Evolution, Research on Technological Innovation, Management and Policy*, R. A. Burgelman and H. Chesbrough, eds. Amsterdam and Oxford: Elsevier Science.

La Porta, R., F. Lopez-de-Silanes, A. Shleifer, and R. W. Vishny. (1998). Law and finance. *Journal of Political Economy* 106:113-155.

Lerner, J., and A. Schoar. (2005). Does legal reinforcement affect financial transactions? The contractual channel in private equity. *Quarterly Journal of Economics* 120(1):223-246.

Mäkelä, M. (2004). Essays on Crossborder Venture Capital: A Grounded Theory Approach. Helsinki University of Technology, Institute of Strategy and International Business Doctoral Dissertation Series 2004/1.

Mäkelä, M., and M. V. J. Maula. (2006). Cross-border venture capital and new venture internationalization: An isomorphism perspective. *Venture Capital: An International Journal of Entrepreneurial Finance* 7(3):227-257.

Manigart, S. (1994). The founding rates of venture capital firms in three European countries. *Journal of Business Venturing* 9(6):525-541.

Manigart S., H. J. Sapienza, and W. Vermeir. (1996). Venture capitalist governance and value-added in four countries. *Journal of Business Venturing* 11(6):439-469.

Megginson, W. L. (2004). Toward a global model of venture capital. *Journal of Applied Corporate Finance* 16(1):8-26.

Murray, G. C., and J. Lott. (1995). Do UK venture capitalists have a bias against investment in new technology-based firms? *Research Policy* 24(2):283-299.

Murray, G. C., and R. Marriott. (1998). Why has the investment performance of technology-specialist, European venture capital funds been so poor? *Research Policy* 27:947-976.

Pagano, M., A. A. Roell, and J. Zechner. (2002). The geography of equity listing: Why do companies list abroad? *Journal of Finance* 57(6):2651-2694.

Pruthi, S., M. Wright, and A. Lockett. (2003). Do foreign and domestic venture capital firms differ in their monitoring of investees? *Asia Pacific Journal of Management* 20:175-204.

Rock, E. B. (2001). Greenhorns, yankees and cosmopolitans: Venture capital, IPO's, foreign firms and the U.S. market. *Theoretical Inquiries in Law* 2(2):1-35.

Romain, A., and B. van Pottlesberghe. (2004, April). The Determinants of Venture Capital: Panel Data Analysis of 16 OECD Countries. Unpublished Working Paper.

Schumpeter, J. A. (1939). *Business Cycles: A Theoretical, Historical and Statistical Analysis of the Capitalist Process* (abridged with an introduction by Rendigs Fels). New York: McGraw-Hill.

Sheahan, M. (2005). Is SOX to blame for IPO decline? What's next? *Venture Capital Journal* (November). Available at http://www.ventureeconomics.com/vcj/protected/1122124957218.html.

Sorenson, O., and T. E. Stuart. (2001). Syndication networks and the spatial distribution of venture capital investments. *American Journal of Sociology* 106:1546-1588.

Stinchcombe, A. E. (1965). Social structure and organizations. Pp. 142-193 in *Handbook of Organizations*, J. G. March, ed. Chicago, IL: Rand-McNally.

Stuart, T. E., and O. Sorenson. (2003). Liquidity events and the geographic distribution of entrepreneurial activity. *Administrative Science Quarterly* 48:175-201.

U.S. Government Accountability Office. (2006, September). *Offshoring U.S. Semiconductor and Software Industries Increasingly Produce in China and India.* Publication GAO 06-423.

Wilson, J. W. (1985). *The New Venturers: Inside the High-Stakes World of Venture Capital.* Menlo Park, CA: Addison-Wesley.

Wong, P. K. (2005). Venture capital funding of Asian entrepreneurs in Silicon Valley: A longitudinal analysis. Paper presented at the Conference on Greater China Capital Market for Innovation & Entrepreneurship, Taipei, Taiwan, May 16-17.

Wright, M., S. Pruthi, and A. Lockett. (2005). International venture capital research: From cross-country comparisons to crossing borders. *International Journal of Management Reviews* 7(September): 135-161.

Zalan, T. (2004). The secret multinationals of the new millennium: Internationalization of private equity firms. *Journal of International Business Studies Literature Review.*

Zero2IPO Research Center. (2008). US$3.25B VC Investment Touches New High (January 23). http://www.zero2ipo.com.hk/china_this_week/detail.asp?id=5753

Zero2IPO Venture Capital Research Center. (2005). China Venture Firms Newly Raised US $4 Billion in 2005; Internet Sector Most Popular to VC Firms. (December 15).

10

Financial Services

RAVI ARON

University of Southern California

INTRODUCTION

The financial services industry is situated at the intersection of twin forces of disruption: technology and globalization. Financial services, perhaps more than any other industry, is buffeted by the forces of change that technology has unleashed and the intensity of competition that globalization has brought in its wake. The nature of work in financial services is such that the input, work in process, and output are all information. There are no heavy and unwieldy manufactured items that need to be carefully managed in transit, there is no need for the maturing of third-party logistics that will delay the entry of global players, there are no issues of establishing warehouse and distribution centers to handle inventory that will deter offshore competitors, and the traditional advantages of scale and incumbency that established players enjoy can be nullified by an intelligent business model that takes no more than a few weeks—if that—to become a business. While business experience—understanding of a market's needs and its willingness to pay for services—is an important prerequisite, the role of physical infrastructure as a barrier to competition is greatly lessened.

Consider the example of ING Direct to illustrate how easily incumbency, scale, and local presence can be neutralized by technology. The high-value-added business models of the full-service banks that dominated the retail financial service industry in the United States should have remained inviolate—after all, they had local presence, scale of operations, and a formidable network of branches that straddled the retail markets from coast to coast. These banks were challenged in their backyards by a European bank, ING Inc., that decided to bypass the battle of the branches and reach out to retail market segments directly through nontraditional distribution channels. ING Direct is a straight-to-consumer banking service that has a very focused product line and one that operates at a significantly lower cost structure than competitors (traditional retail banks). ING Direct does not rely on the traditional channels of distribution—physical branches; instead it distributes its products using the Internet and call centers. As a recent Harvard Business School Case showed,[1] "ING Direct is able to offer depositors higher interest rates on its savings accounts, dispense with fees and service charges, and still make money by reinvesting depositors' funds in longer term assets" (Gary, 2004). A research report showed that ING Direct was able to walk away with $17 billion in direct deposits without an extended banking network (Forrester, 2004). In this parable if the large incumbent retail banks were the Goliaths then technology was the sling that ING's David used to take on the entrenched larger players.

Just as technology releases disruptive change in hitherto stable industries, globalization accelerates that speed of change and amplifies the gains and losses that result. Global free trade agreements and the integration of several regions into trading blocks—such as the European Union (EU), ASEAN, and NAFTA—have opened the financial services markets for several countries to foreign players even as the large established banks and insurance companies must face competition from Asian rivals such as HSBC. If necessity is the mother of invention then competition is the mother of innovation. And competition has unleashed innovations in the financial services industry in mature markets such as the United States and the EU.

A recent research report by the management consulting firm of McKinsey & Company attributed the resurgence of European financial services firms to innovation in a variety of sectors when faced with intense competition from local and foreign players and deregulation (Gary, 2004). The report identifies two different kinds of innovation: the development of new products and services, and the design of innovative ways in which firms produce and deliver these processes. Europe, according to McKinsey & Company, has led innovation in creating derivative products and has also innovated in other product classes such as cash securities. Leading product innovators include BNP Paribas and Société Générale in equity derivatives, and independent advisory houses such as Rothschild. A second kind of innovation—producing and delivering financial services in new ways—including the use of technology in retail financial services and the adoption of global operating models has also resulted in European banks becoming centers of innovation. In other recent research reports, McKinsey & Company identifies that the drivers of success for offshore firms in the Indian and Chinese financial services markets would be the ability to deliver financial services through new and more efficient channels (in India) and to create value-bearing partnerships in China (McKinsey, 2005a,b).

---

[1]As the case makes clear, ING's business model is a clear success in that it enjoys supranormal profits.

Recent reports in the business press have dwelt at length on the idea of "high-end" and "low-end" work. Some of these reports seem to suggest that some work, such as call center work, is routine and can be seen as "low-end," whereas other work, such as financial analytics and investment banking research, is "high-end." The lay reader may well take away the impression that in the global sourcing and delivery of financial services there is a clear understanding of what constitutes low- and high-end work and that corporations in the United States and Europe innovate while their service providers in Asia and other low-wage regimes provide routine low-end services at low cost. There is a widespread assumption in media coverage of these trends that the so-called high-end work is complex whereas the low-end work—as typified by a tech support call center—is essentially low-complexity work. As we show later in this chapter, this assumption both is wrong and reflects a poor understanding of the subjective nature of complexity. To understand how global sourcing of services drives innovation and how the nature and extent of the innovation is itself determined by the subjective perception of complexity, it is necessary to first understand the globalization of the financial services industry.

The Scope and Growth of Offshore Activities in Financial Services

The phenomenon of globalization of financial services has both supply and market (demand) side implications. Large multinational financial services corporations such as American Express, Citigroup, and HSBC Inc. have established a significant sourcing and market presence in Asia and Europe. These companies have a retail, corporate, and investment banking presence in Indian and Chinese markets. In addition to these traditional financial services, investment and risk-based financial services are also offered by multinational VCs and private equity firms. To compete against local companies, the aforementioned firms have to constantly innovate and deliver services to niche segments such as urban retail customers in India.

Citibank launched its web-based corporate banking products in 2001 to capture a significant chunk of the market for cash management and trade services. The bank then rolled out this product widely in the Asia-Pacific region (Malaysia, Australia, Singapore, and Hong Kong) and later into India and parts of Europe.[2] The bank found a new way of distributing corporate banking services via the web based on a flexible electronic product that linked the bank's front office to the clients' back offices. The web-based banking services that the bank launched in Asia and the United States were based on creating a flexible suite of corporate banking products that would connect their clients to the bank's service delivery portals, provide a set of tools to the clients to automate and manage their accounts

---

[2]Citibank's e-Business Strategy for Global Corporate Banking, HBS Case Study, HKU197, 2002, Center for Asian Business Cases, University of Hong Kong.

receivable and payables, and offer advanced treasury services to clients ranging from SMEs and MNCs.[3] The idea of creating an electronic channel that would offer a flexible suite of financial services that would—to quote the bank—connect, extend, and transform the operations of the clients was an idea that was tried in the highly competitive Asian markets of the Far East (Singapore, Malaysia) and then rolled out with variants in other parts of the world. As the technological capabilities offered by the Internet matured, the diffusion of the innovation in other markets such as Europe, India, and China was also accompanied by accommodations of local banking customs, market contexts, and numerous custom features that tailored the innovation to local market conditions.[4] Thus, the expansion of multinational financial services firms into global markets has resulted in the globalization of innovation. Our definition of the globalization of innovation is one where the innovation that takes place is global in nature—not that the output (e.g., the applications) of an innovation process (perhaps done in a single home country) is used globally.

On the sourcing side too there have been numerous innovations in the use of technology as well in new managerial initiatives. OfficeTiger,[5] an American business process outsourcing (BPO) firm with offices in India, Sri Lanka, and the Philippines, has been able to serve offshore clients in projects that often require near-real-time collaboration on highly judgment-intensive processes (i.e., processes require experts such as accountants and financial analysts to intervene and exercise judgment) executed by financial service professionals (Aron and Singh, 2005). A third of OfficeTiger's deadlines are an hour or shorter. The firm has been able to deploy a combination of technology and a collaborative interorganizational managerial structure called the Program Office in order to deliver highly strategic and value-added services to its clients.[6]

Thus, global sourcing of financial services too has left an "innovation footprint" in various parts of the world. There is considerable divergence in the definition of what constitutes offshore production of services and how these are categorized (by work type [i.e., call center] as opposed to vertical industry [i.e., financial services]). There is also considerable variance in the sources of data and the sampling methods used by various firms. Rather than go by any single

---

[3]Citibank's e-Business Strategy for Global Corporate Banking, HBS Case Study, HKU197, 2002, Center for Asian Business Cases, University of Hong Kong.

[4]The details of customization and localization of the innovation require a detailed exposition beyond the scope of this chapter. We refer the interested reader to a business case developed by the Center for Asian Business Cases, University of Hong Kong, entitled "Citibank's e-Business Strategy for Global Corporate Banking" HKU197, 2002.

[5]For an insightful discussion of the business challenges faced by the firm OfficeTiger, we refer the reader to an article in *The New Yorker*, July 5, 2004, titled "The Best Job in Town" by Katherine Foo.

[6]We refer the reader to the *Harvard Business Review* article "Getting Offshoring Right" in December 2005; see issue of the journal for further details.

estimate of this phenomenon, we find that the meta-research done by the firm Pipal Research summarizes the different estimates best. Figures 1 and 2 provide summaries of the estimates. There is wide variance in the estimates of outsourcing because the term is defined differently by different researchers. Some conflate outsourcing with offshoring (i.e., production of services and goods in a different country than the one in which the final output is consumed) while others group all offshoring under outsourcing.[7] BPO is the practice of sourcing services (processes) from a third-party provider of these services. BPO could be an offshore or onshore phenomenon; indeed there is nothing in the nature of BPO that restricts it to a particular location. In popular usage, the term BPO is frequently used to mean offshoring.

Many of these reports include both offshoring and outsourcing activities, whereas some of these reports exclude offshoring of business processes to a subsidiary of the parent corporation (offshoring but on outsourcing). For instance, the Gartner estimate of USD 173 billion includes all the processes in Tables 1 and 2 only to the extent that they are outsourced to a third-party provider, whereas the estimates by IDC (2005) in Figure 1 include processes that are offshored even if they are not outsourced.

Political sensitivity and tensions with labor advocacy groups and organized labor surround the phenomenon of offshore outsourcing. As a result it is not possible to collect data exhaustively from offshore firms on what processes have been offshored and how many agents are employed in executing these processes. We have—as part of our larger research agenda—identified the different kinds of services (and subprocesses) that are offshored in the domain of financial services. Table 1 provides a list of major financial services offshored while Table 2 provides a breakdown of these major financial services to the level of subprocesses.[8] There exist no rigorous estimates that can quantify the extent of offshore outsourcing (or even just outsourcing for that matter) by process type (work type) in financial services in terms of either dollar volumes or of the FTE[9] involved in producing the services.[10] What Tables 1 and 2 provide, however, is a fine-grained analysis of what kinds of financial service-related tasks are offshored.

As we discussed earlier, the advent of BPO also led to the globalization of innovation in financial services. Other examples of such innovation include the

---

[7]When a corporation, say, in the United Kingdom locates its back office in India and sources back office processes through a fully owned subsidiary, it is offshoring but not outsourcing.

[8]From a joint research project between The Wharton School and Gartner in 2003-2006.

[9]Full-time equivalent, a measure of the number of employees involved in executing a process.

[10]There are two reasons why this is not possible: As mentioned earlier, there is intense political sensitivity that surrounds offshore outsourcing and, therefore, senior executives are reluctant to provide any estimates that will result in media attention being focused on their firms. Second, measuring the economic significance of activities at the subprocess level requires the deployment of resources that represents a prohibitively high level of cost and effort, especially given the rapid growth of offshoring, which in turn creates a rapidly changing market that is not easily surveyed.

Source	Market Size	Date/Comment
Gartner	$50 billion (2007)	• 2007/ The market (Offshore Outsourcing) opportunity is expected to go up to 50 billion dollars by 2007. It has predicted that by 2007, total global offshore spending on IT services will reach 50 billion dollars
Gartner	$173 billion in 2007, of which $24.23 billion would be outsourced to offshore contractors	• Gartner: $173 billion in 2007, of which $24.23 billion would be outsourced to offshore contractors • Gartner does not incorporate animation, medical or other (legal) transcription services, GIS, market research, data search, research and development, network consultancy and other non-business processes in its estimates on the ITES market size and potential
IDC	$300 billion (2004)	
NASSCOM-McKinsey report.	$300 billion (addressable market)	• Of which $110 billion will be offshored by 2010
IDC	$382.5 billion (as of Oct. 2005)	• Human resources, procurement, finance & accounting, customer service, logistics, sales & marketing, product engineering, and training
IDC Report	$35,887 billion (2005)	• As of Oct. 2004/ Worldwide Finance and Accounting BPO
Goldman Sachs	$200 – $250 million (2005)	• Feb 2003
Gartner	$234 million (2005) $67.64 billion (HR) $23 billion (F&A)	• Feb 2003. This implies that HR outsourcing accounts for 28-29% of the global BPO market in the period 2000-2005, thus making it the most dominant of BPO solutions required by global customers.

FIGURE 1 The size of the offshore outsourcing market.

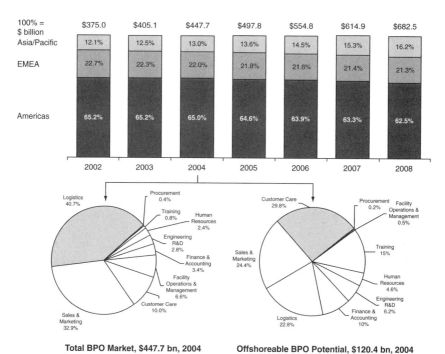

Total BPO Market, $447.7 bn, 2004   **Offshoreable BPO Potential, $120.4 bn, 2004**

FIGURE 2 Growth of total BPO market and offshoreable BPO size.

TABLE 1 Finance and Accounting Processes

Account management (transaction reconciliation, other asset and liability account administration)
Accounts payable
Accounts receivable
Actuarial analysis
Asset pricing research: equity research and fixed income asset research
Billing
Cash flow analysis (including forecasting and related services)
Cost accounting and analysis
Electronic payments (in retail and institutional financial services)
Financial statement analysis
General ledger
Management accounting
Preparing asset schedules
Risk analytics
Tax accounting and analysis (individual and corporate)
Tax management
Treasury and cash management
Underwriting
Yield analysis

TABLE 2 Financial Services Subprocesses

Actuarial (insurance)
All data processing (banking and insurance)
Automated clearing house and/or electronic funds transfer
Business and corporate credit card issuing and processing
Cash management
Check processing and imaging
Claims processing
Core banking
Credit card issuing: retail financial services
Credit card merchant processing: retail financial services
Data warehousing (retail and corporate banking, insurance)
Liability and asset account management (deposits, savings accounts, certificates of deposits, loans, etc.)
Electronic checks
Electronic invoice presentment and payment
Exception processing and automation
Internet banking
Internet lending and asset creation
Investment and portfolio management (insurance)
Investments and trading
Mortgage banking (both loan origination and subsequent account services)
Pension administration (insurance)
Policy servicing and administration (insurance)
Policy underwriting (insurance)
Primary business applications (insurance)
Trade services (as necessary: cargo and business insurance)
Trust (as necessary: retirement and custodial plan trust and administration
Vehicle financing (vehicle leasing and lending)

use of LEAN techniques by Wipro Technologies in India for the production of processes such as financial accounting, management of accounts receivable, and payables. Wipro Technologies was able to cut the time required to execute these processes by between 12 and 47 percent while increasing the quality of output by between 7 and 29 percent.[11] We studied other firms such as Pipal Research Inc. and i-Flex Inc. that provide innovative services similar to those provided by OfficeTiger.[12] Many of the drivers of offshore production of processes are also the drivers of innovation at those locations where these processes were sourced. To understand the globalization of innovation, it is necessary to understand the offshoring of business processes. The advent of GE's BPO initiative in India in

---

[11]From a paper titled "The Use of LEAN Techniques in Offshore Outsourcing," forthcoming (2007).

[12]Pipal Research's Manoj Jain: "We Have Created a Spot Market for Research," *Knowledge @ Wharton*, see http://knowledge.wharton.upenn.edu/india/article.cfm?articleid=4155; "How Some BPO Providers Seek to Build and Protect Their Turf," *Knowledge @ Wharton*, see http://knowledge.wharton.upenn.edu/article.cfm?articleid=1101.

1996 started a spurt of offshore outsourcing activity in the region that has since grown rapidly.[13] There are two reasons for the growth of offshore production of financial services: competitive pressures of a maturing industry and technological enabling factors. Indeed, these are also the factors that drive the globalization of innovation in financial services. Let us begin by analyzing the competitive pressures.

COMPETITION AND PRODUCT MATURITY

The evolution of any product or service is characterized by three phases. In the first phase, when the product or service is initiated, the challenge is to simply make it work and deliver the core benefits associated with the product. We term this the *product creation phase*. There are numerous examples of products in financial services that went through the product creation phase when they were introduced. In retail financial services, the now ubiquitous ATM provides a good example. The ATM was introduced for the first time in 1939 in New York City by the City Bank of New York and was withdrawn soon after for lack of adoption by consumers.[14] It was reintroduced in 1967 in London and has since become a standard feature of all retail financial service products. When the ATM was first introduced it lacked ease of use and was not connected to the savings accounts of the consumers. For one thing there were no mainframe computers then and it was not possible to store account information and support transactions. And for another, without computer networks that could support multiple machines in different locations, the ATM could be used only at the physical location where the bank was situated. The advent of mainframes as the enablers of retail financial services meant that the ATM could be deployed remotely and connected to the accounts. This in turn allowed banks to deliver several financial services through the ATM and consumer adoption followed.

The second phase in the evolution of products and services is the *integration phase*, where the objective is to integrate the product to work seamlessly with the firm's existing products. As the ATM evolved into the integration phase it was extended to offer a suite of services that went beyond dispensing cash to include deposits, balance enquiry, money transfer, and so on. As the industry matures and all other firms offer similar products and differentiation becomes increasingly more costly, the firm in the final phase outsources several components associated with the product. We term this the *orchestration phase*, where firms outsource (often to offshore entities) several aspects of production, delivery, and postpurchase service. Most banks now are connected to ATM networks run by

---

[13]"The Little Start-Up That Could: A Conversation with Raman Roy, Father of Indian BPO," *Knowledge @ Wharton*, see http://knowledge.wharton.upenn.edu/article.cfm?articleid=794.

[14]MIT Series on Inventor of the Week: Luther George Simjian, see http://web.mit.edu/invent/iow/simjian.html.

third parties. In the orchestration phase, ownership of ATM networks is no longer a source of competitive advantage because ATM services are bought by banks from third-party firms that provide connectivity to ATM networks. Indeed, major ATM networks are now managed by third-party firms that are not in the banking business, thus turning the ATM network into a "banking utility" comparable to electric power and other commoditized services. Many wealth management and personal financial services (PFS) firms often offer portfolios of services that include offerings of other firms.

Financial services are natural candidates for outsourcing, in the production and delivery of financial services, input, output, work in process—are all information. Most products and services in the financial services industry are mature and as a consequence there is considerable competition among firms that offer all of these services. Almost all retail banks offer the spectrum of services ranging from savings and checking accounts through money market accounts and PFS including wealth management services. As a result there is considerable pressure on firms both to differentiate themselves by providing a higher quality of service—more accurate resolution of transactions, quicker balancing of asset accounts, better customer service, faster turnaround on loan and credit card applications—and to lower the costs of production of services. As a result firms have been forced to relocate the production of several services to wage regimes where skilled (white collar) labor can be sourced more cost-effectively. Financial service firms have offshored processes both to third-party providers[15] and to their own subsidiaries offshore—known as *captive centers* (or simply "captives")—that execute these processes offshore. The phenomenon of offshoring that we witness is an evolution of financial service products where firms have entered the orchestration phase.

As we can see, the business case for offshore production of financial services was chiefly due to the maturation of products and the fact that differentiation of products had become costly. This alone was not enough to result in firms moving offshore. The technological platforms and business infrastructure that would enable such a move were not in place before the advent of the Internet and its wide and ubiquitous adoption in countries like India and China. In 1996 Raman Roy set up a captive center for General Electric (GE) to process transactions and deliver support services.[16] Roy first demonstrated his ability to deliver high-quality financial services out of India as a part of American Express's back-office team and then demonstrated it again with GE. After his success with GE he set up Spectramind Inc. as a third-party offshore financial service provider for MNCs. As the success of Roy's Spectramind received attention from the business and trade press, several Western companies started sourcing services from India

---

[15]As mentioned earlier, this is often referred to as BPO.

[16]"The Little Start-Up That Could: A Conversation with Raman Roy, Father of Indian BPO, Part 2," *Knowledge @ Wharton*, see http://knowledge.wharton.upenn.edu/article.cfm?articleid=798.

and other third-party services providers and IT majors started offering these services.[17]

Since then the BPO industry in India has grown rapidly. Corporations in the United States and United Kingdom could transfer raw data across the Internet and Indian companies would process the data according to prespecified rules and transmit information back to the clients. The service could range from simple transaction processing to finance and accounting and as the industry matured it would grow to include financial analytics and research. For all of this to happen two factors were necessary: (1) the availability of the Internet as the physical medium of connection and (2) the availability of firms that would provide these services, because not all corporations could run their own captive centers and the presence of third-party firms—such as Spectramind—was necessary to scale up the offshoring trend.[18] In the period from 1999 to 2002 several third-party providers of BPO services came into being (principally) in India but also in countries such as China and the Philippines. By the end of 2004 there were several third-party providers of these services as well as captive centers in several regions of the world including Sri Lanka, Jamaica, and the Philippines.[19] The combination of competitive pressures in a maturing industry and the presence of technological factors of enablement made it possible for corporations to migrate the production of some financial services to offshore providers.

Offshore production of financial services is often seen as being analogous to the production of goods offshore. The production of physical goods offshore was facilitated by the emergence of large and cost-effective labor pools in Asia and parts of Latin America. Specialized supply chains emerged in industries such as apparel, auto components, consumer electronics, and toys, which enabled firms to set up offshore operations to service markets in the advanced industrialized countries, thereby leveraging wage differentials for factory labor. While there are similarities between the two forms of production, it would be wrong to assume that the production of services is chiefly about white collar unskilled labor, such as seen in call centers, which offers attractive wage arbitrage possibilities.

It is useful here to make a distinction between the production of financial services at a lower cost made possible by an attractive wage arbitrage to be found in migrating to low-labor-cost countries and the sourcing of innovative solutions, processes, and management techniques from providers offshore. The production of financial services can be characterized by lower costs, productivity gains

---

[17]"The Little Start-Up That Could: A Conversation with Raman Roy, Father of Indian BPO, Part 2," *Knowledge @ Wharton*, see http://knowledge.wharton.upenn.edu/article.cfm?articleid=798; "In a Global Economy, Competition Among BPO Rivals Heats Up," *Knowledge @ Wharton*, see http://knowledge.wharton.upenn.edu/article.cfm?articleid=642.

[18]"The Little Start-Up That Could: A Conversation with Raman Roy, Father of Indian BPO, Part 2," *Knowledge @ Wharton*, see http://knowledge.wharton.upenn.edu/article.cfm?articleid=798.

[19]"Move Over, India: The Shifting Geography of Offshore Outsourcing Creates New Challengers," *Knowledge @ Wharton*, see http://knowledge.wharton.upenn.edu/article.cfm?articleid=1100.

(sometimes), and continuous improvements in efficiency over a period of time. These services produce modest gains in cost and quality ranging anywhere from 1 to 10 percent over a period of 3 to 4 years.

In contrast to these incremental improvements, the execution of processes can be characterized by order-of-magnitude gains in both costs and quality over the same period of time. Usually, such gains result not from incremental improvements but from radical redesign of elements of work. Such redesign can include some of the following: process redesign, radical changes in the workflows and information flows associated with process production, automation in production leading to less human intervention, new information and knowledge transfer mechanisms that make it possible to offshore processes that were hitherto thought of as being too risky to decouple from where they are consumed (e.g., United States, United Kingdom), and the deployment of managerial practices that allow managers to monitor and influence work remotely. Some features of such innovations include the following: order-of-magnitude gains (as opposed to steady and incremental gains in costs and quality), gains realized over a compressed time period (as opposed to small incremental gains), use of new techniques (as opposed to improvements to existing methods), and simultaneous improvements to aspects of production that are thought to be substitutes (such as cost of production and quality of output; in the absence of innovation, lowering the cost of production substantially would result in lowering the output quality).

Certain financial services such as bond pricing research, equity research, asset pricing research in general, cash flow analysis, financial statement analysis, tax management, and modeling are now executed offshore. While the production of these financial services has been characterized by highly innovative practices, some so-called low-end services have proven to be difficult to migrate offshore and have been characterized at best by gains from wage arbitrage and small incremental improvements in efficiency. To understand why some complex services have been successfully migrated to offshore entities and have, within a short period of migration, resulted in innovations in the production and delivery of these services while other seemingly simple services could not be migrated offshore due to a high degree of operational errors, it is necessary to understand the subjective nature of process complexity that underlies the production of services.

The lay person might well think that the work involved in tasks such as asset pricing, tax analysis, financial analytics, and modeling is complex whereas the work involved in tasks such as customer service and selling credit card products is relatively simple. In fact, our research into the nature of complexity points to the contrary. There is considerable divergence between managers in different countries in their ratings of what is complex and what is not and as a result there are resulting differences between countries on the locus of innovation in the production and delivery of financial services. Executives in India and Singapore tend to rate the complexity of different kinds of work very differently from executives in the United States and United Kingdom.[20] To understand how the different regions of the world forge very different kinds of innovation it is necessary to investigate the subjective nature of complexity and how offshoring may well set up a *mechanism for complexity arbitrage*.

THE SUBJECTIVE NATURE OF COMPLEXITY

We surveyed executives of BPO firms and captive centers and their clients and users[21] respectively and sought their views on what kinds of tasks they deemed to be "complex" and which ones not. Executives from BPOs and captive centers were mostly located in India, Singapore, and Mauritius while executives from their clients (and users) were located in the United States and United Kingdom. We showed executives an identical set of processes. As a measure of complexity we asked executives from the four countries to rate an identical set of processes on the level of complexity involved in executing the processes.[22] Only those executives that managed the execution and delivery of a process were asked to rate the complexity of the process.[23] Each process was rated by four executives (one for each region) and some raters rated more than a single process.[24] They rated processes on a Likert scale of 1 to 7, where the highest level of complexity corresponded with a rating of 7 and the lowest complexity level was rated at 1. The results of the survey are shown in Table 3.

It is clear that while the executives of the client firms in the United States and United Kingdom are in broad agreement on process complexity—just as Indian and Singaporean executives are in agreement with each other—they differ to a significant degree from their Indian and Singaporean counterparts. For instance, U.S. and U.K. executives rate process[25] P8 as a low-complexity process while their Indian and Singaporean counterparts rate the process as being of relatively

---

[20]Based on a survey of BPOs and captive centers undertaken by the author and his doctoral students.

[21]For captive centers the term "client" is less accurate as the users of the service and the providers belong to the same firm. Instead of clients we call them users.

[22]As a measure of complexity we asked executives how difficult it was to ensure that a particular process could be executed at an acceptable level of quality, at a required scale, reliably, and predictably. It is clear that implicit in this definition of complexity is the difficulty associated with staffing the process and the costs incurred in retaining and training the workforce to reach the adequate levels of quality.

[23]In all there were 37 processes for which we were able to obtain comparable data for 28 processes. Each process was rated by an executive from each of four countries: the United States, United Kingdom, India, and Singapore. The 28 processes were rated by 112 raters.

[24]The same manager often managed two or three different processes and provided a rating for each.

[25]The companies that provided us with the data have placed strict restrictions on us whereby we are not allowed to mention the names of processes or firms or reveal any information that may lead to the disclosure of the names of the companies or their clients. We therefore use generic names for processes such as "P1", "P2", etc.

TABLE 3 Subjective Perceptions of Process Complexity

Processes	Indian BPO Firms	Singaporean BPO Firms	UK Firms	US Firms
P1	1	2	3	2
P2	3	4	2	3
P3	2	3	5	6
P4	3	4	5	5
P5	3	2	7	7
P6	4	5	4	3
P7	7	6	1	2
P8	6	5	2	2
P9	3	4	5	4
P10	5	4	3	2
P11	3	2	7	7
P12	3	2	5	6
P13	2	3	5	6
P14	5	5	3	4
P15	5	4	3	2
P16	3	4	5	4
P17	1	1	7	7
P18	6	7	2	1
P19	3	4	4	4
P20	6	7	2	1
P21	6	6	2	2
P22	1	3	5	7
P23	5	4	3	3
P24	7	5	1	2
P25	2	2	6	6
P26	6	5	4	7
P27	2	1	7	6
P28	3	3	5	4

higher complexity. Similarly, executives in India and Singapore tend to rate process P5 as a low-complexity process while executives in the United States and United Kingdom rate it as being of relatively higher complexity. Table 4 provides a summary of the levels of convergence and divergence of opinions, between executives from the buyer and provider countries, on process complexity. The coefficient of correlation in the ratings of process complexity between the four regions is furnished in Table 4.

The preceding findings can be explained when we analyze the factors that cause the divergence of opinions between the four regions. We found[26] that when a process involved computational work, algorithmic execution, quantitative analysis, numerical analysis, or large dataset handling and analysis, the executives from the United States and United Kingdom tended to rate such processes as being highly complex, while their Indian and Singaporean counterparts rated

---

[26]In interviews with executives that responded to the survey.

TABLE 4 Correlation Coefficients of Process Complexity Ratings

Correlations	India	Singapore	United Kingdom	United States
India	1.000			
Singapore	0.831	1.000		
t-stat	7.61099815			
P-value	4.4461E-08			
United	−0.760	−0.813	1.000	
Kingdom	−5.9608215	−7.112491		
t-stat	2.7238E-06	1.4903E-07		
P-value				
United States	−0.644	−0.720	0.861	1.000
t-stat	−4.2870909	−5.2887715	8.61462229	
P-value	0.0002204	1.5708E-05	4.2947E-09	

such work as being of low complexity. Similarly, when the work involved in executing these processes required interpretation, judgment, understanding of the market context, understanding the client's business context, communication, persuasion, or disambiguation,[27] the executives in India and Singapore rated the processes as being highly complex while their Western counterparts rated such work as being of low complexity. Why do executives in India and Singapore find the latter kind of work complex? The answer to this question also has to with difference between domain expertise and domain experience and the difference in turn explains the very different kinds of innovation in financial services that we see in Western and Asian economies.

DOMAIN EXPERTISE AND DOMAIN EXPERIENCE

Domain expertise refers to an expert's understanding of a body of knowledge and ability to use a set of rules to execute some tasks in a given domain. For instance, an employee in a captive center in India may have acquired the expertise needed to do equity research and forecast a stock price window for a basket of stocks. The techniques needed to do this can be learned from a doctoral course in asset pricing and require no experience on the part of the agent actually working in an investment bank. Similarly an agent working for a BPO in China could well forecast cash flows and create scenario models for a client firm without ever having worked for a corporate bank. The mastery of a set of techniques and un-

---

[27]Sometimes it is necessary for an agent that is executing the process to use personal understanding of the business context of the process to remove ambiguities that may arise in executing the process.

derstanding of rules needed to perform such work is termed *domain expertise* in this context. On the other hand, in order to design a new financial product or offer an asset of a specific risk category to an investor it is necessary that the manager have an understanding of the nuances of the market that cannot be codified or captured in algorithms or cannot be easily transmitted to offshore providers via documentation. A private equity investor needs to understand the business of the client as well as be able to judge the quality of the management of the company in which he plans to invest. Similarly a wealth management specialist needs to have a clear understanding of the social and financial context of a high-net-worth individual as well as her personal goals in order to offer her a wealth management solution. These skills cannot be acquired without actually working in the business environment in which they are required. Employees of financial service firms in developed economies have the domain experience needed to execute these tasks. The tasks that require interpretation, communication, persuasion, understanding of the business context, and so forth very often need domain experience to execute. Those tasks that require computation, quantitative analysis, algorithmic execution, statistical analysis, and so forth can often be captured in terms of rules and expressed as techniques that can be learned. To execute such tasks it is enough for the agent to have domain expertise—there is no need for domain experience. It is indeed this difference in the nature of the two sets of tasks that drives the divergence in perceptions of complexity. This difference is one of the principal reasons[28] for the divergence in the approach to innovation adopted by managers in different parts of the world. The locus of innovation is different for the different regions in the world.

THE LOCUS OF INNOVATION

To launch new innovative features in financial products or entirely new categories of financial products it is necessary for the designers of these features (or products) to have considerable domain experience. It is not possible to launch new features or services for a segment of the market without knowing the business context of the market segment. On the other hand, it is possible to improve how some back-office support processes work without having the same level of domain experience. As an example, consider a credit card-based financial service that a company may offer to the merchants that accept its credit card. It is necessary for the company to know the needs of the merchants and have an estimate for their willingness to pay for the service, their preferred channel of service delivery, and the compliance procedures that they must follow. However, this company's offshore captive center or BPO provider need not have the same level

---

[28]Other factors too are important. Factors such as compliance regime in different countries, the economic systems, the extent of participation in the economy permitted by national governments, and the sociocultural beliefs of the consumers are also important. These factors can be thought of as exogenous or can be seen as parameters set by nature and constitute systemic variables.

of understanding of the market in order to make some of the supporting back-office processes work well. The offshore provider may be able to intelligently automate some segments of the process, may reengineer some information flows, and may bring experts to resolve transactions that are otherwise difficult to reconcile. Thus, he can innovate with processes to be able to better support the new innovative features that the product carries.

Product innovation requires domain experience while process innovation requires only domain expertise. Executives in Asia often focus on process innovation, leaving product innovation to their Western counterparts. Product and process innovations often have very different implications and market dynamics associated with them. We explore these two kinds of innovation in the section that follows.

PRODUCT AND PROCESS INNOVATIONS

The principal difference between Asian and Western executives in the kinds of innovation that they bring to the market has to do with process and product innovations. While Asian executives[29]—especially in the financial services industry—are far more likely to bring to market process innovation, their Western counterparts focus almost exclusively on product innovations as a means of creating competitive advantage. This difference in focus of managerial priorities is because of the lack of domain experience on the part of the providers and is not due to any other factor. As we mentioned earlier, product innovation requires considerable domain experience, making it a daunting challenge for providers of offshore services to release innovations for the Western markets.[30] In addition to this, there are crucial differences between product and process innovation that explain why the offshore providers and their Western clients specialize in different chunks of the value chain. Our studies of the offshore financial services industry since 2000 indicate that there is what amounts to a rough division of innovation labor—Western corporations increasingly look to product innovations to boost growth while their service providers (and captive centers) in Asia support them with process innovations.

Product Innovation

Product innovation in financial services is about adding new features to existing products or offering new products to the market. It is almost always customer

---

[29]It does not matter whether these Asian executives are employed by Western or Asian firms. Their lack of domain experience does not go away irrespective of who employs them.

[30]Of course for non-Western markets this is less of a problem. In Bangladesh, for instance, Mohammed Yunus was able to formulate a microcredit mechanism via the Grameen Bank and serve an underserved segment of the market.

facing.[31] Product innovation usually takes several months of R&D effort (including market research, research into product feature mix, pricing schemes, and the kinds of delivery channel to be deployed). It is usually piloted first to a few market segments; once it reaches a stable state, it is rolled out in scale. Examples of such innovation include the introduction of web-based banking and web-based insurance delivery. However, once product innovation is introduced into the market, the extent to which the innovation has succeeded takes a far shorter time to measure and it can be measured through relatively well-understood and unambiguous measures of product performance such as extent of consumer adoption, market share, profitability indices (including measures of gross and net margins), and customer satisfaction scores.

Product innovation nevertheless has an interesting "dark side"—it often adds to the complexity of operations. Each new suite of features or product offerings brings greater complexity in operations and more intense demands on the company's service infrastructure. As an example consider the following: a bank adds a new feature—the ability to apply online for a loan and consolidate all asset balances into a single-view account at the level of each customer. This feature places considerable strain both on the back-office operations and on the technology infrastructure needed to support the feature. The underlying information systems have to be modified to be able to store information in customer-centric accounts as opposed to the legacy product-based storage of transactions. It will not suffice if the information is aggregated in real time based on customer requests from multiple accounts. As the number of customer requests increase, this real-time aggregation is bound to break down under the load of processing requests. The entire database has to be redesigned and applications have to be rewritten. If the technology were substantial, the human operations challenges tend to be even more daunting. The bank would have historically supported customer requests based on product type (banking related, credit card-related, etc.), while now it has to transition to a system where the customer's requests have to be processed with a minimal set of interactions, both with the customer and between the bank's back-office processors. The bank's back-office staff are usually product-level experts that specialize by product type. Thus, a mortgage loan service staff will have little understanding on how to service a question related to credit card interest rate and the bank's new feature of relationship banking may flounder due to a sharp and significant increase in operational complexity. Adding new features and products often results in placing a strain on the organization's service operations amplifying the levels of complexity. This is where process in-

---

[31]Customer-facing activities are those that involve direct interaction with end customers, have a direct impact on the end customer experience, or both. Thus, a new product or service that is offered to the market is a customer-facing initiative, as opposed to sourcing a process from an overseas provider that will be subsumed into a service by a corporation. In the second case a corporation builds the process (sourced offshore) into its service but the offshore provider does not directly interact with the end customers.

novation—especially by offshore providers—plays a significant role as a complement to product innovation.

Process Innovation

Unlike product innovation, process innovation is not market-facing in its direct impact. Process innovation is about making significant changes to how tasks are executed, changing information flows and workflows to achieve radical simplification of process execution in order to achieve greater process efficiency and more effective support of the company's market objectives. Process innovation supports product innovation by reducing complexity in operations, making information work and its output more predictable, and improving the alignment between a company's service and operations and its market objectives.

In the preceding example of the financial services corporation that introduced innovative features into the market, the offshore provider of back-office processes to whom the firm outsourced these processes was able to radically reengineer processes by changing both the process architecture—consisting of information flows and workflow—and the skill sets of the agents that did the work. The company reengineered the entire back-office process from end to end to dispense with product-based specialists and instead moved to a model of deploying agents that had a 360-degree view of customer accounts and could process all kinds of products. To be able to support this process architecture the company needed back-office workers of a higher skill level than before. Under the changed scheme the agents had to be able to understand standard loans, split-coupon rate loans, mortgages, and credit lines. They hired agents with the necessary skill sets and trained them additionally as needed.[32]

In another example of process innovation supporting product innovation, a large financial services MNC launched a new financial loan product to its corporate customers whereby the company would offer a series of small tranche loans to corporations for funding working capital requirements. The loan amounts would be priced, negotiated, and delivered to the corporation within 24 hours notwithstanding the complexity of the deal. Furthermore, the company offered to price each tranche based on the risk level associated with the tranche as opposed to a fixed-rate coupon for the whole loan. This was especially attractive to corporate customers who believed that there would be a reduction in risk as the business moved forward. To make this possible, the company had to radically redesign its operations. It offshored the loan-processing tasks, including application management, credit risk assessment, and structuring the loan, first to a captive center that broke up the operations into modules. Thus, each set of related tasks

---

[32]"'Smart Growth': Innovating to Meet the Needs of the Market without Feeding the Beast of Complexity," *Knowledge @ Wharton*, see http://knowledge.wharton.upenn.edu/article.cfm?articleid=1585&CFID=2650244&CFTOKEN=97771378#.

would be carried by a modular operations unit team. Whenever a new customer entered the market a new operational silo could be assembled in real time to service that customer. The corporate loan-processing division was radically reorganized into modular units that specialized in a particular task.[33] For each new customer an assembly line of services would be created in real time by plugging in the modularized components.

The offshore firm was able to do this in very short time as it could recruit employees with the specialized skills needed to achieve this within a much shorter period than was possible in the United States.[34] Second, in order to get this design correct, the company had to carry some slack labor in the early months before it could reach a stable state of operations. This was not a problem in the offshore location because of the far lower wage costs. The firm's redesigned, modularized business processes are shown in Figure 3.

Product and process innovations differ in many aspects (as seen in Table 5), which we highlight in this section. *Process innovation often complements product innovation and the two kinds of innovations have a lock-and-key structure*[35] *of complementarily.* They also differ in some important ways; process innovation invariably takes far less time to deploy than product innovation. However, the metrics that measure the effectiveness of process innovation are rarely (if ever) direct—they tend to measure the impact of product innovation such as process output quality and customer satisfaction with service. However, these measures are compounded by a number of internal factors and are at best proxies for measuring process innovation effectiveness. Second, it takes much more time to judge the effectiveness of process innovations. Unlike the case of product innovation, where critical success factors such as market share and gross and net margins often provide an accurate picture of the success of the innovation, in the case of process innovation not only are there few (if any) direct metrics, there is also considerable lag between introducing the innovation and measuring its effectiveness. The absence of direct measures means that it is necessary to capture measures of a variety of indirect proxies and to compare their change over a period of time to get some idea of how successful the innovation has been. Thus, while process and product innovations are complementary in the financial services industry,

---

[33]The reader should note the direct contrast with the preceding example, where such specialized units were broken down into more versatile operational divisions. What should be clear is that there is no canonical way of engineering process innovation and very different solutions would work in slightly different situations.

[34]Mainly because it was located in a country with a very large labor pool with the necessary skills.

[35]The idea of a lock-and-key structure is used to illustrate how one kind of innovation holds the key to the success of another kind. New product introduction tends to snarl up the operational processes of a firm. Process innovation opens up operational silos, eases information flow, and reengineers workflow, thereby reducing complexity and allowing the benefits of product innovation to be delivered to end customers.

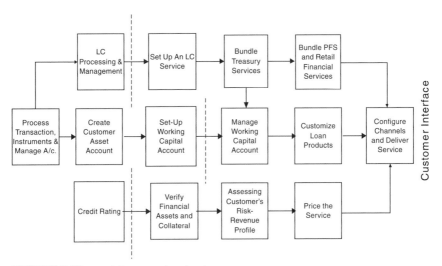

FIGURE 3 The modular operational unit.

TABLE 5 Product Versus Process Innovation

Innovation Factor	Product Innovation	Process Innovation
Locus of innovation	Market facing	Inside the firm, directed at back-office and service operations
Principal goal	Competitive advantage, higher market share and revenues	Operational efficiency, redesign of operations
Principal impact	Increases operational complexity	Lowering of complexity, increases service effectiveness
Gestation time	Comparatively long, often preceded by elaborate (several months of) R&D and market research efforts	Short; time lag between design and deployment is minimal
Metrics to measure effectiveness	Direct metrics, including market share, and gross and net margins	Only indirect proxies exist for measuring the effectiveness of the innovation
Time for market validation	Comparatively short; within a couple of quarters of introducing the innovation, CSFs can be measured	Much longer; it takes several periods of observation of indirect proxies for success to arrive at some estimate of the innovation's effectiveness
Prerequisite for innovation	Domain experience and understanding of market and business context	Domain expertise, ability to manage complexity and deal with rapid changes in business volume
IP assets	Proprietary in nature; companies guard these carefully	Rarely, if ever, proprietary, companies share these in a nonzero fashion
Contract structure	Coordination between firms, clearly defined boundaries and control structures	Collaboration between firms and between departments within firms; blurring of organizational boundaries

it is clear that they are very dissimilar. Table 5 contrasts process and product innovations.

Given the aforementioned nature of product and process innovation, are corporations willing to source from offshore providers? If so, what is the degree of acceptance of such sourcing mechanisms?

SOURCING INNOVATIONS OFFSHORE

We surveyed the views of senior executives of Fortune 1000 corporations on their willingness to source innovations from offshore providers (see Tables 6 and 7).[36] While the survey was not restricted to corporations in the financial services industry alone, we captured data about the industry to which respondents belonged and were able to extract the responses from the executives of the financial services industry.[37] We found that senior executives in corporations are prepared to source innovations from offshore providers as long as they have a means of dealing with the risk inherent in such a relationship.

In response to the question, "Can you partner with an offshore provider of innovation?" most executives (68 percent surveyed[38]) expressed their willingness to source innovations from offshore outsourcing firms. Only a small minority (13 percent) expressed the view that it was strategically risky to source innovations from offshore firms. It is particularly interesting to note that there is little or no concern expressed about the ownership of the offshore provider. Of those managers that were open to sourcing from offshore providers only a minority wanted to have full ownership control of the offshore provider. About 56 percent of these managers were open to sourcing from a firm that could become a strategic partner and another 15 percent or so were satisfied with additional control over outcomes offered by a joint venture with the offshore firm.

Product innovation often results in the creation of IP assets that corporations try to protect. Large investment banks have proprietary program trading algorithms that they go to great lengths to safeguard. Similarly corporate banks have developed risk assessment and management systems for different kinds of businesses that are protected as valuable IP assets. As opposed to this, process innovation rarely results in the creation of sensitive IP assets. The gains of process innovation in financial services can be shared in a non-zero-sum fashion between the offshore provider and its Western clients.

A final contrast between product and process innovation that we wish to comment on is that product innovation usually involves tight control via contract

---

[36]The survey was conducted by Knowledge@Wharton in 2006. About 75 percent of respondents worked either in financial services firms or were executing processes that belonged to the finance and accounting category.

[37]This is a subset of a larger set of responses.

[38]Sum of the first three rows, which represent the willngness to source innovations from offshore providers under different conditions.

TABLE 6 Sourcing Innovations from Abroad

Can you partner with an offshore provider of innovation?	Number of Responses	Percentage
Yes, if it is our fully owned subsidiary	12	7%
Yes, if we can establish a JV with an overseas firm	27	15%
Yes, if we can exert some measure of operational control over the overseas firm and make it a strategic partner	101	56%
No. We cannot risk sourcing innovative products and processes from an overseas firm	23	13%
Other please specify	16	9%
Total responses	179	

TABLE 7 Product and Process Innovations and IP Assets

Product innovation usually results in valuable intellectual property (IP) assets that need to be protected		
Strongly disagree	3	2%
Somewhat disagree	3	2%
Neither agree nor disagree	7	6%
Somewhat agree	51	40%
Strongly agree	62	49%
Total	126	
Process innovation, rarely, if ever, results in proprietary IP assets		
Strongly disagree	11	9%
Somewhat disagree	17	13%
Neither agree nor disagree	19	15%
Somewhat agree	48	38%
Strongly agree	31	25%
Total	126	
The company that brings an innovative product to the market often takes ownership of the innovation		
Strongly disagree	1	1%
Somewhat disagree	8	6%
Neither agree nor disagree	11	9%
Somewhat agree	59	47%
Strongly agree	47	37%
Total	126	
Gains from process innovation are often shared in a zero-sum manner		
Strongly disagree	31	25%
Somewhat disagree	33	26%
Neither agree nor disagree	49	39%
Somewhat agree	11	9%
Strongly agree	2	2%
Total	126	

by one firm, due in part to the nature of the proprietary IP assets, which takes ownership of the innovation and takes it to the market. As opposed to this, process innovation works through collaboration. Some of the most successful examples of offshore process innovations feature close collaboration between the client in the United States/United Kingdom and the provider located in India/Sri Lanka/ China. The work that occurs on OfficeTiger's premises involves a high degree of collaboration between the clients in the United States and United Kingdom and OfficeTiger's Asian offices. Sometimes the same project may be broken up in real time so that one chunk of it is handled in Sri Lanka while another is executed out of the Philippines even as a third piece is executed in Chennai in India. OfficeTiger's managers work with the client, monitor the progress of work in three countries in real time, and stitch together the output so as to create a seamless delivery vehicle for the client.

This kind of collaboration raises the question of governance: what is the optimal governance structure for these offshore innovation centers? Is it necessary for corporations to own (via a captive center) the innovations or will they be able to source these innovative solutions from a third-party provider? It turns out that the emergence of a new governance structure made possible by advances in information and communication technologies is particularly well suited to sourcing innovations from offshore providers. Before we investigate this governance structure we will first discuss the different governance options open for sourcing services from offshore providers. We term this collection of options the *governance spectrum*.

THE GOVERNANCE SPECTRUM

Economists, following in the central tradition of Ronald Coase in 1933 (Coase, 1937) have long posited that the two principal ways of coordinating work are through an organization or the market. These two solutions—organization and market—that straddle opposite ends of the means of governing the production of work, map onto the "make or buy" options that executives face. The make option corresponds to doing the work in-house while the buy option is about outsourcing work to a third party. In offshore outsourcing in particular there are some interesting options that have emerged in recent years. Figure 4 provides a schematic representation of the spectrum of governance solutions.

Build operate transfer (BOT) and run to outsource (RTO) are temporally separated variants of the make and buy options. In several offshoring deals in the financial services domain we saw both BOT and RTO arrangements. In the BOT model, a third party runs an offshoring center for the client that after a period of time, is bought by the client and made into an in-house offering. Ownership in an RTO moves in the opposite direction: the client sells an in-house facility to a third-party provider after a prespecified period of time. A key point that is worth noting in this context is that offshore sourcing of financial services deals are par-

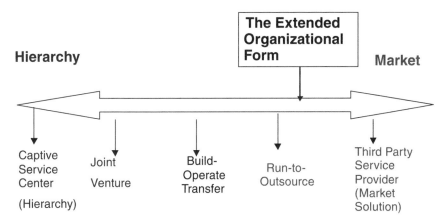

FIGURE 4 The governance spectrum.

ticularly well suited for the governance structures—BOT, RTO, and the extended organizational form (EOF)—that we discuss here. Furthermore, we believe that the sourcing of innovations in financial services—as opposed to the sourcing of services that are chiefly about labor arbitrage[39]—requires some special features of governance, which we term the EOF and which we discuss later in this section.

In cases of both BOT and RTO, the transfer of ownership as well as the handover of assets is easier for financial services firms for a variety of reasons. First, the input, output, and work in process are all information. There are no messy transfers of semifinished inventory nor is there transfer of raw materials and commodities that are factors of production (such as oil or coal). As a result there are few if any issues that have to do with valuing tangible assets that enter production. Second, it is possible for buyers to exert considerable operational control over suppliers via technological monitoring mechanisms and to integrate the work of the offshore supplier closely with the buyer's firm. Investment banks and corporate banks have used fine-grained monitoring mechanisms to keep a close watch on the production process taking place offshore. Due to the continuous proximity enabled by technology, the transfer of control from suppliers to buyers (or the other way around) becomes a lot easier than in the case of offshore production of other products and services. Finally, financial services firms have a culture of measurement and use of metrics to track the critical success factors of their major initiatives (whether in their end-customer markets or having to do with offshore suppliers). As a result, both the buyer and the supplier have a reasonably clear idea of the health of the operation at the time of transfer of control.

---

[39]Lower wages for comparable skills in Asia and parts of Eastern Europe.

These three factors—monitoring, transfer of control, and metrics—taken together allow offshore financial services centers to work as if they are an extension of a client's organization and to be able to sense and respond to the client's context. For the sourcing of process innovations it is essential that the service provider be allowed the leeway to experiment and arrive at innovative solutions to problems. Second, too rigid a controlling structure results in the service provider being reduced to execute against specifications and not try to formulate process innovations that may complement the client's product innovations and address the problems of complexity. The optimal governance structure that allows both the sourcing of process innovations and delivers on contractual terms is a hybrid governance structure that has some measure of flexibility built into the contract while giving the client sufficient monitoring capabilities. We call this hybrid governance structure, which has features of both market-based governance and a hierarchy, the *extended organizational form*. This is a governance structure that is seen predominantly in the financial services industry for the aforementioned reasons.

EMERGENCE OF THE EXTENDED ORGANIZATIONAL FORM

The extended organizational form (EOF) is a hybrid governance structure that brings together some elements of market and hierarchy. The strength of market-based governance (or sourcing processes from a third-party provider) is that it enforces the discipline of cost containment. Third-party firms have to compete with each other, and in order to lower costs they specialize or acquire scale economies and contain costs. However, the weakness of this mode of production is that the incentives of the supplier of services and the buyer are not aligned. The supplier would prefer to minimize the effort at any given price while the buyer would like to induce the highest possible effort from the supplier at the lowest possible price, thus setting up the classic problem of moral hazard.[40] The use of contracts to mitigate moral hazard often involves wasteful expenditure involved in monitoring work, inspecting output, enforcing penalties, and so forth, which economists often refer to as transaction costs associated with market-based governance (Coase, 1937). As opposed to market-based governance the firm could produce these services in-house through a captive center in an offshore location. This would largely eliminate the misalignment of incentives and provide the corporation control over its production facilities offshore through the device of organizational hierarchy. However, the corporation would lose the benefits of scale and specialization and the reduction in costs due to competition that a market solution would offer.

---

[40]The situation that arises when the incentives of the supplier and buyer diverge. When buyers cannot fully observe the effort level of the supplier, the supplier's incentives would be to cut costs (make a suboptimal effort).

A solution to this problem of bringing together the strengths of the two forms of governance is the EOF. It affords the clients both more control over the supplier of services and lower costs of production. Some examples of the EOF include Pipal Research, a firm that provides expertise-driven financial analytics and research services to corporations in the United States and the EU. The company is headquartered in Chicago and has offices in India, China, and London. Once the client and provider establish a contractual relationship, the client works directly with researchers in India and China and the client's managers can direct Pipal Research's associates as well as collaborate with them in research.

An example of the EOF can be seen in the operation of OfficeTiger, a U.S.-based company with operations in several countries including India, Philippines, Sri Lanka, and Switzerland. The firm faces deadlines for work that range from one hour to a month. As a result of facing short deadlines, OfficeTiger does not have the luxury of negotiating arm's-length contracts each time a client commissions a project. The corporation has established a managerial control mechanisms called the "Program Office" by which its managers work closely with managers of the clients. Furthermore, the company has created one of the most comprehensive management information systems that we have studied, called T-Tracks, which it deploys through extranets at the client's site. Client managers as well as OfficeTiger's managers can track the progress of work and slice and dice the work by project, by team, by shift, and even by individual agent in real time. The use of a very fine-grained monitoring system combined with a system of metrics that allows every aspect of work to be tracked and monitored allows both OfficeTiger and clients to execute projects of considerable complexity.[41] In fact OfficeTiger's agents become extensions of the client's organization—an example of the EOF—which in turn allows the client to migrate the production of strategically significant services such as financial analytics, equity research, and market research to OfficeTiger's production teams.

Another example of this phenomenon is the company EquinoxCorp.[42] EquinoxCorp provides services across the mortgage finance value chain that range from simple loan processing to predictive analytics leading to customer profitability gradient and retention. The company has created a network of partnerships along the mortgage processing value chain so that it is able to offer predictive analysis and customer retention services based on the use of a proprietary platform that captures data from several stages of the mortgage financing process.[43] Corporations such as EquinoxCorp and OfficeTiger are networks that span several corporations. Their ability to offer innovative solutions results from process

---

[41]"How Some BPO Providers Seek to Build and Protect Their Turf," *Knowledge @ Wharton*, see http://knowledge.wharton.upenn.edu/article.cfm?articleid=1101.

[42]Recently acquired by i-Flex Solutions a subsidiary of Oracle Inc.

[43]"How Some BPO Providers Seek to Build and Protect Their Turf," *Knowledge @ Wharton*, see http://knowledge.wharton.upenn.edu/article.cfm?articleid=1101.

innovations that they forge across the value chain in the production and delivery of several products ranging from mortgage financing in retail financial services through tax analysis and financial analytics in corporate banking to strategic market analysis, forecasts, and research in investment banking.

To be able to deliver process innovations and calibrate their needs to the context of the clients and their changing market needs, companies such as OfficeTiger and Pipal Research require both considerable flexibility to try and experiment with their solutions as well as the ability to collaborate in real time with their clients. The EOF allows this flexibility as well as establishes deep collaborative links that span joint technology platforms, interorganization information systems, extranet-based tools, and managerial mechanisms (such as the Program Office) that allow the firms to deliver innovations to clients offshore. In the survey (mentioned earlier) we asked executives what forms of governance might be best suited to source different kinds of services and products. Their responses are shown in Figure 5.[44] It is clear in the assessment of these executives that for sourcing innovations the form of governance structure that works best is the "sense and adjust" collaborative networks.

The EOF is a recent phenomenon. The strength of this form of governance emerges from the combination of fine-grained monitoring enabled by technology and the flexibility of work enabled by removal of restrictions on experimentation. In our survey of business process offshoring starting in 1999, we found few if any cases of EOF in the first few years. It is only in the past 5 years or so that this organizational structure has become relatively more prevalent. The EOF, as we saw earlier, is a particularly effective tool for mitigating risk and for aligning the interests of the buyer and the provider of services, especially in financial services where the provider's ability to innovate and create value for the buyer is significant (*Economist*, 2006). Indeed our past research also points in this direction: we find the EOF being deployed in domains where close collaboration between the buyer and the provider of financial services is necessary to transform sourcing via contracts into collaborative networks that can deliver innovations.

CONCLUSION

The diffusion of innovations in financial services across the regions of the world has been propelled by the twin forces of market participation by financial services firms and by the sourcing of process innovations that emerged as a result of global sourcing of financial services. A key factor that drives the division of labor in bringing innovations to market is the divergence in perceptions of process complexity. As we showed earlier, process complexity is subjective and the divergence in views of complexity often stem from the fact that many executives in Asia and Eastern Europe have domain expertise but domain experience is needed

---

[44]Based on 102 responses.

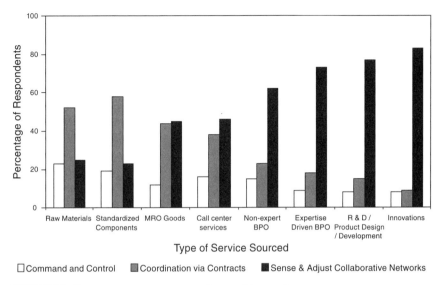

FIGURE 5 Governance structures, sourcing, and innovations.

to deliver innovative products. Thus, service providers in Asia and elsewhere have often restricted themselves to process innovations while their counterparts in the United States and EU have been able to deliver product innovations.

The strength of this model, however, is the complementarity between process and product innovations. Process and product innovations have a lock-and-key structure, where process innovations hold the key to resolving the problem of complexity that is unleashed by the introduction of new products. Finally, the sourcing of innovations in financial services requires forms of governance that are very different from the contractual, command-and-control mechanisms used for the procurement of more routine and highly codifiable forms of work such as call centers and data processing. For sourcing innovations from offshore providers it is necessary to be able establish a collaborative mechanism that allows the Western clients to monitor their offshore service providers in fine-grained fashion, which in turn allows them to give these firms the leeway to experiment and arrive at innovative process-based solutions. Second, this form of governance also establishes deep collaborative linkages between firms, thus enabling the client and the buyer to come up with highly customized solutions in relatively short time frames.

Where do we expect to see these trends lead? The cost of bandwidth has been falling over the past decade even as the reach of the Internet has been increasing. Interoperability between computing platforms has made it relatively easy to transport vast datasets between firms that are on very different platforms. The emergence of collaborative technological platforms such as Wikis and Web services (that can connect disparate systems over the Internet) as well as structured blogs will deepen the collaborative capabilities of corporations. Increasingly companies are able to monitor the working of the offshore providers at finer and finer levels of detail thanks to technological advances. These forces will result in greater global sourcing of innovations from various regions and will expand the sourcing footprint of corporations to include regions other than China and India. Brazil, Sri Lanka, Philippines, the Republic of South Africa, and Poland could well become important providers of financial services innovations in the near future. International trade trends are also making it easier for corporations to compete in offshore markets. Free-trade agreements between the United States and Singapore and between the United States and Australia, as well as the integration of Eastern European countries into the EU, have all resulted in increased market participation in global markets by financial services firms the world over. Technology has brought instant transparency to business practices and product strategies adopted by firms in different regions and markets. As a firm introduces a new product or new features of an existing product, it will rapidly get copied by all other firms and will become a standard feature of the next generation of products. Firms will thus have to compete constantly for profits and market share. Competition is perhaps the most powerful driving force of innovation. As competition intensifies in world markets, we will surely see greater innovation in these markets.

ACKNOWLEDGMENTS

We wish to thank the Mack Center for Technological Innovation and the Fishman Davidson Center for Service and Operations Management at the Wharton School, University of Pennsylvania, for supporting this research.

REFERENCES

Aron, R., and J. Singh. (2005). Getting offshoring right. *Harvard Business Review* December: 135-143.
Coase, R. (1937). The nature of the firm. *Economica* 4:386-405.
Economist. (2006). Partners in wealth. January 19.
Emloh, David von, and Yi Wang. (2005b). Competing for China's credit card market. *McKinsey.* Available at http://www.mckinseyquarterly.com/competing_for_chinas_credit_card_market_1713.
Ensor, B. (2004). How ING DIRECT stole $17 billion in bank deposits. *Forrester.* Available at http://www.forrester.com/Research/Document/Excerpt/0,7211,41632,00.html.
Gary, L. (2004). Taking disruption to the bank, Strategy & Innovation, September-October 2004.
Gupta, Rajat K. (2005). Fulfilling India's promise. *McKinsey Quarterly Special Edition.* Available at http://www.mckinseyquarterly.com/public_sector/economic_policy_fulfilling_indias_promise_1673.

Roxburgh, Charles. (2006). Industry comment: The outlook for European corporate and investment banking. *McKinsey Quarterly* Web Edition. Available at http://www.mckinseyquarterly.
com/industry_comment_the_outlook_for_european_corporate_and_investment_banking_1838.
Segupta, J., and Renny Thomas. (2005a).What Indian consumers want from banks. *McKinsey.* Available at http://www.mckinseyquarterly.com/what _indian_consumers_want_from_banks_1662.